J.W. Deacon BSc, PhD
Institute of Cell & Molecular Biology
University of Edinburgh

Modern Mycology

Third edition

http://helios.bto.ed.ac.uk/bto/course-info/third-yr/micro3m/profiles.htm

b

**Blackwell
Science**

© 1980, 1984, 1997 by
Blackwell Science Ltd
Editorial Offices:
Osney Mead, Oxford OX2 0EL
25 John Street, London WC1N 2BL
23 Ainslie Place, Edinburgh EH3 6AJ
350 Main Street, Malden
 MA 02148 5018, USA
54 University Street, Carlton
 Victoria 3053, Australia
10, rue Casimir Delavigne
 75006 Paris, France

Other Editorial Offices:
Blackwell Wissenschafts-Verlag GmbH
Kurfürstendamm 57
10707 Berlin, Germany

Blackwell Science KK
MG Kodenmacho Building
7–10 Kodenmacho Nihombashi
Chuo-ku, Tokyo 104, Japan

The right of the Author to be
identified as the Author of this Work
has been asserted in accordance
with the Copyright, Designs and
Patents Act 1988.

First published 1980
Japanese edition 1980
Second edition 1984
Spanish edition 1985
Third edition 1997
Reprinted 1998 (twice), 2000

Set by Excel Typesetters, Hong Kong
Printed and bound in the United Kingdom
at the University Press, Cambridge

The Blackwell Science logo is a
trade mark of Blackwell Science Ltd,
registered at the United Kingdom
Trade Marks Registry

DISTRIBUTORS

Marston Book Services Ltd
PO Box 269
Abingdon, Oxon OX14 4YN
(*Orders*: Tel: 01235 465500
        Fax: 01235 465555)

USA
Blackwell Science, Inc.
Commerce Place
350 Main Street
Malden, MA 02148 5018
(*Orders*: Tel: 800 759 6102
              781 388 8250
        Fax: 781 388 8255)

Canada
Login Brothers Book Company
324 Saulteaux Crescent
Winnipeg, Manitoba R3J 3T2
(*Orders*: Tel: 204 837-2987)

Australia
Blackwell Science Pty Ltd
54 University Street
Carlton, Victoria 3053
(*Orders*: Tel: 3 9347 0300
        Fax: 3 9347 5001)

A catalogue record for this title
is available from the British Library

ISBN 0-632-03077-1

Library of Congress
Cataloging-in-Publication Data

Deacon, J.W.
  Modern Mycology/J.W. Deacon.–3rd ed.
     p.   cm.
  Rev. ed. of: Introduction to modern mycology.
  2nd ed. 1984.
  Includes bibliographical references and index.
  ISBN 0-632-03077-1
  1. Mycology.   2. Fungi.
  I. Deacon, J.W. Introduction to modern mycology.
  II. Title.
QK603.D4   1997
579.5–dc21   97-3738
                                              CIP

For further information on
Blackwell Science, visit our website:
www.blackwell-science.com

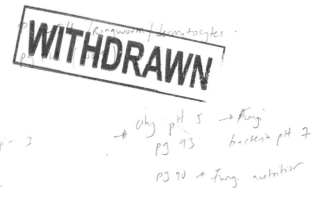

Contents

Preface

This text is intended as an introduction to fungi for microbiologists, botanists and biologists in general. It is the successor to *Introduction to Modern Mycology* (1980, 1984) and follows the same format but has been updated thoroughly to incorporate new information and to reflect new approaches in the subject. The emphasis is on the behaviour, physiology, activities and practical significance of fungi — a group of organisms distinct from all others. The book contains extensive sections on the fungal pathogens of plants and animals, including humans, and on the use of fungi as biological control agents of pests and pathogens.

I am grateful to the many people who have allowed me to reproduce material in this book, and particularly to colleagues around the world, acknowledged in the captions, who supplied some of the original photographs. My task in updating the text was made much easier by the availability of many recent reviews, not least the Proceedings of the Fifth International Mycological Congress, held in 1994, and published in *Canadian Journal of Botany* **73** (Suppl.). Lastly, it is my pleasure to acknowledge a personal debt to Professors Noel Robertson and the late Denis Garrett, who fostered my early interest in mycology.

Jim Deacon
Edinburgh
October 1996

Chapter 1

Introduction

Fungi are a unique group of organisms, different from all others in their behaviour and cellular organization. The uniqueness of fungi is a prominent feature of this book, which adopts a functional rather than a taxonomic approach. In the first part of the book, we deal with the growth, physiology, behaviour and genetics of fungi, including their roles in biotechnology. In the second part we cover the main activities of fungi—as decomposers of organic matter, spoilage agents, plant pathogens, pathogens of humans and biological control agents of pests and pathogens. A final chapter is devoted to the major ways of preventing and controlling fungal growth, as this presents a major challenge in modern mycology. However, to place all this in perspective, we begin with an overview of the fungi and an outline of the major fungal groups.

Towards a definition of fungi

In order to arrive at a definition of fungi and to see their place among organisms as a whole, we must look briefly at the history of classifying organisms. The earliest classifications assigned all organisms to either plants or animals, and so fungi were included in the plant kingdom because they are immotile and they have cell walls. It was only in 1866 that this increasingly unworkable system was changed. Then, Haeckel, a disciple of Darwin, proposed a third kingdom—the protista or **protists**—for all the microscopic life forms, including algae, bacteria, protozoa and fungi. The next major change came with the development of electron microscopy in the 1950s, when the fundamental difference in cellular organization between bacteria and other organisms was apparent. The

extreme simplicity of bacterial cells, lacking a distinct nucleus, led to their separation as **prokaryotes**, while all other organisms were termed **eukaryotes** (*eu*=true; *karyos*=kernel, nucleus). Whittaker (1969) then argued that fungi are different from the other eukaryotes in many fundamental respects, so he assigned them to a separate kingdom, although he recognized that the organisms grouped as fungi at that time were **polyphyletic**, derived from more than one ancestor.

The separation of fungi from other organisms heralded the **Five Kingdom** approach to classifying organisms—an approach that some people still use. It recognizes two **domains**: (i) the prokaryote domain for bacteria; and (ii) the eukaryote domain for the kingdoms of plants, animals, fungi and protists. However, recent research on deoxyribonucleic acid (DNA) sequence analysis and comparative biochemistry has placed increasing strains on this system. In particular, it is clear that some bacteria are so different from others that they must be separated as a third domain—the archaebacteria, now termed the **Archaea**; although they lack a nucleus, they have several features that resemble those of eukaryotes. In addition to this, DNA sequence analysis suggests that there are many small groups, as well as the large groups, of eukaryotes that have diverged from one another. By such methods it is possible to begin to construct a **universal phylogenetic tree**, although this will be a long task and doubtless will undergo various modifications on the way. Figure 1.1 shows an example of such a tree, proposed by Woess (1994). For our present purposes, we see from Fig. 1.1 that fungi represent one of the ultimate branches of the **eucarya domain**, evolutionarily closest to present-day plants and animals, but distinct from them.

1

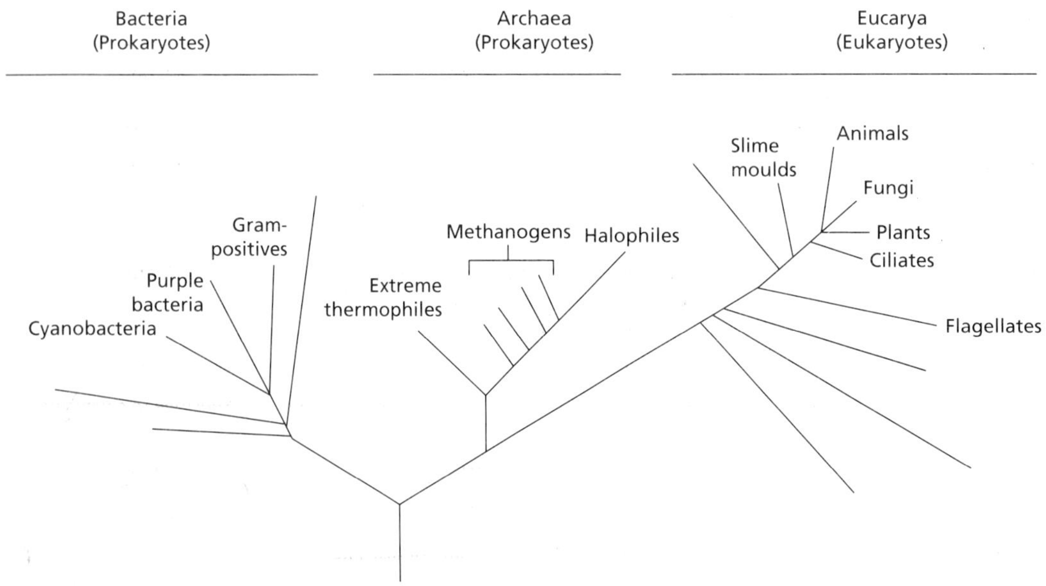

Fig. 1.1 Universal phylogenetic tree, showing the three domains: Bacteria, Archaea and Eucarya, based on a diagram in Woess (1994), but showing only some of the major groups of organisms. The tree is based on nucleotide sequences of the DNA that codes for the RNA in the small subunit of ribosomes. The root of the tree is based on comparison of the genes encoding translation elongation factors (specific proteins involved in protein synthesis).

The fungi and fungus-like organisms

It should be noted that the 'fungal branch' in Fig. 1.1 refers to a 'core' group of fungi which typically have chitin in their walls. Mycologists often refer to these as the true fungi because they are a clearly defined group, characterized by a combination of cellular and biochemical features, summarized in Table 1.1. We deal with the functional significance of many of these points in later chapters because they affect, for example, our ability to control pathogenic fungi in a plant or a human host. But, it is worth noting here that the 'true fungi' are more like animals than plants.

Table 1.2 gives an outline of the organisms that can be considered as fungi in a broader sense. It shows that at least three groups of organisms have traditionally been studied by mycologists.

1 An assemblage loosely termed **slime moulds**, which are evolutionarily distinct from one another. They grow as wall-less protoplasmic stages, often engulfing bacteria and other food particles by phagocytosis. They resemble fungi only insofar as they produce walled spores, but even their walls have a predominance of galactosamine rather than chitin (a polymer based on glucosamine). In many ways these organisms are closest to protozoa. We will say little more about them, but their features are summarized on p. 14 and in Fig. 1.4.

2 A group of fungus-like organisms, the **Oomycota**, which have cellulose in their walls and many other plant-like features (Table 1.1). DNA sequence analysis shows that these organisms are closely related to diatoms and brown algae in a group (kingdom) termed the **Stramenopila** (stramenopiles). Nevertheless, the oomycota have evolved a typical fungal lifestyle and several of them are important plant pathogens. It is appropriate to regard them as fungi in a broad sense.

3 A core of true fungi, as mentioned earlier, typified by having chitin in their walls. This group is subdivided as shown in Table 1.2 and will be discussed later. All of them seem to have a common ancestor, perhaps similar to the present-day choanoflagellates, which are unicellular nonphotosynthetic protozoa with a single flagellum surrounded by a buccal cavity.

Table 1.1 Some major characteristics of the chitin-walled (true) fungi in comparison with animals and oomycota.

Character	True fungi (chapter reference)	Animals	Oomycota
Growth habit	Hyphal, tip growth (3)	Not hyphal	Hyphal, tip growth
Nutrition	Heterotrophic, absorptive (5)	Heterotrophic, ingestive	Heterotrophic, absorptive
Cell wall	Chitinous (2)	Chitin in insect exoskeleton	Cellulose
Nuclei	Haploid, membrane persists during division; spindle pole bodies do not have centriolar arrangement (2)	Typically diploid; typical centrioles	Diploid; typical centrioles
Histones	Histone 2B-like animals		Like plants
Microtubules	Sensitive to benzimidazoles and griseofulvin (14)	Sensitive to colchicine	Sensitive to colchicine
Lysine	Synthesized by AAA pathway (6)	Not synthesized	DAP pathway
Golgi cisternae	Unstacked, tubular (2)	Stacked	Stacked
Mitochondria	Plate- or disc-like cisternae (2)	Plate or disc-like	Tubular
Translocable carbohydrates	Polyols, trehalose (6)	Trehalose in insects	Glucose, etc., as plants
Storage compounds	Glycogen, lipids, trehalose (6)	Glycogen, lipids, trehalose in some	Mycolaminarin
Mitochondrial codon usage	UGA codes for tryptophan	For tryptophan	For chain termination
Sterols	Ergosterol (2,14)	Cholesterol	Plant sterols

AAA, α-aminoadipic acid; DAP, diaminopimelic acid.

In this book we focus on the walled fungi of groups (2) and (3) above. These share four major features that lead to a broad definition of fungi.

1 All fungi are **eukaryotic**. In other words, they have membrane-bound nuclei containing several chromosomes, and a range of membrane-bound cytoplasmic organelles (mitochondria, vacuoles, etc.). Other characteristics of eukaryotes include: cytoplasmic streaming, DNA that contains non-coding regions termed introns, membranes that typically contain sterols, and ribosomes of the 80S type in contrast to the 70S ribosomes of bacteria ('S' refers to Svedberg units, a measure of the sedimentation rate during centrifugation).

2 Fungi typically are filamentous, the individual filaments being termed **hyphae** (singular — hypha). The hyphae grow only at their extreme tips, so fungi exhibit **apical growth** in contrast to the intercalary growth of most other filamentous organisms. Hyphae branch successively behind their tips, the resulting network of hyphae being termed a **mycelium**.

3 Fungi are **heterotrophs** (chemo-organotrophs). In other words, they need pre-formed organic compounds as energy sources and as carbon sources for cellular synthesis, in contrast to autotrophic plants and some bacteria. The cell wall prevents fungi from engulfing food by phagocytosis. Instead, they absorb simple, soluble nutrients through the wall and cell membrane, and these simple nutrients may be released from more complex polymers by enzymes (depolymerases) that are secreted into the external environment.

4 Fungi reproduce by both sexual and asexual means, producing **spores** as the end-product. Fungal spores vary enormously in shape, size and other properties, related to their various roles in dispersal or dormant survival.

Thus, we can define fungi, albeit in indigestible terms, as eukaryotic, characteristically mycelial

Table 1.2 The major groups of fungi and fungus-like organisms.

---

1 *True fungi* with walls typically containing chitin, and many other characteristic cellular and biochemical features

(a) *Chytridiomycota*. Typically unicellular, or primitive chains of cells, attached to a food base by tapering rhizoids; sexual reproduction by fusion of **motile gametes**; asexual reproduction by cytoplasmic cleavage in a **sporangium**, producing motile, **uniflagellate zoospores**

(b) *Zygomycota*. Typically hyphae without cross-walls (**aseptate**); sexual reproduction by fusion of sex organs (**gametangia**) leading to thick-walled resting spores (**zygospores**); asexual reproduction by cytoplasmic cleavage in a **sporangium**, producing **non-motile spores**

(c) *Ascomycota*. Hyphae with cross-walls (**septa**) or yeasts; sexual reproduction by fusion of modified hyphae (or yeasts), sometimes by fusion of a 'male' spore (**spermatium**) with a 'female' receptive hypha (**trichogyne**), leading to development of an **ascus** containing **ascospores**; asexual reproduction as in deuteromycota (see below)

(d) *Deuteromycota*. Hyphae (with septa) or yeasts; sexual reproduction absent, rare or unknown; asexual spores (**conidia**) formed in various ways from hyphae but never by cytoplasmic cleavage in a sporangium

(e) *Basidiomycota*. Hyphae (with dolipore septum) or yeasts; asexual spores rare in most groups; sexual reproduction by fusion of compatible hyphae, leading ultimately to production of **basidiospores** on **basidia**, sometimes on or in a large fruiting body (e.g. toadstool)

2 *Organisms that resemble fungi* in behaviour and lifestyle but with cellulose-based walls and cellular and biochemical features resembling those of plants

Oomycota. Hyphae aseptate; asexual reproduction by formation of motile, **biflagellate zoospores** in a sporangium; sexual reproduction by fusion of a 'male' sex organ (**antheridium**) with 'female' sex organ (**oogonium**), leading to production of thick-walled resting spores (**oospores**)

3 *Organisms with some fungus-like features*, but with a naked, protoplasmic somatic stage, perhaps most closely related to protozoa

(a) *Acrasids* and *dictyostelids* (sometimes termed **cellular slime moulds**). Amoeboid organisms that phagocytose bacteria and other food particles; they aggregate to form a fungus-like fruiting body that releases dry, air-borne spores

(b) *Myxomycota* (plasmodial slime moulds). Grow as a network of protoplasm (the **plasmodium**) that engulfs bacteria and other food particles; at the onset of starvation they form fruiting bodies that release dry, air-borne spores

(c) *Plasmodiophorids*. Obligate intracellular parasites of fungi, algae and higher plants; exist as naked plasmodia in the host cells; form highly persistent thick-walled resting spores

---

heterotrophs, with absorptive nutrition and reproduction by spores.

Growth forms of fungi

There are three major growth forms of fungi (Fig. 1.2). Most fungi are mycelial, with a network of hyphae, and colloquially they are termed 'moulds'. Some of the more primitive fungi, such as the chytridiomycota (see later), often have single rounded cells or dichotomously branched chains of cells, attached to a food source by tapering **rhizoids.** Some other fungi grow as unicellular **yeasts** and produce daughter cells either by budding (e.g. *Saccharomyces cerevisiae*) or by binary fission (e.g. *Schizosaccharomyces pombe*). These different growth forms can be related to habitat. For example, yeasts are common in moist environments that are rich in simple, soluble nutrients, and some fungi can even alternate between a yeast and a mycelial phase in response to environmental conditions. Such **dimorphic** fungi (with two shapes) include some of the important pathogens of humans and higher animals (see Chapter 13); they grow as yeasts for spread in water films or in body fluids, but as hyphae for invasion of the

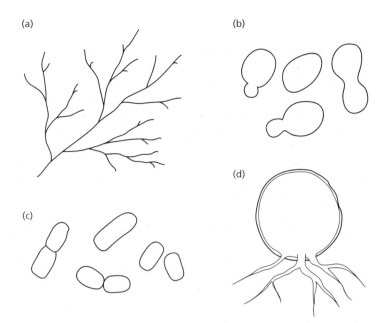

Fig. 1.2 Growth forms of fungi. (a) Mycelial; (b) budding yeast; (c) fission yeast; (d) chytridiaceous growth form with rhizoids. Not drawn to scale.

tissues. *Candida albicans* is a classic example of this. It is common as a yeast on the mucosal membranes of humans, where it does little damage; but in response to some conditions the yeast cells can produce hyphae which invade the mucosa and cause significant problems. Some of the other opportunistic pathogens of humans grow as hyphae in culture or on organic matter at 20–25°C, but as yeasts at 37°C.

Activities of fungi

All fungi require organic nutrients as energy sources and for cellular synthesis, but a broad distinction can be made according to how the nutrients are obtained—by growth as **parasites** (or **symbionts**) on another organism (a host), or by growth as **saprotrophs** (saprophytes) on non-living materials. Here we review briefly the parasitic and saprotrophic activities of fungi, but we discuss them in detail in Chapters 10–13.

Fungal parasites of plants

A parasite can be defined as an organism that obtains some or all of its nutrient requirements from the living tissues of another organism with which it lives in intimate association. Therefore, parasitism is a lifestyle and it involves specific adaptations. A distinction can be made according to how the parasite obtains its nutrients: **necrotrophic** parasites (*necros* = death; *trophy* = feeding) kill the host tissues as part of the feeding process — for example by producing toxins or degradative enzymes; **biotrophic** parasites (*bios* = life) feed from living cells without killing them, often by means of special nutrient-absorbing structures. We discuss this important distinction in Chapters 11 and 12.

A parasite that causes disease is termed a **pathogen**, and in this respect the fungi are pre-eminent because they cause more than 70% of all the major crop diseases. For example, potato blight is caused by one of the cellulose-walled fungi, *Phytophthora infestans*. This disease devastated the potato crops of Ireland in the 1840s, leading to the starvation of up to one million people and large-scale emigration to the rest of Europe and the USA. Even today the control of *P. infestans* and its close relatives, the downy mildew fungi, accounts for about 15% of world fungicide sales, and there is widespread concern because *P. infestans* has developed resistance to some of the fungicides that have kept it under control in recent years (see Chapter

14). Other notable pathogens of crop plants are the rust, smut, powdery mildew and vascular wilt fungi, discussed in Chapter 12. Fungal pathogens also have environmental impact when introduced accidentally into new regions. For example, Dutch elm disease, caused by *Ophiostoma novo-ulmi* and related *Ophiostoma* spp. (see Chapter 9), has destroyed most of the elm trees in Britain and Western Europe in the last 30 years, and in North America earlier this century. Chestnut blight caused by *Cryphonectria parasitica* (see Chapter 8) has devastated the native American chestnut population in the USA, and *Phytophthora cinnamomi* has damaged large areas of indigenous eucalypt forest in Australia (see Chapter 12).

There is, however, another side of the coin because some parasites can be beneficial — they provide advantages that outweigh the costs to the host. Two important examples of this are seen in **lichens** and **mycorrhizas**, where fungi live in balanced association with a photosynthetic partner. Lichens are remarkable because the fungus and its partner (a green alga or a cyanobacterium) form a composite organism. The fungus produces a tissue around the photosynthetic cells, protecting them and supplying them with mineral nutrients absorbed from the underlying substratum (a rock, tree trunk, etc.); the photosynthetic partner provides the fungus with organic carbon and, in the case of cyanobacteria, organic nitrogen fixed from atmospheric nitrogen. Lichens grow in some of the most inhospitable environments on earth, where they have a major role at the base of a food chain. During their long evolutionary association the lichen partners have become so closely attuned that the most common algae of lichens (*Trebouxia* spp.) probably do not have a free-living existence, and there are thousands of fungi that are found only in lichen partnerships. In some forms of lichen there is a mechanism for ensuring the dispersal of the partnership to new environments, by propagules such as **soredia** which consist of groups of algal cells enmeshed in fungal hyphae.

Mycorrhizas are intimate associations between fungi and the roots or other underground organs of plants. They are less conspicuous than lichens but are found on more than 70% of all plant species in agricultural and natural environments, and they can be essential for plant nutrition. Typically,

the fungus obtains its organic carbon from the host and, in return, benefits the plant by increasing the uptake of mineral nutrients such as phosphorus or nitrogen from soil. Mycorrhizal fungi also can protect plants from pathogenic attack, from drought or from potentially toxic heavy metals in some environments. The major mycorrhizal associations are discussed in detail in Chapter 12, but it is worth mentioning here that mycorrhizas date back to the earliest land plants (Lewis, 1987) and these associations might even have been essential for the colonization of land by plants.

The many adaptations of fungi for parasitism of plants are discussed in Chapters 4, 9 and 12. Here we will note just one further point: the hyphal apex has remarkable penetrating power, which enables hyphae to invade even intact plant surfaces. This is why fungi are so important as plant parasites, in contrast to bacteria which often have to invade through wounds or natural openings.

Fungal parasites of humans

In contrast to the many fungi that parasitize plants, relatively few infect humans or other warm-blooded animals. Nevertheless, the fungal pathogens of humans pose a significant threat to immunocompromised individuals and to transplant patients. The fungus *Pneumocystis carinii* is now the single most common cause of death of acquired immune deficiency syndrome (AIDS) patients in North America and Western Europe. The incidence of such infections has highlighted the scarcity of drugs that can be used to control one eukaryote in the tissues of another. Only a handful of chemicals are sufficiently specific to control fungi without affecting the tissues of humans; this contrasts with antibacterial agents which can be targeted at the many cellular differences between prokaryotes and eukaryotes (see Chapter 14).

Most of the fungi that infect humans are opportunistic parasites which grow more commonly as saprotrophs in soil, composts, bird excreta, etc., but can infect through wounds or when air-borne spores enter the lungs. Examples include *Aspergillus fumigatus*, *Histoplasma capsulatum*, *Blastomyces dermatitidis* and *Coccidioides immitis* (see Chapter 13). A different type of behaviour is shown by *C. albicans* which is a common and harmless commensal on the mucosal membranes of the mouth, gut and vagina of healthy individuals but can invade locally, causing 'thrush' in infants, vaginitis of pregnant women and stomatitis in up to 50% of people who wear dentures. In extreme cases it can grow systemically and cause death after colonizing the blood from an intravenous catheter to which the yeast cells adhere. Another type of behaviour is found in the **dermatophytes** or 'ringworm' fungi, which grow on the skin, nails and hair, causing athlete's foot and similar superficial diseases. The dermatophytes are not life-threatening, but are estimated to occur on 40% of people worldwide, especially in the developing countries. They are also common on wild and domesticated animals, which can be a source of infection of humans. They grow on the dead keratinized tissues rather than by invading living cells, but they cause irritation and inflammation which can lead to secondary infections by bacteria. All these fungal pathogens of humans are discussed in Chapter 13.

Fungal parasites as biological control agents

Fungi parasitize many types of host, including other fungi (**mycoparasites**; see Chapter 11), insects (**entomopathogens**; see Chapter 13) and nematodes (**nematophagous fungi**; see Chapter 13). In the past, such fungi might have been regarded as curiosities, but now are recognized as significant population regulators of their hosts and as potential biological control (biocontrol) agents of major pests or plant pathogens. We discuss biocontrol at many points in this book, notably in Chapters 11 and 14.

Fungal saprotrophs

Fungal saprotrophs have major roles in decomposition of organic matter and thus in the recycling of nutrients in natural and agricultural environments. Indeed, there are few naturally occurring organic compounds that cannot be utilized by one fungus or another — from the simplest organic compound, methane, which is utilized by some yeasts, to the most complex polymers such as lignin which is degraded by some wood-rotting fungi. However, the breakdown of cellulose by fungi is perhaps most important on a global scale because cellulose accounts for about 40% of all plant structural material — the most abundant natural polymer on earth. Polymer breakdown is achieved by the release of extracellular enzymes and is intimately related to the hyphal growth form of fungi (see Chapter 5). But, different fungi are adept at degrading different types of polymer, so fungal saprotrophs often grow in complex, mixed communities reflecting their different enzymic capabilities (see Chapter 10).

Decomposer fungi cause spoilage of useful products such as foods, structural timbers, leather, canvas, books, etc. These undesirable activities are grouped under the term **biodeterioration**, as opposed to **biodegradation** for the more general saprotrophic activities of fungi. The dry-rot fungus, *Serpula lacrymans*, is a major cause of timber decay in buildings (see Chapter 4). The 'sooty moulds' such as *Aureobasidium pullulans* (Fig. 1.3), *Alternaria* and *Cladosporium* spp. are common on bathroom and kitchen walls, where they utilize the soluble cellulose gels that are used as stabilizers in emulsion paints or as wallpaper pastes; these fungi cause discolouration because of their darkly pigmented spores and hyphae. The 'kerosene fungus' *Amorphotheca resinae* causes significant problems by growing on the long-chain hydrocarbons in aviation fuel, machine lubricants, etc. Often, the control of these spoilage fungi is difficult and costly, because the individual fungi are adapted to tolerate extreme environmental conditions (see Chapter 7). For example, fungi are particularly notable for their ability to grow at low water availability: almost all fungi can grow below the 'permanent wilting point' of plants, and some *Aspergillus* spp. that cause grain spoilage can grow in conditions of extreme water-stress. These fungi not only cause the rotting of foodstuffs but also can produce potent **mycotoxins** such as the aflatoxins, which are among the most potent known carcinogens and have been implicated in liver cancer of humans. These and other mycotoxins are among the vast range of **secondary metabolites** of fungi, discussed in Chapter 6.

Fungi in biotechnology

Fungi have many traditional roles in biotechnology, but also some novel roles and there is major scope for their future commercial development

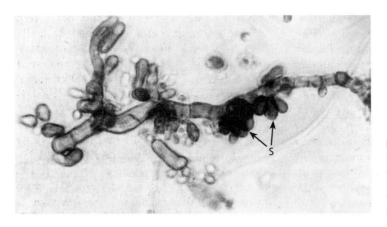

Fig. 1.3 *Aureobasidium pullulans* (deuteromycota) in a scraping from a kitchen wall. This fungus has melanized hyphae that readily fragment and that produce clusters of bud-like spores (S) at the septa.

(Wainwright, 1992). Some of these roles are outlined here.

Foods and food flavourings

Currently, about 1.5 million tonnes of edible mushrooms are produced commercially each year around the world. Much of this involves the common mushroom *Agaricus bisporus* (also known as *A. brunnescens*) discussed in Chapters 4 and 10, but *Lentinula edodes* (grown on logs) and *Volvariella volvacea* (on rice straw) are major mushroom crops in Japan and South-East Asia. In a major new development, mycelium of the fungus *Fusarium graminearum* has been grown in fermenter vessels and marketed as a novel food termed 'Quorn' (Trinci, 1992). It is now widely available in British supermarkets, both as meat-like chunks and in various oven-ready meals. Quorn 'mycoprotein' is the one significant product to have emerged from the efforts over many years to produce 'single-cell

protein' for human consumption but, ironically, the costs involved in the technology and to ensure the strict safety of the product have meant that it is appropriate only for diet-conscious western consumers rather than for the protein-deficient developing world. In nutritional composition Quorn compares favourably with meat (Table 1.3) because it has a high-protein content but low-fat content and absence of cholesterol — fungi have different sterols from those of animals (see Chapter 2). We discuss the production process of Quorn in Chapter 3.

Fungi are used to produce several traditional foods and beverages, including alcoholic drinks (ethanol from the yeast *Saccharomyces cerevisiae*) and bread where yeast produces carbon dioxide for raising the dough. *Penicillium roquefortii* is used in the later stages of production of the blue-veined cheeses such as Stilton and Roquefort, to which it imparts a characteristic flavour. *P. camembertii* is used to produce the soft cheeses such as Camem-

Table 1.3 Nutritional composition of Quorn mycoprotein, compared with traditional protein sources. (Data from Trinci, 1992.)

	Units	Quorn	Cheddar cheese	Raw chicken	Raw lean beef	Fresh cod
Protein	g 100 g^{-1}	12.2	26.0	20.5	20.3	17.4
Dietary fibre	g 100 g^{-1}	5.1	0	0	0	0
Total fats	g 100 g^{-1}	2.9	33.5	4.3	4.6	0.7
Fat ratio	Polyunsaturated: saturated	2.5	0.2	0.5	0.1	2.2
Cholesterol	mg 100 g^{-1}	0	70	69	59	50
Energy	kJ 100 g^{-1}	334	1697	506	514	318

bert and bries; it grows on the cheese surface, forming the 'crust' and produces proteases which progressively degrade the cheese to give the soft consistency. Less well known but more significant is the role of fungi in the fermentation of traditional foods in developing countries. For example, *Rhizopus oligosporus* is used to convert cooked soybeans to a nutritious staple food in Indonesia, the product being called **tempeh**. This involves only a short (24–36 h) incubation time, during which the fungus degrades some of the fat and also a trypsin inhibitor in soybeans, so that the naturally high-protein content of this crop is more readily available in the diet, and a 'flatulence factor' also is removed. The food termed **gari** is part of the staple diet in southern Nigeria; it is produced from the high-yielding root crop, cassava, perhaps better known in its processed form, tapioca. Raw cassava contains a toxic cyanogenic glycoside termed linamarin, which is removed during a prolonged and largely uncontrolled fermentation in village communities. Much of this process involves bacteria, but the fungus *Geotrichum candidum* gives the product its desired flavour (see Ekundayo, 1980).

With the increasing popularity of 'convenience' foods in western societies, fungi are set to play a new role as sources of natural flavour and odour components. The mycelia of many fungi produce such compounds in fermenter vessels, and several companies have patented them. But, some of these compounds could present a challenge even to the most imaginative marketing agencies. When a taste panel sampled a range of mycelial metabolites the terms used to describe them included 'fruity', 'mushroom', 'medicinal', 'sweat' and 'rabbit-burrow' (Gallois *et al.*, 1990).

Fungal metabolites

The metabolites of fungi can be grouped into two broad categories: (i) **primary metabolites** which are the end-products of the common metabolic pathways (intermediary metabolism); and (ii) **secondary metabolites** which are a diverse range of compounds formed by specific pathways of particular organisms (see Chapters 3 and 6). Both types of product accumulate in the culture medium when growth is restricted by some factor but when the biochemical machinery continues to operate, like the engine of a car taken out of gear. This is done purposefully in commercial conditions. For example, ethanol accumulates as a metabolic end-product when yeast is grown in a sugar-rich medium favouring metabolism but in anaerobic conditions that limit growth, and this is the basis of the alcoholic drinks industry. Similarly, some fungi produce large amounts of organic acids from sugars when their growth is limited by low pH. *Aspergillus niger* and *A. wentii* are used in this way to produce citric acid for the soft drinks industry, and *A. terreus* is used for production of itaconic acid, from which polymers are made in the manufacture of paints, adhesives, etc. In conditions of normal pH, *A. niger* is also used to produce gluconic acid (as a dietary supplement) by the direct enzymic oxidation of glucose supplied as the substrate. Many other primary metabolites can be produced by fungal fermentations, examples being glycerol, mannitol, isocitric acid and α-ketoglutaric acid, although the use of fungi to produce them commercially may not be economically feasible if they can be produced more cheaply by chemical synthesis.

A vast range of secondary metabolites are produced by fungi and are described by Turner (1971) and Turner and Aldridge (1983). Many of them have no commercial value, but others are extremely important. A classic example is **penicillin** (actually a group of structurally related compounds; see Chapter 6), produced commercially from strains of *Penicillium chrysogenum* but originally discovered as a metabolite of *P. notatum*. This discovery literally changed the course of medicine and has saved millions of lives. Other antibiotics produced by fungi include the **cephalosporins** (structurally similar to penicillin, but now produced commercially by the filamentous bacteria, actinomycetes), **griseofulvin** (from the fungus *P. griseofulvum*) which is used to treat dermatophyte infections of humans (see Chapter 14) and **fusidic acid** (from various fungi) which is used to control staphylococci that have become resistant to penicillin. Several other fungal secondary metabolites have practical applications in horticulture, medicine or research (Table 1.4). For example, **cyclosporin** from *Trichoderma polysporum* is used as an immunosuppressant to prevent organ rejection in transplant surgery. The antibiotic **gliotoxin** (from *T. virens*) has potential for biocontrol of plant

Table 1.4 Fungal secondary metabolites produced commercially for pharmaceutical, agricultural and research uses.

Usage	Product	Fungal source	Application
Medicine	Penicillins	*Penicillium chrysogenum*	Antibacterial
	Cephalosporins	*Cephalosporium acremonium*	Antibacterial
	Griseofulvin	*P. griseofulvum*	Antifungal
	Fusidin	*Fusidium coccineum*	Antibacterial
	Cyclosporin	*Trichoderma polysporum*	Immunosuppressant
	Ergot alkaloids	*Claviceps purpurea*	Induces labour; migraine treatment
Agriculture	Zearalenone	*Gibberella zeae*	Growth promoter for cattle
	Gibberellins	*G. fujikuroi*	Plant hormones
Research	Gliotoxin	*Trichoderma virens*	Immunosuppressant
	Cytochalasins	*Helminthosporium dermatoideum*, etc.	Anti-actin agents
	Fusicoccin	*Fusicoccum amygdali*	Stomatal opening
	Phalloidin	*Amanita phalloides*	Actin binding
	α-Amanitin	*A. phalloides*	RNA polymerase II inhibitor

pathogens (see Chapter 11), but also is a powerful immunosuppressant and a possible alternative to cyclosporins. Fungi also produce many antiviral and antitumour agents which might be developed commercially (Jong & Donovick, 1989).

In a different context, some of the polysaccharides of fungi have potential commercial value. **Pullulan** is an α-1,4-glucan (polymer of glucose) produced as an extracellular sheath by *Aureobasidium pullulans*, the fungus mentioned earlier as growing on bathroom walls. This polymer is used in Japan to make a film-wrap for foods. A potential new market could develop from the discovery that fungal wall polymers or their partial breakdown products can be powerful elicitors of plant defence responses (see Chapter 12) so they might be used to 'immunize' plants (Hadwiger *et al.*, 1988). For example, the β-glucan fractions from walls of the yeast *Saccharomyces cerevisiae* have this effect. So too does **chitosan**, the de-acetylated form of chitin in fungal cell walls (see Chapters 2 and 6). At present, chitosan is used on a large scale in Japan for clarifying sewage, but the source of this chitosan is crustacean shells. Fungi are an alternative, easily renewable source of this and other polymers.

Enzymes and enzymic conversions

Saprotrophic fungi and some plant-pathogenic fungi produce a range of extracellular enzymes with important commercial roles (Table 1.5). The

Table 1.5 Examples of fungal enzymes produced commercially. (From Wainwright, 1992.)

Enzyme	Fungal source	Application
α-Amylase	*Aspergillus niger, A. oryzae*	Starch conversions
Amyloglucosidase	*A. niger*	Starch syrups, dextrose foods
Pullulanase	*Aureobasidium pullulans*	Debranching of starch
Glucose aerohydrogenase	*A. niger*	Production of gluconic acid
Proteases (acid, neutral, alkaline)	*Aspergillus* spp., etc.	Breakdown of proteins (baking, brewing, etc.)
Invertase	Yeasts	Sucrose conversions
Pectinases	*Aspergillus, Rhizopus*	Clarifying fruit juices
Rennet	*Mucor* spp.	Milk coagulation
Glucose isomerase	*Mucor, Aspergillus*	High-fructose syrups
Lipases	*Mucor, Aspergillus, Penicillium*	Dairy industry, detergents
Hemicellulase	*A. niger*	Baking, gums
Glucose oxidase	*A. niger*	Food processing

pectic enzymes of fungi are used to clarify fruit juices; a fungal **amylase** is used to convert starch to maltose during bread-making, and a fungal rennet is used to coagulate milk for cheese-making. A single fungus, *Aspergillus niger*, accounts for almost 95% of the commercial production of these and other bulk enzymes from fungi, although specific strains of the fungus have been selected for particular purposes. The methanol-utilizing yeasts (*Candida lipolytica*, *Hansenula polymorpha* and *Pichia pastoris*) have potential commercial value because they produce large amounts of **alcohol oxidase**, which could be used as a bleaching agent in detergents. The wood-rotting fungus *Phanaerochaete chrysosporium* is extremely active in degrading lignin; it has the potential to be developed for delignification of agricultural wastes and by-products of the wood-pulping industry, so that the cellulose in these materials could be used as a cheap substrate for production of fuel alcohol by yeasts (see Chapter 10).

In addition to these examples of 'bulk' enzymes, fungi have many internal enzymes and enzymic pathways that can be exploited for bioconversion of compounds such as pharmaceuticals. For example, fungi are used for bioconversion of steroids, because fungal enzymes perform highly specific dehydrogenations, hydroxylations, etc., on the complex aromatic ring systems of steroids. Precursor steroids are fed to a fungus that is held at low nutrient level, either in culture or entrapped on an inert bed, so that the steroid is absorbed by the fungus, transformed and then released into the culture medium from which it can be retrieved.

Heterologous (foreign) gene products

Genetic engineering of fungi, particularly *Saccharomyces cerevisiae*, has developed to the stage where the cells can be used as factories to produce pharmaceutical products, by the introduction of foreign (heterologous) genes. Indeed, yeast was used to produce the first vaccine from a genetically engineered organism that was approved for human use. It is the vaccine against hepatitis B, produced by engineering the gene for a virus coat antigen (hepatitis B surface antigen, HBsAg) into the yeast genome. There are several advantages in using yeast to synthesize such products. For example, *S. cerevisiae* is already grown on a large

industrial scale so that companies are familiar with its cultural conditions. Also, the genetics and molecular genetics of yeast are well researched (see Chapter 8) and yeast has a well-characterized secretory system (see Chapter 2) for exporting gene products into a culture medium from which the products can be harvested and purified. Examples of heterologous gene products that have been produced experimentally from yeast include epidermal growth factor (involved in wound healing), atrial natriuretic factor (for management of hypertension), interferons (with antiviral and antitumour activity) and α-1-antitrypsin (for potential relief from emphysema). The developments in fungal molecular genetics that have made these applications possible are discussed in Chapter 8. There are, however, disadvantages in using *S. cerevisiae*. In particular, this fungus is genetically quite different from other fungi and other eukaryotes, including its use of different codons for some amino acids, so that it does not always correctly read the introduced genes. For this reason, attention has switched to some other fungi, such as the fission yeast *Schizosaccharomyces* and the filamentous fungus *Aspergillus nidulans*.

The major taxonomic groups of fungi

The taxonomy of fungi is in a state of flux. Traditionally, the major groups of fungi and the relationships between them have been based on comparative morphology and the developmental patterns of the sexual reproductive structures. Now these relationships are being reassessed by nucleic acid sequence analysis, with special emphasis on the nuclear DNA that codes for the small ribonucleic acid (RNA) subunit of the ribosome. The information is still patchy because some fungal groups have received much more attention than others, and in any case there is debate about the best way to compare the data. This subject is discussed in detail by Alexopoulos *et al.* (1996).

Meanwhile we need a workable classification for practical purposes — one that distinguishes at least the major groups, without trying to rank them in a formal hierarchy. In the past, these groups have been given various names (Table 1.6). Here we will use the names adopted by Alexopoulos *et al.* (1996) and partly following those in *The Mycota* (Wessels & Meinhardt, 1994). The

Table 1.6 Comparison of names that have been applied to the major fungal groups.

Current name	Past names	
Oomycota	Oomycetes	
Chytridiomycota	Chytridiomycetes	} Mastigomycotina, Phycomycetes
Zygomycota	Zygomycetes	
Ascomycota	Ascomycetes, Ascomycotina	
Deuteromycota	Deuteromycetes, Deuteromycotina, Fungi imperfecti	
Basidiomycota	Basidiomycetes, Basidiomycotina	

Table 1.6 Comparison of names that have been applied to the major fungal groups.

This table is provided for comparison with names that might be found in older texts.

major groups are **Chytridiomycota, Zygomycota, Ascomycota, Deuteromycota, Basidiomycota** and **Oomycota**, and these will be used as lower case, common names: **chytridiomycota, zygomycota, ascomycota, deuteromycota, basidiomycota** and **oomycota**. The wall-less slime moulds and similar organisms will be referred to as the **myxomycota**, **acrasids**, **dictyostelids** and **plasmodiophorids**.

The features of these major fungal groups are outlined below and summarized on pp. 14–27 and in Figs 1.4–1.10.

Somatic stages

To some extent, the fungal groups can be distinguished by their somatic (non-reproductive) stages. The chytridiomycota typically grow as single large cells or a rudimentary mycelium, often attached to the food source by thin, tapering rhizoids. The other fungi grow as either hyphae or yeasts, but in the case of hyphae we can distinguish those that have cross-walls (**septa,** singular septum) at fairly regular intervals (ascomycota, deuteromycota and basidiomycota) and those that typically lack septa (zygomycota and oomycota). This distinction is coupled with several behavioural and ecological features, as will be seen in later chapters. It also provides a first step towards placing any fungus in a group, because septa are seen easily with a compound microscope. The type of septum can differ between groups, although this requires the use of an electron microscope: in ascomycota and deuteromycota the septa usually have a wide central pore through which cytoplasm and even nuclei can move along the hyphae, whereas in basidiomycota the pore is usually narrower and prevents the passage of major organelles (see Chapter 2). In addition, some (but not all) basidiomycota have **clamp connections** at the septa. These are small, backwards-projecting branches that arise just in front of each septum and fuse with the hypha immediately behind the septum; they serve to regulate the nuclear distribution in the hyphae (see Chapter 4, Fig. 4.15).

There is an important functional distinction between septate and aseptate fungi, because the hyphae of septate fungi can fuse with one another by localized breakdown of the walls at points of contact. This is termed **anastomosis.** It occurs in the older regions of a fungal colony and it enables the hyphal contents to be withdrawn from parts of the colony and mobilized to other parts for the building of complex, differentiated structures such as toadstools (see Chapter 4). The somatic hyphae of aseptate fungi do not anastomose, so they can produce only small differentiated structures from the resources of individual hyphae.

Asexual reproductive stages

Asexual spores are the main dispersal agents of fungi (see Chapter 9). In chytridiomycota, zygomycota and oomycota the asexual spores are formed in a **sporangium**—a large multinucleate cell in which the cytoplasm is cleaved around individual nuclei to form the spores, termed **sporangiospores.** These spores are either non-motile (zygomycota) or they can swim by means of one (chytridiomycota) or two flagella (oomycota). The asexual spores of ascomycota, deuteromycota and basidiomycota are termed **conidia** (singular conidium). They are never formed in a sporangium and never have flagella, and they are produced in various ways (by budding, fragmentation, etc.) from modified somatic hyphae (see Chapter 4).

Sexual stages

Sexual reproduction in fungi is rather complex, so we consider only the main features at this stage. The 'lower fungi' (chytridiomycota, zygomycota, oomycota) usually have well-defined sex organs, described on pp. 16–21. The 'higher fungi' (ascomycota and basidiomycota) usually have inconspicuous sex organs, and in some cases mating is achieved by the fusion of somatic hyphae. The behaviour of deuteromycota is considered later. In any case, sexual reproduction tends to occur at the onset of unfavourable conditions for growth, and it results in the production of thick-walled spores or other structures that serve for dormant survival. This function can be even more important than the role of sexual reproduction in generating variation. In fact, some fungi are self-fertile (**homothallic**) and undergo sexual reproduction on a single colony; others are self-sterile (**heterothallic**) and require two colonies of different **mating types** for sexual reproduction, but even some of these fungi can revert to homothallism if a suitable partner is not available.

The sexual behaviour of the basidiomycota deserves special mention here. As shown in Fig. 1.10, the hyphae that develop from spores have a single nucleus in each hyphal compartment, so these hyphae are termed **monokaryons** (i.e. with one nuclear type). A monokaryon can fuse with another monokaryon of a different mating type, and all the subsequent hyphae develop with two nuclei in each compartment, one from each parent. The fungus can then grow for most of its life as a **dikaryon** (with two nuclear types). At a later stage, in response to specific environmental conditions, these dikaryotic hyphae will produce a complex fruitbody (**basidiocarp**) such as a mushroom or toadstool, and this will contain special cells termed **basidia** (singular basidium) in which the two nuclei fuse, followed by meiosis to produce four sexual spores (**basidiospores**). When the basidiospores germinate they give rise to monokaryons again. In essence, therefore, the basidiomycota have a single mating event, which involves fusion of the normal somatic hyphae, and

they delay the sexual process until the paired nuclei have been multiplied many times and a complex fruiting body can be built. Some of the ascomycota show a more limited development of this type—sexual fusion occurs late in the life cycle and leads to a localized development of dikaryotic hyphae which will produce the sexual cells (**asci**; singular ascus) in which meiosis occurs; meanwhile, some of the other hyphae develop around these dikaryotic hyphae to produce a fruitbody, termed the **ascocarp**. The sexual systems of fungi are considered in more detail in Chapter 8, where we discuss fungal genetics.

The special case of deuteromycota

The deuteromycota is an artificial grouping, used for convenience to accommodate a large number of fungi that seldom or never produce sexual stages. However, in all other respects these fungi are like the ascomycota, and DNA sequence analysis confirms that this is their natural home. We need to discuss the deuteromycota here because they include some of the most common and important fungi, such as *Penicillium*, *Aspergillus*, *Trichoderma* and *Fusarium* (see Fig. 1.9). These generic names are termed **form-genera**, because they refer literally to the form (shape) of the asexual sporing stages that we commonly see, although in recent years there has been greater emphasis on the developmental processes that lead to these forms. In any case, problems can arise when a sexual stage is discovered in one of these fungi. The rules of fungal nomenclature dictate that a name must be given to the sexual stage, and this name takes precedence over the 'asexual' name. So, at a trivial level, a single fungus can be known by more than one name; some common examples of this are given on p. 24. Fungi that reproduce by asexual means can be extremely variable because they form clonal populations, and it is of interest that some of the fungi that are most common and important in terms of their environmental and industrial roles seem largely to have abandoned sexual reproduction in their natural environments.

Slime moulds and other organisms distantly related to fungi (Fig. 1.4)

Four groups of organisms with naked, protoplasmic stages have traditionally been allied to the fungi: the plasmodial slime moulds (**myxomycota**), the cellular slime moulds comprising the acrasids (**acrasiomycota**) and dictyostelids (**dictyosteliomycota**), and the plasmodiophorids (**plasmodiophoromycota**). They have affinities with some of the protozoa (Barr, 1992). They resemble fungi in the fact that they produce walled, wind-dispersed spores, but these spores have a predominance of galactosamine polymers in the walls, unlike those of fungi.

Myxomycota

Myxomycota have a multinucleate network of protoplasm (the **plasmodium**) which engulfs bacteria and other food particles by phagocytosis. They are common on moist rotting wood and similar organic substrata where bacteria are abundant. Typically, the whole plasmodium converts to fruiting structures (sporangia) in appropriate conditions (nutrient depletion and light), and the sporangia release many haploid spores that are wind-dispersed. The spores germinate to produce either **myxamoebae** or flagellate swarmers, which fuse in pairs, and the resulting diploid cell grows into a plasmodium. *Physarum polycephalum* has been studied intensively by morphogeneticists because it can be maintained on defined (bacterium-free) media. Details of the myxomycota can be found in Martin and Alexopoulos (1969).

Acrasids and dictyostelids

Acrasids and dictyostelids are biologically similar to one another but clearly separable by nucleotide sequence comparisons. They are unicellular amoeboid organisms that engulf bacteria and other food particles by phagocytosis. They are common in moist organic-rich soil, leaf litter, animal dung, etc. Details of the group are given by Raper (1984). *Dictyostelium discoideum* has been studied intensively as a model of primitive differentiation systems and of chemotaxis (Armitage & Lackie, 1990). At the onset of starvation, the myxamoebae aggregate in response to pulses of cyclic adenosine monophosphate (cAMP) released from the cells, and form a multicellular mass termed the **grex**. This migrates and then converts to a fruit-body, the **sorocarp**, which consists of a cellulosic stalk and a globular head containing walled spores. After dispersal, the spores germinate to produce myxamoebae.

Plasmodiophorids

Plasmodiophorids grow as obligate intracellular parasites of plants, algae or fungi; they cannot be grown in culture in the absence of a host. The most important organism in this group is *Plasmodiophora brassicae*, which causes the clubroot disease of cruciferous crops (cabbage, etc.) discussed in Chapter 12. A similar organism, *Spongospora subterranea* causes powdery scab of potato tubers. However, the most common are *Polymyxa* spp., which are found as symptomless parasites in the roots of many plants and are vectors of some important plant viruses (see Chapter 9). All these organisms have thick-walled resting spores which persist for many years in soil. The resting spores germinate to release a zoospore which then locates a host by chemotaxis, encysts on the host surface and injects the protoplast into the host. The protoplast of *P. brassicae* develops into a small primary plasmodium in a root hair. Zoospores are released from the primary plasmodia and fuse to form diploids which infect the root cortical cells and develop into large secondary plasmodia. Meiosis is thought to occur in these plasmodia, producing resting spores which enter the soil when the root decays. Details of the group as a whole can be found in Karling (1968) and Buczacki (1983).

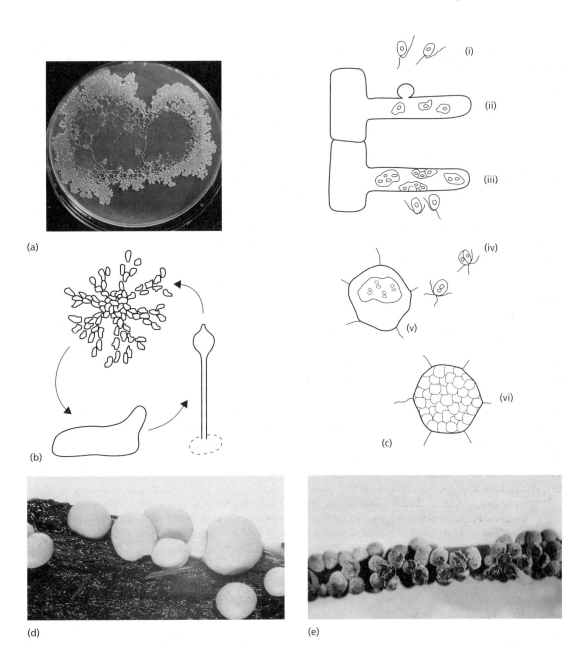

Fig. 1.4 Slime moulds and other wall-less organisms. (a) A naked plasmodium of *Physarum polycephalum* (myxomycota) growing on an agar plate. (b) *Dictyostelium discoideum* (cellular slime mould), showing aggregation of myxamoebae to form a multicellular grex which then converts into a stalked fruitbody. Spores released from the fruitbody germinate to release myxamoebae. (c) Proposed stages in the infection cycle of *Plasmodiophora brassicae* (plasmodiophorid): (i), uninucleate zoospores; (ii), zoospores encyst on a root hair and release a protoplast into the host cell; (iii), the protoplasts grow to produce small primary plasmodia, which convert into sporangia and release zoospores; (iv) the zoospores fuse in pairs; (v), the fused zoospores infect root cortical cells and develop into large secondary plasmodia; (vi), at maturity the secondary plasmodia convert into resting spores which are released into soil. The resting spores germinate eventually to release haploid zoospores which repeat the cycle. (d) Immature sporangia of a slime mould (myxomycota) on rotting wood; each is about 1 cm in diameter. (e) Mature sporangia of the slime mould *Physarum cinereum* on a grass blade; some of the sporangia have opened to reveal a mass of darker spores.

Oomycota—the cellulose-walled fungi
(Fig. 1.5)

Examples: *Phytophthora, Pythium, Saprolegnia.*

Main distinguishing features

Somatic stage

Typically a mycelium lacking septa; diploid; with a wall of cellulose and other glucans (polymers of glucose) and many cellular features different from other fungi (see Table 1.1).

Asexual reproduction

By diploid **zoospores** with an anterior tinsel-type flagellum and a posterior whiplash flagellum, formed by cytoplasmic cleavage in a **sporangium**. In some members (e.g. downy mildew pathogens) the sporangia are wind-dispersed and germinate by a hypha rather than zoospores.

Sexual reproduction

By the production of a male sex organ (**antheridium**) and a female sex organ (**oogonium**). Nuclei pass from the male to female organ through a fertilization tube, then fuse in pairs to form diploid nuclei. One or more thick-walled **oospores** then develop in the oogonium. They serve in dormant survival then germinate to produce either diploid hyphae or a sporangium that releases diploid zoospores.

Ecology and significance

The oomycota contains at least four distinctive subgroups. The **Saprolegniales** (water moulds) includes species of *Achlya* and *Saprolegnia* that are common saprotrophs in freshwater habitats, but *S. diclina* and *S. parasitica* are important pathogens of salmonid fish (Stuart & Fuller, 1968). The closely related *Aphanomyces* spp. are aggressive root pathogens (e.g. *A. euteiches* on pea and beans, *A. cochlioides* on spinach), but *A. astaci* is a pathogen of crayfish and has virtually eliminated the native crayfish from much of Europe. The **Leptomitales** is a small group of aquatic fungi, such as *Leptomitus lacteus* which is common in sewage-polluted

waters. This group has several peculiar features, such as fermentative rather than respiratory metabolism (see Chapter 6); they need sulphur-containing amino acids because they cannot use inorganic sulphur sources, and they have a mixture of chitin and cellulose in the cell walls. The **Lagenidales** is a small group of common symptomless parasites of plant roots (e.g. *Lagena radicicola*; Macfarlane, 1970) or parasites of algae, fungi or invertebrates (e.g. *Lagenidium giganteum* on nematodes, mosquito larvae, etc.). Often they cannot be grown in culture in the absence of a host. The **Peronosporales** is the most important group, because it contains many serious plant pathogens. *Pythium* spp. cause seedling diseases and attack the feeder roots of almost all crop plants (see Chapter 12), but a few *Pythium* spp. parasitize other fungi and have potential for biocontrol of plant diseases (see Chapter 11). *Phytophthora* spp. also are plant pathogens, with notable examples such as *P. infestans* which causes potato blight and *P. cinnamomi* which causes root rot of pines, eucalypts, fruit trees and many other plants. Also in this group are the **downy mildew** pathogens (e.g. *Bremia lactucae* on lettuce, *Plasmopara viticola* on grapevines) which are obligate biotrophic parasites (see Chapter 12). Many of them produce wind-dispersed sporangia that can germinate by a hypha to infect the host leaves, rather than by producing zoospores. They are the most 'terrestrial' of the oomycota.

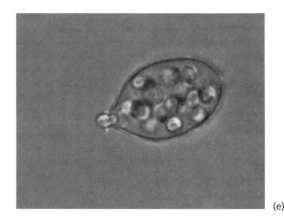

(e)

Fig. 1.5 *Continued.* (e) Zoospores in a detached sporangium of *Phytophthora parasitica*, showing one zoospore escaping through the dissolved tip of the sporangium.

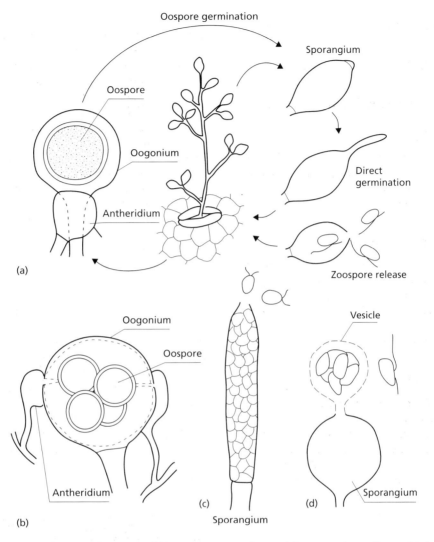

Fig. 1.5 Characteristic features of oomycota. (a) *Phytophthora infestans* (potato blight), showing stages in the infection cycle. When strains of opposite mating type meet in the host tissues they produce antheridia and oogonia; fertilization leads to the production of a thick-walled, resting oospore. The oospore germinates to produce a hyphal outgrowth which produces lemon-shaped sporangia. The sporangia are wind-dispersed and land on host leaves where they germinate either by producing a germ-tube (direct germination) or by releasing laterally biflagellate zoospores which initiate infection. Once established in the host tissues, the fungus produces sporangiophores through the stomatal openings and the sporangia are dispersed to infect other leaves. (b) Sexual reproduction by *Saprolegnia*; several oospores are formed in each oogonium. (c) Asexual reproduction in *Saprolegnia*. The sporangium develops as a terminal swelling on a hypha; at maturity it releases primary zoospores (anteriorly biflagellate) which encyst rapidly and the cysts then release secondary zoospores (laterally biflagellate) which are dispersed to new sites. (d) Asexual reproduction in *Pythium*, where a sporangium discharges its contents into a thin membranous vesicle; the laterally biflagellate zoospores are cleaved in this and then released for dispersal.

Chytridiomycota (Fig. 1.6)

Examples: *Allomyces, Olpidium, Rhizophlyctis.*

Main distinguishing features

Somatic stage

Typically unicellular or dichotomously branched chains of cells with tapering rhizoids for anchorage or absorption. Usually haploid, but some (e.g. *Allomyces*) alternate between haploid and diploid phases. Wall composed of chitin and glucans.

Asexual reproduction

By zoospores with a posterior whiplash flagellum, formed by cytoplasmic cleavage in a sporangium. In some cases the whole body (thallus) converts into a sporangium (e.g. *Rhizophlyctis*), in others the sporangia form on part of the thallus (e.g. *Allomyces*).

Sexual reproduction

Usually by fusion of motile male and female gametes, but sometimes only the male gamete is motile. The fusion product (zygote) can be a resting spore; alternatively (e.g. *Allomyces*) it grows into a diploid somatic generation that later produces resting sporangia (Fig. 1.6). Details are unknown for most of the group.

Ecology and significance

The chytridiomycota are divided into the predominantly unicellular types (Chytridiales) and the blastocladians (Blastocladiales) with usually branched chains of cells attached to a substrate by rhizoids (e.g. *Allomyces*). Many are saprotrophs in freshwater environments or wet soils, and they depend on zoospores for locating suitable substrata by chemotaxis (see Chapter 9). They accumulate on 'baits' such as pollen grains, insect exoskeletons, etc., in these environments. Some species have marked enzymatic activities: *Rhi-zophlyctis rosea* is strongly cellulolytic in soil, and *Chytriomyces hyalinus* strongly degrades chitin. *Olpidium* spp. are common symptomless parasites of plant roots, and can be vectors of plant viruses (see Chapter 9). A few species are serious plant pathogens, the most notable being *Synchytrium endobioticum* which causes potato wart disease. Several species parasitize algae or other fungi; e.g. *Rozella allomycis* is an obligate parasite of *Allomyces. Catenaria anguillulae* parasitizes nematodes and fungal resting spores in soil (see Plate 13.2, opposite p. 210), and the *Coelomomyces* spp. parasitize mosquito larvae and are of interest as biocontrol agents (see Chapter 13). A few chytrid-like organisms are obligately anaerobic inhabitants of the rumen, where they contribute to the digestion of herbage (see Chapter 7); these species are also unusual in having several flagella on their zoospores. Overall, the chytridiomycota is an inconspicuous and little-known group of fungi, but it has many diverse and important roles in natural habitats. Details of the group are given in Fuller and Jaworski (1987).

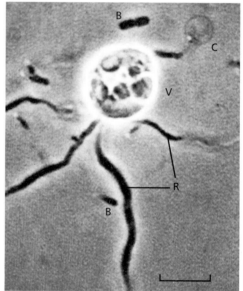

Fig. 1.6 *Continued.*

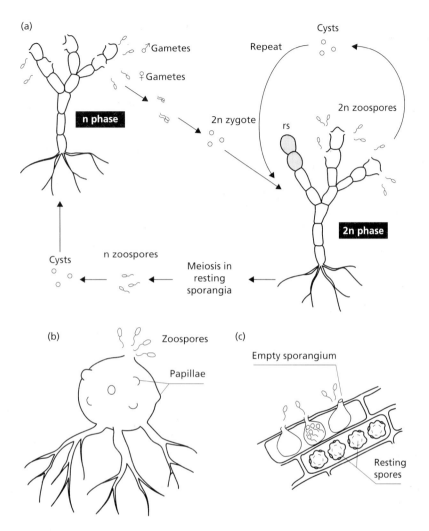

Fig. 1.6 Characteristic features of Chytridiomycota.
(a) Life cycle of *Allomyces*. The haploid somatic thallus
produces male and female gametangia that release
motile gametes. These fuse in pairs then encyst to form
diploid zygotes which germinate to produce a diploid
thallus. The diploid thallus produces sporangia which
release diploid zoospores to repeat the diploid phase. In
adverse conditions it produces thick-walled resting
sporangia (rs) which eventually germinate to release
haploid zoospores which give rise to haploid thalli.
(b) *Rhizophlyctis rosea*, a species that has a single large
somatic cell with rhizoids; the whole cell content is
cleaved to form zoospores which are released through

papillae. (c) *Olpidium brassicae* in a cabbage root. The
fungus grows as a naked protoplast in the root cells. At
maturity, the protoplasts convert into sporangia that
produce exit tubes through the host cell wall and release
zoospores; these spores encyst on a host root and the cyst
germinates to release a protoplast into the host. In
unfavourable conditions the thallus can convert into
thick-walled resting spores. (d, *facing page*) *Catenaria
anguillulae* zoospore cyst (C) which germinated to
produce a large lipid-filled vesicle (V), then rhizoids (R)
developed from this. Cells of the bacterium *Bacillus* (B)
are also seen. Bar = 10 µm.

Zygomycota (Fig. 1.7)

Examples: *Mucor, Pilobolus, Erynia, Piptocephalis.*

Main distinguishing features

Somatic stage

Typically mycelial and non-septate, but septa are present in some subgroups; some species have both hyphal and yeast forms (dimorphic); some grow as protoplasts in insect hosts; haploid; wall of chitin, chitosan and polyglucosamine (see Chapter 2).

Asexual reproduction

Asexual reproduction is typically by non-motile spores formed in a **sporangium**, mounted on a **sporangiophore**. The sporangium is often large and globose (e.g. *Mucor*), with a wall that breaks down at maturity to release dry (wind-dispersed) or wet spores. Some genera (e.g. *Piptocephalis*) produce narrow sporangia with linearly arranged spores (**merosporangia**); some (e.g. *Thamnidium*) produce rounded, few-spored sporangia (**sporangioles**) as well as normal sporangia. In one group the sporangium is not cleaved into spores but is released as a single spore.

Sexual reproduction

By fusion of sex organs (**gametangia**) formed at the tips of specialized aerial hyphae (**zygophores**), under the influence of volatile pheromones (see Chapter 4). Fusion of gametangia leads to the production of a thick-walled, resting spore, the **zygospore**. This germinates after meiosis to form a hypha or a sporangium.

Ecology and significance

The best-known subgroup of zygomycota is the **Mucorales** (e.g. *Mucor*) which contains many common saprotrophs of soil and animal dung. Typically, these fungi exploit simple, soluble nutrients, but some *Mortierella* spp. degrade chitin, and *Rhizopus stolonifer* is important in rotting of soft fruits such as strawberries. A few members (e.g. *Dimargaris* and *Dispira* spp.) are biotrophic para-

sites of other zygomycota. Another important group, the **Glomales** contains the six genera of arbuscular mycorrhizal fungi (*Glomus, Acaulospora*, etc.) that occur on roots of plants worldwide and facilitate phosphorus uptake from soil (see Chapter 12). These fungi show primitive features, little changed since they evolved 350–460 million years ago. None of them can be grown in pure culture, but the related *Endogone* spp. (which form ectomycorrhizas with trees) can be grown in culture. Another group, the **Entomophthorales** includes many insect pathogens (e.g. *Entomophthora muscae* on house-fly; *Erynia neoaphidis* on aphids) of potential use in biocontrol (see Chapter 13); they seem to grow as naked protoplasts in the insect host. Other species of this group occur in the gut and faeces of amphibians and reptiles (e.g. *Basidiobolus ranarum*; see Chapter 3). *Basidiobolus* and the related *Conidiobolus coronatus* can infect humans and other animals, typically through nasal lesions. The *Entomopthorales* as a whole is characterized by sporangia that are dispersed and function as single spores. The **Zoopagales** is a small group that produces merosporangia; it includes some biotrophic parasites of other zygomycota (e.g. *Piptocephalis*; see Chapter 11) and some parasites of nematodes (*Cystopage, Stylopage*; see Chapter 13) and other small animals. The **Trichomycetes** are a peculiar, morphologically distinct group of fungi that attach to the gut wall of aquatic arthropods or the aquatic larval stages of terrestrial insects, and probably feed on the gut contents.

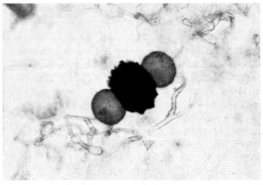

(e)

Fig. 1.7 *Continued.* (e) *Rhizopus sexualis* zygospore (about 100 μm) with suspensors.

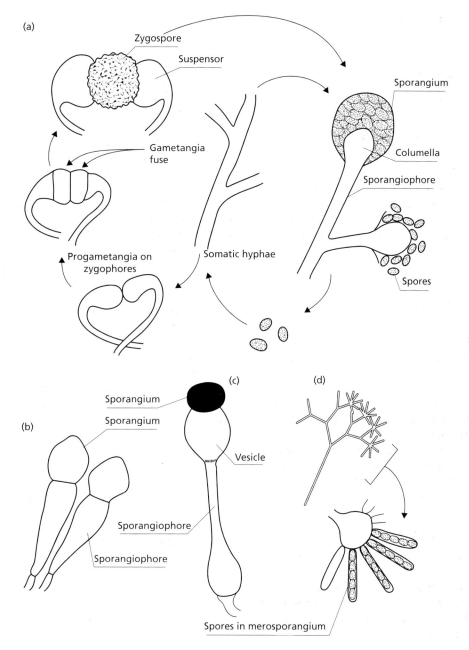

Fig. 1.7 Characteristic features of Zygomycota. (a) Life cycle of *Mucor*. Asexual reproduction occurs by the production of sporangiophores with sporangia at their tips. The sporangium can have a central columella and spores are released by breakdown of the sporangium wall. They germinate to form new somatic hyphae. Sexual reproduction occurs when colonies of opposite mating types produce aerial branches (zygophores) which grow towards each other and produce progametangia at their tips. Septation in the progametangia leads to the production of terminal gametangia which fuse to form a thick-walled resting

spore, the zygospore. The subterminal parts of the progametangia can swell to produce suspensors. Meiosis occurs in the zygospore and it germinates to produce a sporangiophore or a hypha. (b) The sporangia of the insect-pathogenic Entomophthorales (e.g. *Entomophthora*, *Erynia*) are released as single cells and function as dispersal spores. (c) *Pilobolus*, in which the sporangium is borne on a swollen vesicle which ruptures to shoot the sporangium free during dispersal (see Chapter 9). (d) *Piptocephalis*, showing a branched sporangiophore at the tips of which are many merosporangia, each containing a few spores.

21

Ascomycota (Fig. 1.8)

Examples: *Neurospora, Eurotium, Ascobolus, Saccharomyces.*

Main distinguishing features

Somatic stage

Usually a mycelium with simple septa, sometimes yeasts; haploid, although some yeasts have alternating haploid and diploid phases; wall of chitin with glucans, but the yeasts have glucans and mannans (see Chapter 2).

Asexual reproduction

Non-motile spores (conidia) formed in various ways but never by cleavage in a sporangium; described under deuteromycota (see pp. 24–25), which are the 'asexual' stages of ascomycota.

Sexual reproduction

By fusion of somatic hyphae or yeasts, or of a 'male' spore (a **spermatium** or normal conidium) with a female receptive hypha (**trichogyne**). This leads to production of one or more **asci** in which nuclei fuse to form a diploid nucleus which undergoes meiosis; a further mitotic division typically leads to the production of eight ascospores in the ascus. The asci can be naked (e.g. yeasts) or enclosed in a fruiting body (**ascocarp**) such as a closed **cleistothecium** (e.g. *Eurotium*), a flask-shaped **perithecium** (e.g. *Neurospora*), a disc- or cup-shaped **apothecium** (e.g. *Ascobolus*) or a hyphal tissue, the **pseudothecium**, which morphologically resembles a perithecium.

Ecology and significance

The ascomycota is a large and important group, divided into three sub-groups. The **archiascomycetes** include the fission yeasts (*Schizosaccharomyces* spp.) which are found in sugar-rich environments and have become important model organisms in studies of the cell cycle (see Chapter 3). This group also includes the plant pathogen *Taphrina deformans*, which causes leaf-curl of peach and apricot trees, and *Pneumocystis carinii* which causes a virulent pneumonia in AIDS patients (see Chapter 13). The **ascomycetous yeasts** include *Saccharomyces* and *Pichia* spp., which are common on fruit surfaces. *S. cerevisiae* is used in bread-making and for alcoholic drinks (although the beer yeast is a separate species, called *S. carlsbergensis* or *S. pastorianus*). Molecular genetical methods for *Saccharomyces* and *Schizosaccharomyces* are highly advanced, and these fungi can now be used to produce many foreign proteins (see Chapter 8). The **mycelial ascomycota** are common saprotrophs that degrade cellulose and other structural polymers; examples include *Chaetomium* in soil and in composts, *Xylaria* and *Hypoxylon* which degrade wood, *Sordaria* and *Ascobolus* which are common on the dung of herbivores, and *Lulworthia* on wood in estuarine environments. Some ascomycota form mycorrhizas with forest trees (e.g. *Tuber* spp., the truffles); some parasitize humans (e.g. the dermatophytes *Arthroderma* spp.; see Chapter 13). Many are important plant pathogens, including *Ophiostoma novo-ulmi* (Dutch elm disease; see Chapter 9), *Cryphonectria parasitica* (chestnut blight; see Chapter 8), *Claviceps purpurea* (ergot of cereals) and the powdery mildew pathogens, such as *Erysiphe graminis* on cereals and *Sphaerotheca pannosa* on roses (see Chapter 12). Ascomycota are the fungal partners in an estimated 96% of lichens; about 14 000 fungi are involved in these cases and very few of these have been found growing in a free-living condition.

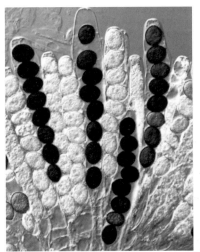

(e)

Fig. 1.8 *Continued.*

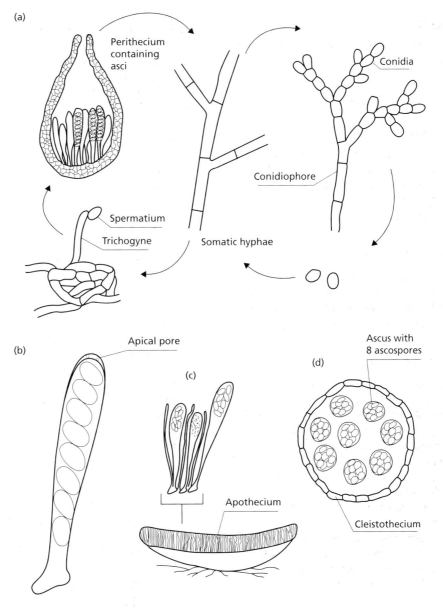

Fig. 1.8 Characteristic features of Ascomycota. (a) Life cycle of *Neurospora crassa*. The somatic hyphae produce conidiophores with branched chains of conidia. After dispersal these germinate to form new somatic hyphae. Sexual reproduction occurs by fusion of a 'male' spore (spermatium) with a female receptive hypha, the trichogyne. Then a perithecium develops around the fertilized hyphae, which produce asci. Nuclear fusion in each ascus is followed by meiosis, then mitosis to produce eight haploid ascospores. At maturity the asci elongate up the perithecium neck and release the ascospores, which germinate to form somatic hyphae. (b) Detail of ascus structure, showing the row of eight ascospores and the apical pore through which the ascospores are shot when mature. (c) *Ascobolus*, an example of a fungus with a cup-shaped ascocarp, termed an apothecium, in which the asci are packed together with sterile hairs (paraphyses). (d) *Eurotium*, the sexual stage of some *Aspergillus* spp., which contains asci within a closed ascocarp, the cleistothecium. The ascospores are released when the ascus and cleistothecium walls break down. (e, *facing page*) Asci containing ascospores (about 20 μm) in a crushed perithecium of *Sordaria macrospora*. (Courtesy of N. Read, from Read N.D. & Lord K.M. (1991) *Experimental Mycology*, **15**, 132–9.)

Deuteromycota (Fig. 1.9)

Examples: *Alternaria, Aspergillus, Aureobasidium, Cladosporium, Geotrichum, Humicola, Penicillium, Gloeosporium, Pesotum, Phomopsis.*

Main distinguishing features

Somatic stage

As for the ascomycota (see pp. 22–23).

Asexual reproduction

By conidia formed in various ways. For example, by budding (*Aureobasidium, Cladosporium*), ballooning of a hyphal tip (e.g. *Humicola*), hyphal fragmentation (e.g. *Geotrichum*), extrusion from a flask-shaped cell termed a **phialide** (*Aspergillus, Penicillium*), and so on. The spore-bearing hyphae (**conidiophores**) may be simple, branched or aggregated into a stalk termed a synnema or **coremium** (e.g. *Pesotum*); also, they may arise from a pad of tissue (an **acervulus** as in *Gloeosporium*) or be enclosed in a flask-shaped **pycnidium** (e.g. *Phomopsis*).

Sexual reproduction

Absent, rare or unknown; where present, it is typical of ascomycota and is given a generic name, termed a **teleomorph**. However, often the name of the asexual stage (**anamorph**) is retained for common usage. Table 1.7 shows some common examples.

Ecology and significance

The deuteromycota is an artificial grouping, based on the absence or rarity of sexual stages. It is commonly divided into the yeast-like **blastomycetes**, the **hyphomycetes** which have simple, hypha-like conidiophores, and the **coelomycetes** which produce conidia on an acervulus or in a pycnidium. All are common, and important in different ways. The yeasts include *Candida albicans* (diploid), which is a common commensal and

Table 1.7 The generic names for some of the sexual stages (known as teleomorphs) and asexual stages (anamorphs) of fungi that are frequently seen only in the asexual form.

Anamorph	Teleomorph
Aspergillus	Eurotium Emericella
Penicillium	Eupenicillium Talaromyces
Trichoderma	Hypocrea
Fusarium	Gibberella Nectria
Microsporum Trichophyton	Arthroderma
Blastomyces Histoplasma	Ajellomyces
Rhizoctonia	Ceratobasidium Thanatephorus Waitea

occasional pathogen of humans (see Chapter 13). The yeast-like *Geotrichum candidum* is a common spoilage fungus of dairy products; in culture it grows as hyphae which readily fragment into cells behind the colony margin. The hyphomycetes include many common saprotrophs of soil (e.g. *Trichoderma, Penicillium, Gliocladium, Fusarium*) or plant surfaces (*Cladosporium, Alternaria* and *Aureobasidium*). Also in this group are many spoilage fungi of cereal grains and other foodstuffs (*Penicillium, Aspergillus* and *Fusarium* spp.) in which they produce mycotoxins (see Chapter 6). Several are major plant pathogens, including the vascular wilt fungi *Verticillium dahliae* and *Fusarium oxysporum* (see Chapter 12) and the *Fusarium* spp. that cause root rots and foot rots; others are parasites of insects (e.g. *Metarhizium, Beauveria*) or nematodes (e.g. *Arthrobotrys*), discussed in Chapter 13. The coelomycetes include several seed-borne and leaf-spot pathogens such as *Colletotrichum, Phoma* and *Ascochyta* spp. Industrially, the deuteromycota are widely used for production of antibiotics (penicillin, griseofulvin), enzymes (especially from *Aspergillus niger*) and many other compounds (see Tables 1.4 & 1.5, p. 10).

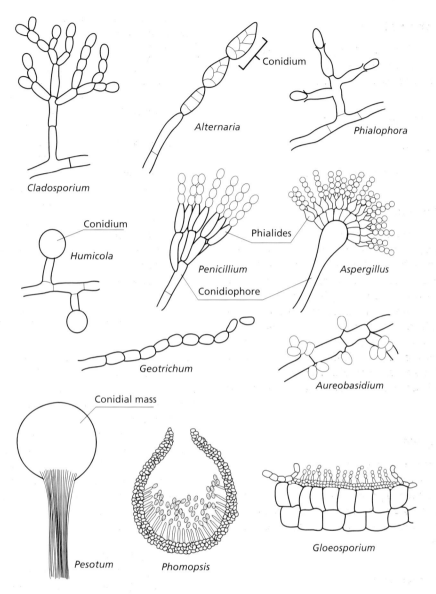

Fig. 1.9 Characteristic features of Deuteromycota. Some representative methods of conidium production are shown: by a budding process at the ends of conidiophores in *Cladosporium* or near the septa of hyphae of *Aureobasidium*; from a small apical 'pore' in *Alternaria*; by extrusion from flask-shaped phialides in *Phialophora*, *Aspergillus* and *Penicillium*; by hyphal fragmentation in *Geotrichum*; and by blowing-out of hyphal tips in *Humicola*. The conidiophore of *Penicillium* is branched in a brush-like manner, with phialides at the branch tips, whereas in *Aspergillus* the conidiophore swells into a terminal vesicle that bears the phialides. In *Pesotum* a large head of conidia is produced in a mucilaginous mass at the tips of aggregated conidiophores (a coremium). In *Phomopsis* the conidia are released in a mucilaginous mass which oozes from the neck of a flask-shaped pycnidium. In *Gloeosporium* (growing within fruit or leaf tissues) the host cuticle is ruptured and the conidia develop on conidiophores that are produced from a stromatic mass of hyphae (acervulus).

Basidiomycota (Fig. 1.10)

Examples: *Agaricus, Ganoderma, Lycoperdon, Puccinia.*

Main distinguishing features

Somatic stage

Typically mycelial with a complex **dolipore septum** (see Chapter 2) and sometimes with **clamp connections** (see Fig. 4.15, p. 86); some species grow as yeasts. Wall of chitin and glucans, but chitin and mannans in the yeasts; haploid but the main somatic stage often has paired nuclei and is termed a **dikaryon**.

Asexual reproduction

By conidia, but uncommon in the toadstool- and bracket-producers. The rusts can form two types of asexual spore, termed uredospores and aeciospores (Fig. 1.10).

Sexual reproduction

Fusion of somatic hyphae of two compatible strains termed **monokaryons** (with one nuclear type) is followed by nuclear division and migration so that each hyphal compartment contains a pair of nuclei of different types (a **dikaryon**). The fungus can grow for most of its life in this state, but in response to environmental triggers it forms a fruitbody (**basidiocarp**, e.g. a toadstool). Basidia develop on or in this; then the pair of nuclei fuse in each basidium to form a diploid nucleus. Meiosis follows and the four haploid daughter nuclei migrate into basidiospores which develop from the basidia. The rust fungi differ from this in having sex organs and sometimes several spore stages, shown in Fig. 1.10.

Ecology and significance

The basidiomycota contains several subgroups but the distinctions between them are not always clear. One clear group, the **rust fungi** (Uredinales), contains some of the most economically important pathogens of crop plants (e.g. *Puccinia graminis* on wheat, *Uromyces appendiculatus* on bean), dis-

cussed in Chapter 12. Another important group of pathogens, the **smut fungi** (Ustilaginales) are unusual because they grow only as yeasts (monokaryons) in culture but as mycelia (dikaryons) in their hosts. Other yeast forms, but of different affinity, include the pink yeast *Sporobolomyces roseus*, which is a common saprotroph on senescent plant leaves, and *Coccidioides immitis* which is a significant pathogen of immunocompromised people (see Chapter 13). Many of the basidiomycota produce large fruiting bodies such as puffballs (e.g. *Lycoperdon*), toadstools (e.g. *Agaricus*), woody brackets (e.g. *Ganoderma*) or flat crusts on the soil surface (e.g. *Rhizoctonia*) or rotting wood (e.g. *Stereum*). Most of these 'higher' basidiomycota are saprotrophs that degrade polymers such as cellulose, hemicelluloses and lignin. They are found in composts (e.g. *Coprinus*), leaf-litter (e.g. *Mycena*), the thatch of dead leaf sheaths in old grasslands (e.g. *Marasmius oreades*, which produces fairy rings) and in wood where they cause major decay (e.g. *Serpula lacrymans*, dry rot), discussed in Chapter 10. Some basidiomycota cause major tree diseases; e.g. *Armillaria mellea* in hardwood stands, *Heterobasidion annosum* in conifers (see Chapter 11), *Rigidoporus lignosus* in rubber plantations, *Poria* root rot of tea plantations. Yet, other species form mycorrhizas with forest trees, examples being *Amanita, Russula, Cortinarius, Boletus*, etc. (see Chapter 12). A few basidiomycota (*Agaricus bisporus, Volvariella volvacea, Lentinula edodes*) are grown commercially as mushroom crops; others are likely to be commercialized as fermenter cultures for production of flavour and odour components.

(e)

Fig 1.10 *Continued.*

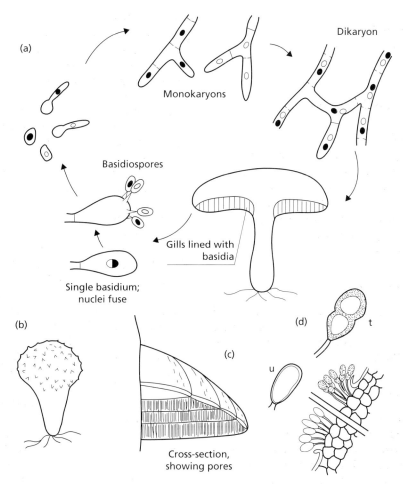

Fig. 1.10 Characteristic features of Basidiomycota. (a) Life cycle of *Agaricus*. Hyphae of two monokaryons fuse to form a dikaryon, with one nucleus of each parental type in each hyphal compartment. The dikaryon ultimately produces a fruitbody (toadstool), the gills or pores of which are lined with basidia. The nuclei fuse in each basidium, then the diploid nucleus undergoes meiosis and the four haploid nuclei enter the developing basidiospores. Basidiospores germinate to produce somatic hyphae (monokaryons). (b) Puffball fruitbody of *Lycoperdon*; the basidiospores are released inside this and dry to a powdery mass; they are dispersed through a pore at the tip of the puffball (see Chapter 9). (c) Large, woody fruitbody of *Ganoderma* on a tree trunk; this fruitbody forms annual zones of pores, lined with basidia. (d) Sporing pustules of a rust fungus, *Puccinia graminis*, on a cereal stem. Early in the season the fungus produces dikaryotic **uredospores** (u) for dispersal; later it produces thick-walled dikaryotic **teliospores** (t) for dormant survival. During germination of the teliospores the nuclei fuse and the diploid nucleus undergoes meiosis; a short hypha (probasidium) is produced and it gives rise to four haploid basidiospores. These are dispersed and infect the leaves of barberry bushes (*Berberis* spp.) in which they produce monokaryotic hyphae. Sex organs are produced on barberry, leading to the fusion of two monokaryons, then dikaryotic spores (**aeciospores**) are produced and these can reinfect a cereal. Thus, *P. graminis* is a rust fungus with an obligatory alternation of hosts (wheat and barberry). (e, *facing page*) *Puccinia graminis* on a barberry leaf; transverse section of an aecial cup containing aeciospores.

References

Alexopoulos, C.J., Mims, C.W. & Blackwell, M. (1996) *Introductory Mycology*, 4th edn. John Wiley, New York.

Armitage, J.P. & Lackie, J.M. (eds) (1990) *Biology of the Chemotactic Response*. Society for General Microbiology Symposium 46. Cambridge University Press, Cambridge.

Barr, D.S. (1992) Evolution and kingdoms of organisms from the perspective of a mycologist. *Mycologia*, **84**, 1–11.

Buczacki, S.T. (1983) *Zoosporic Plant Pathogens: a Modern Perspective*. Academic Press, London & New York.

Ekundayo, J.A. (1980) An appraisal of advances in biotechnology in Central Africa. In: *Fungal Biotechnology* (eds J.E. Smith, D.R. Berry & B. Kristiansen), pp. 243–71. Academic Press, London.

Fuller, M.F. & Jaworski, A. (eds) (1987) *Zoosporic Fungi in Teaching and Research*. Southeastern Publishing Corporation, Athens, GA.

Gallois, A., Gross, B., Langlois, D., Spinnler, H-E. & Brunerie, P. (1990) Influence of culture conditions on flavour compounds of 29 lignolytic basidiomycota. *Mycological Research*, **94**, 494–504.

Hadwiger, L.A., Chiang, C., Victory, S. & Horovitz, D. (1988) The molecular biology of chitosan in plant–pathogen interactions and its application to agriculture. In: *Chitin and Chitosan: Sources, Chemistry, Biochemistry, Physical Properties and Applications* (eds G. Skhak, B.T. Anthonsen & P. Sandford), pp. 119–38. Elsevier Applied Sciences, Amsterdam.

Jong, S-C. & Donovick, R. (1989) Antitumor and antiviral substances from fungi. *Advances in Applied Microbiology*, **34**, 183–262.

Karling, J.S. (1968) *The Plasmodiophorales*. Hafner Publishing Co., New York.

Lewis, D.H. (1987) Evolutionary aspects of mutualistic associations between fungi and photosynthetic organisms. In: *Evolutionary Biology of the Fungi* (eds A.D.M. Rayner, C.M. Brasier & D. Moore), pp. 161–78. Cambridge University Press, Cambridge.

Macfarlane, I. (1970) *Lagena radicicola* and *Rhizophydium graminis*, two common and neglected fungi. *Transactions of the British Mycological Society*, **55**, 113–16.

Martin, G.W. & Alexopoulos, C.J. (1969) *The Myxomycota*. Iowa University Press, Iowa City.

Raper, K.B. (1984) *The Dictyostelids*. Princeton University Press, Princeton, NJ.

Stuart, M.R. & Fuller, H.T. (1968) Mycological aspects of diseased Atlantic salmon. *Nature*, **217**, 90–2.

Trinci, A.P.J. (1992) Myco-protein: a twenty-year overnight success story. *Mycological Research*, **96**, 1–13.

Turner, W.B. (1971) *Fungal Metabolites*. Academic Press, London.

Turner, W.B. & Aldridge, D.C. (1983) *Fungal Metabolites. II*. Academic Press, London.

Wainwright, M. (1992) *An Introduction to Fungal Biotechnology*. Wiley, Chichester.

Wessels, J.G.H. & Meinhardt, F. (eds) (1994) *The Mycota. I. Growth, Differentiation and Sexuality*. Springer-Verlag, Berlin. (Six further volumes will appear in the series, *The Mycota*.)

Whittaker, R.H. (1969) New concepts in kingdoms of organisms. *Science*, **163**, 150–60.

Woese, C.R. (1994) There must be a prokaryote somewhere: microbiology's search for itself. *Microbiological Reviews*, **58**, 1–9.

Chapter 2
Structure and ultrastructure

Fungi have a unique structure and organization, central to their mechanism of growth and to many of their activities. In this chapter we begin with an overview of fungal structure and then consider some of the more interesting and important features in more detail. A lot of the material we will cover has come from recent advances in electron microscopy and in the use of specific cytochemical techniques that enable subcellular components to be visualized even in living fungal hyphae.

General structure: the hypha

The hypha is essentially a tube with a rigid wall, containing a moving slug of protoplasm (Fig. 2.1). It is of indeterminate length but often has a fairly constant diameter, ranging from 1 to 2 µm to 30 µm or more (usually 5–10 µm), depending on the species and growth conditions. Hyphae grow only at their tips, where there is a tapered region termed the **extension zone**; this can be up to 30 µm long in the fastest-growing hyphae such as those of *Neurospora crassa* which extend at up to 40 µm min^{-1}. Behind the growing tip, the hypha ages progressively and in the oldest regions it may break down by autolysis or be broken down by other organisms (heterolysis). While the tip is growing, the protoplasm moves continuously from the older regions of the hypha to supply the tip with materials for growth. So, a fungal hypha continuously extends at one end and continuously ages at the other end, drawing the protoplasm forward as it grows.

The hyphae of ascomycota, deuteromycota and basidiomycota have cross-walls (septa) at fairly regular intervals, whereas septa are absent from hyphae of most oomycota and zygomycota, except where they occur as complete walls to isolate old or reproductive regions. Nevertheless, the functional distinction between septate and aseptate fungi is not as great as might be thought, because septa have pores through which the cytoplasm and even the nuclei can migrate towards the growing hyphal tip. Strictly speaking, therefore, septate hyphae do not consist of cells but of interconnected compartments.

The hyphal wall has a complex organization, described later. It is thin at the apex (about 50 nm) but thickens behind this to about 125 nm at 250 µm behind the tip of *Neurospora crassa*. The plasma membrane lies close to the wall and seems to be firmly attached to it in some regions because hyphae can be difficult to plasmolyse.

Hyphae contain a range of membrane-bound organelles such as nuclei, mitochondria, vacuoles, endoplasmic reticulum, a Golgi body or its equivalent and lipid bodies. In this respect fungi are like eukaryotes in general, although some aspects of fungal ultrastructure are different from those of other organisms. For example, the mitochondria of most fungi have cristae that are plate-like and resemble those of most animals, unlike the tubular cristae of plants; only the oomycota and some smaller related groups have tubular cristae. Also, most fungi do not have a typical Golgi body consisting of a stack of membranous cisternae. Instead, they have a 'Golgi equivalent' which appears like a ring of sausages or rounded cisternae in electron micrographs, although the oomycota have a more typical Golgi body. These features can be seen in the electron micrographs in Figs 2.2–2.4.

The distribution of nuclei varies between and even within the fungal groups. The aseptate fungi have many nuclei in a common cytoplasm, so these fungi are **coenocytic** (Fig. 2.3). The septate

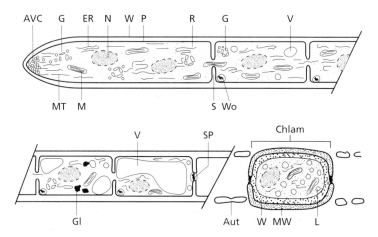

Fig. 2.1 Diagrammatic representation of a fungal hypha, showing an apical region, progressive ageing and vacuolation behind the apex, and autolysis and chlamydospore formation in the oldest region. AVC, apical vesicle cluster; MT, microtubules; G, Golgi body; M, mitochondrion; ER, endoplasmic reticulum; N, nucleus; W, wall; P, plasmalemma; R, ribosomes; S, septum; Wo, Woronin body; V, vacuole; Gl, glycogen; SP, septal plug; Aut, autolysis; MW, melanized wall; L, lipid; Chlam, chlamydospore.

fungi commonly have several nuclei in the apical compartment, but often only one or two in each compartment behind the apex. However, some septate fungi have a regular arrangement of just one nucleus in each compartment (e.g. the eyespot fungus of cereals, *Pseudocercosporella herpotrichoides*, and *Idriella bolleyi*, discussed in Chapter 11). In many basidiomycota, also, there is a regular arrangement of one nucleus in each compartment of the monokaryon, and two in each compartment of the dikaryon (see Chapter 1, Fig. 1.10). This is ensured by a special type of septum, the **dolipore septum** (see later), which has pores too narrow to allow the passage of nuclei, and by a special pattern of branching to produce clamp connections as explained in Chapter 4. The nuclei of fungi are usually small and difficult to see by normal light microscopy because their optical properties are similar to that of the cytoplasm. However, they are seen easily when stained with a fluorescent nuclear dye and observed under a fluorescence microscope (see Fig. 2.13).

Differentiation along the hypha

There is a progressive change in organization of the cytoplasm with distance behind the hyphal tip (see Fig. 2.1). The apical compartment is variable in length but typically is between 150 and 500 µm long. It is densely protoplasmic and lacks conspicuous vacuoles, although it does contain small, tubular vacuoles that can be seen in electron micrographs (see Fig. 2.4) or by special staining methods under a fluorescence microscope (see Fig. 2.12). The apical compartment is rich in organelles, including a dense zone of mitochondria immediately behind the growing apex. However, the extreme tip (the terminal 1–5 µm) is filled with a cluster of small membrane-bound vesicles, the **apical vesicle cluster** (AVC) which play a major role in growth (see Figs 2.2 & 2.4). The individual vesicles are too small to be seen with a light microscope, but early light microscopists noted a distinctive phase-dark structure in the growing tips of ascomycota and deuteromycota. This was termed the **Spitzenkörper** and it is thought to correspond to a central region of the AVC.

Compartments behind the hyphal tip may contain conspicuous rounded vacuoles. These are small at first but become progressively larger in the older compartments, and ultimately they may occupy much of the volume, restricting the nucleus and the rest of the cytoplasm to a thin peripheral zone (see Fig. 2.1). In older regions, also, there can be substantial numbers of lipid bodies which represent energy storage reserves. Various crystalline inclusions also occur in the

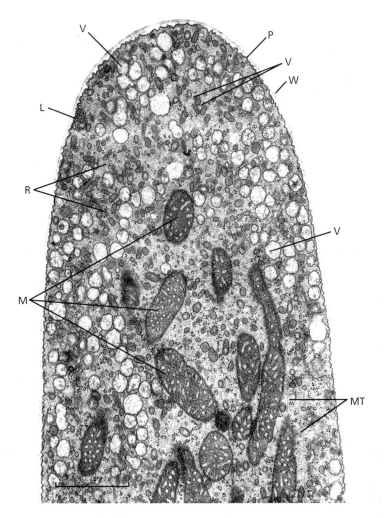

Fig. 2.2 Electron micrograph of the hyphal tip of *Pythium aphanidermatum* (Oomycota), prepared by conventional chemical fixation and stained to reveal internal organelles. The apex contains a cluster of Golgi-derived vesicles of two major classes: large vesicles with electron-lucent contents and small vesicles with electron-dense contents. Mitochondria (M) are abundant in a sub-apical zone; they have tubular cristae typical of Oomycota. Microtubules are also seen, but this is unusual in material prepared by chemical fixation. W, wall; V, vesicles; P, plasmalemma which has an indented appearance; L, lomasome, a proliferation of membranes beneath the plasmalemma, now thought to be an artefact; R, ribosomes; M, mitochondria; MT, microtubules. Bar = 1 µm. (Photograph courtesy of C. Bracker; from Grove & Bracker, 1970.)

older compartments, and structures termed **Woronin bodies** occur just behind each septum. In electron micrographs these bodies are seen to consist of an electron-dense proteinaceous lattice surrounded by a membrane. They serve to plug the septal pores if hyphae are damaged or as hyphal compartments age or undergo differentiation.

In the oldest parts of hyphae the compartments may be empty and even the wall may break down by the actions of lytic enzymes. Usually, by this stage the protoplasm and energy storage reserves have been withdrawn from these compartments and mobilized to other parts of the mycelium to produce survival structures, such as thick-walled **chlamydospores** (resting spores).

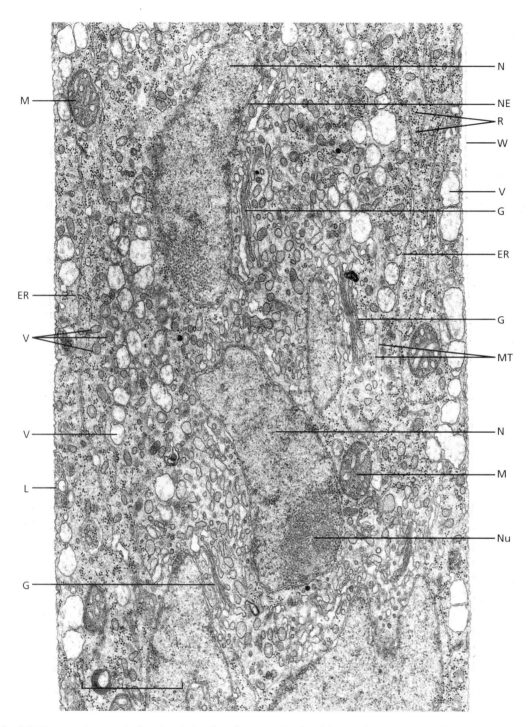

Fig. 2.3 Electron micrograph of a sub-apical region of a hypha of *Pythium aphanidermatum* (Oomycota), representing a region behind the apex shown in Fig. 2.2 and prepared in the same way; the wall (W) is poorly stained. N, nucleus; NE, nuclear envelope; M, mitochondrion; R, ribosomes; V, vesicles; G, Golgi body consisting of stacked cisternae typical of Oomycota; ER, endoplasmic reticulum; MT, microtubules; L, lomasome; Nu, nucleolus. Bar = 1 μm. (Photograph courtesy of C. Bracker; from Grove & Bracker, 1970.)

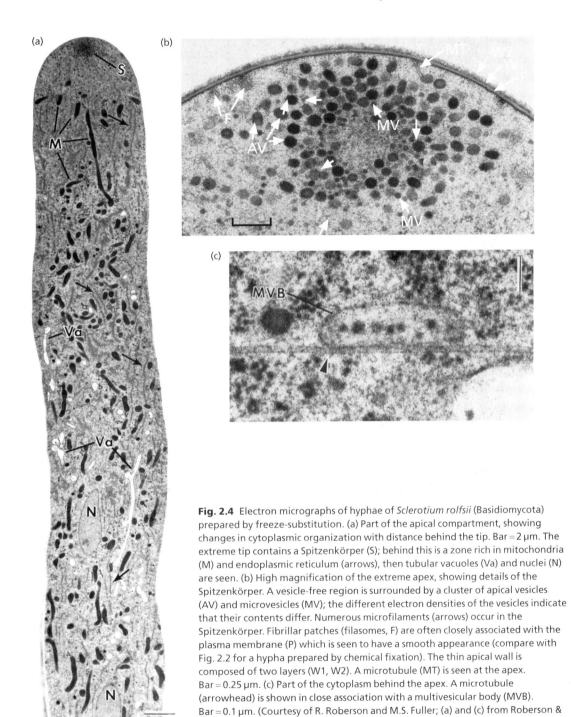

Fig. 2.4 Electron micrographs of hyphae of *Sclerotium rolfsii* (Basidiomycota) prepared by freeze-substitution. (a) Part of the apical compartment, showing changes in cytoplasmic organization with distance behind the tip. Bar = 2 μm. The extreme tip contains a Spitzenkörper (S); behind this is a zone rich in mitochondria (M) and endoplasmic reticulum (arrows), then tubular vacuoles (Va) and nuclei (N) are seen. (b) High magnification of the extreme apex, showing details of the Spitzenkörper. A vesicle-free region is surrounded by a cluster of apical vesicles (AV) and microvesicles (MV); the different electron densities of the vesicles indicate that their contents differ. Numerous microfilaments (arrows) occur in the Spitzenkörper. Fibrillar patches (filasomes, F) are often closely associated with the plasma membrane (P) which is seen to have a smooth appearance (compare with Fig. 2.2 for a hypha prepared by chemical fixation). The thin apical wall is composed of two layers (W1, W2). A microtubule (MT) is seen at the apex. Bar = 0.25 μm. (c) Part of the cytoplasm behind the apex. A microtubule (arrowhead) is shown in close association with a multivesicular body (MVB). Bar = 0.1 μm. (Courtesy of R. Roberson and M.S. Fuller; (a) and (c) from Roberson & Fuller, 1988; (b) from Roberson & Fuller, 1990.) *Continued overleaf.*

(d)

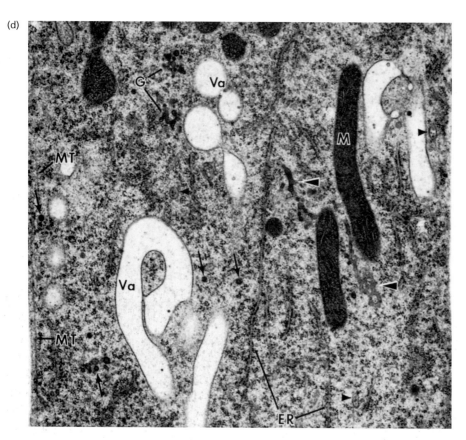

Fig. 2.4 (*continued*) (d) Portion of a sub-apical region of a hypha of *Sclerotium rolfsii*, showing the longitudinal orientation of organelles. The mitochondria (M) have plate-like cristae typical of 'true' fungi (compare with Fig. 2.3). The cytoplasm contains endoplasmic reticulum (ER), tubular vacuoles (Va), microtubules (MT) and Golgi equivalents (G) distinct from the Golgi bodies in Fig. 2.3. Also seen are apical vesicles (arrows), tubular cisternae (large arrowheads) and multivesicular bodies (small arrowheads) like the one in Fig. 2.4(c). (Courtesy of R. Roberson and M.S. Fuller; from Roberson & Fuller, 1988.)

We return to many of these points later in this chapter, but the general picture gives us some idea of the remarkable organization of fungal hyphae. They extend only at the extreme tip where there is a peculiar organization of the cytoplasm; they age progressively behind this point, and they break down or are broken down in their oldest regions. Yet, all the compartments are connected through the septal pores, and in most fungi the cytoplasm streams continuously towards the tip from the older regions, supporting growth at the tip and ensuring that the protoplasm is evacuated from the older zones where nutrients have been depleted from the environment.

The hypha as part of a colony

Up to now we have considered a single hypha, but branches can arise at almost any point behind the apex. As they grow, these first-order branches produce second-order branches, and so on, and they diverge from one another to give the colony its characteristic circular outline (Fig. 2.5). In the older parts of a colony, where nutrients have been depleted from the environment, much narrower branches can arise, and instead of diverging they grow towards one another and fuse (anastomose) by localized breakdown of their walls at the points of contact (Fig. 2.6). This creates a network for the

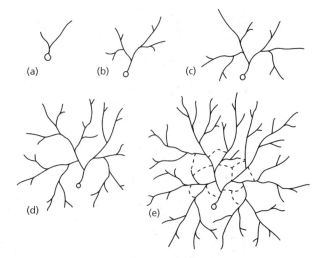

Fig. 2.5 (a–e) Stages in development of a fungal colony from a germinating spore. The broken lines in (e) represent narrow anastomosing hyphae near the colony centre.

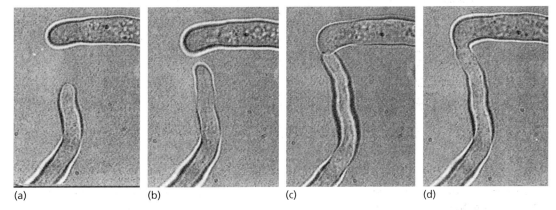

(a)   (b)   (c)   (d)

Fig. 2.6 Anastomosis of two hyphae of *Rhizoctonia solani*, videotaped through a thin film of water agar. (a) The upper hypha (about 10 μm wide) has stopped growing and the lower hypha has orientated towards the other hyphal tip. (b) 6 min later, a small bulge has appeared near the tip of the upper hypha and the lower hypha has reorientated towards this. (c) After 14 min, the hyphal tips have made contact. (d) After 32 min, the hyphal tips have fused by localized breakdown of their walls, establishing cytoplasmic continuity. (Courtesy of P. McCabe.)

remobilization of protoplasm to produce chlamydospores or other, larger differentiated structures (see Chapter 4). However, the anastomosis of hyphae seldom occurs in aseptate fungi, so these can produce only small differentiated structures from the resources of individual hyphae.

The colony often becomes coloured behind the hyaline (colourless) margin. This can be due to pigmentation of the hyphae themselves — for example, the deposition of melanin in hyphal walls. Alternatively, the colour can be caused by pigments released into the culture medium, the red colour of some *Fusarium* spp. being an example of this. Most often, however, the colony colour is a result of spore production behind the colony margin, because the spore walls of different fungi have characteristic pigmentation — black in *Aspergillus niger*, green in many other *Aspergillus* and *Penicillium* spp., pink in *Neurospora crassa*, and so on.

General structure: yeasts

Several fungi grow as budding yeasts and rarely, if ever, as hyphae. Examples of this include *Saccharomyces cerevisiae* (ascomycota), *Candida* spp.

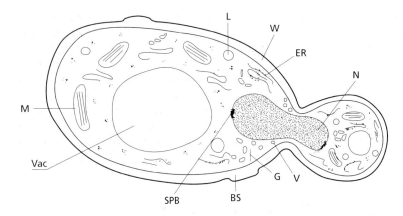

Fig. 2.7 Diagrammatic representation of a budding yeast, *Saccharomyces cerevisiae* (about 5 μm). W, wall; Vac, large central vacuole; BS, bud scar; M, mitochondrion; L, lipid body; G, Golgi; ER, endoplasmic reticulum; V, vesicle; SPB, spindle-pole body, equivalent to a centriole in other eukaryotes; N, nucleus.

(deuteromycota) and *Sporobolomyces roseus* (basidiomycota). Some other fungi are dimorphic: they can change between a hyphal and a yeast form depending on environmental conditions (see Chapter 4). Therefore, yeasts are not fundamentally different from hyphae; they merely represent a different growth form, and the reasons for this will be discussed in later chapters.

The structure of a representative yeast, *S. cerevisiae*, is shown in Fig. 2.7. Each cell has a single nucleus and a typical range of cytoplasmic organelles, including a large, conspicuous vacuole (the only structure that is clearly visible in a yeast cell by normal light microscopy). The cell divides by budding, and during this process the nucleus divides in the mother cell and one of the daughter nuclei enters the bud. Once the bud has reached nearly full size it is separated by the development of a septum. The first stage of this involves the deposition of a plate of chitin between the mother and daughter cell. Then, other wall materials are deposited on both sides of this plate, and the cells separate by enzymic cleavage of the wall between the chitin plate and the daughter cell. This process leaves a **bud scar** on the mother cell and a **birth scar** on the daughter cell. These scars are inconspicuous, but the bud scar is seen clearly if cells are treated with a fluorescent brightener which binds to the chitin plate. In this way, *S. cerevisiae* can be shown to exhibit **multipolar budding** because buds arise from a different point on the cell surface each time, whereas some other yeasts have **bipolar budding** in which the buds always arise from one or other end of the cell. In theory, therefore, the cells of bipolar species are immortal because there is no limit to the number of times they can bud,

whereas cells of multipolar species eventually would exhaust all possible sites—calculated to be up to 100 although only about 40 scars have ever been seen on a single cell. Nevertheless, in practice this distinction is unimportant because in exponentially growing cultures (where one cell produces two in a unit of time, two produce four, and so on) 50% of all the cells will be first-time parents, 25% will be second-time parents, and so on, an infinitesimally small number being very old parents.

Yeasts have become important 'model' organisms for cytological studies, contributing much to our understanding of fungi in general. For example, *S. cerevisiae* shows localized growth similar to that of hyphae, but it occurs at the bud site rather than at a hyphal apex. Small membrane-bound vesicles are seen at this site, just as they are seen in the hyphal tip. It is also of interest that the walls of *S. cerevisiae* have a low-chitin content (usually about 1%) and most or all of this is found in the bud scars. Yet, this fungus shows multipolar budding so it must be able to synthesize chitin almost anywhere on the cell surface, and study of this has helped in the understanding of how wall synthesis can be localized in fungi (see Chapter 3). Of more general relevance, the study of yeasts has contributed to an understanding of the origin of eukaryotes. *S. cerevisiae* produces occasional spontaneous mutants termed 'petites' which are seen as small, slow-growing colonies on agar. Petite mutants are unable to respire because of deficient mitochondria, and this provided the first clear evidence that mitochondria cannot be synthesized *de novo*: at least one mitochondrion must be inherited from a mother cell during budding. It is now known that mitochondria contain a circle of deoxyribonucleic acid (DNA) which codes for some of the mitochondrial structural proteins and some components of the respiratory electron-transport chain (see Chapter 6). Mitochondria also have ribosomes that more closely resemble the 70S ribosomes of bacteria than the 80S ribosomes of the rest of the eukaryotic cell. Evidence such as this led to the **endosymbiotic theory**, that the mitochondria of eukaryotes originated as bacteria that once lived as symbionts within the cells of other prokaryotes and that progressively lost their ability to exist as free-living entities. Nucleic acid sequence comparisons now strongly suggest that mitochondria evolved from a bacterial ancestor close to the present-day purple bacteria (see Chapter 1, Fig. 1.1). In a similar way, the chloroplasts of photosynthetic eukaryotes are thought to have originated as endosymbiotic cyanobacteria (see Fig. 1.1).

Fungal walls

Fungal walls merit detailed consideration because they serve many important roles. They determine the shape of fungi, because removal of the wall by enzymic treatments leads to the release of spherical protoplasts. So, the way in which a fungus grows — whether as hyphae or yeasts — is determined by the wall components and how these are assembled and bonded to one another. The wall is also the interface between a fungus and its environment: it protects against osmotic lysis; it acts as a molecular sieve regulating the passage of large molecules through the wall pore space;

and if the wall contains pigments such as melanin it can protect the cells against ultraviolet radiation or the lytic enzymes of other organisms. Furthermore, the wall can have several physiological roles. It can have binding sites for enzymes, such as invertase (which degrades sucrose to glucose and fructose) and β-glucosidase (which degrades cellobiose to glucose in the final stages of cellulose breakdown), and it can have antigenic properties that mediate the interactions of fungi with other organisms.

The wall components

Chemical analysis of fungal walls reveals a predominance of polysaccharides, lesser amounts of proteins and still smaller amounts of lipids. The types of polysaccharide are found to differ between the major fungal groups, as shown in Table 2.1. The chytridiomycota, ascomycota, deuteromycota and basidiomycota typically have chitin and glucans (polymers of glucose) as their major wall polysaccharides. Chitin consists of long, straight chains of β-1,4-linked N-acetylglucosamine residues (see Chapter 6), whereas the fungal glucans are branched polymers, consisting mainly of β-1,3-linked backbones to which short side-chains are linked by β-1,6 bonds. The zygomycota typically have a mixture of chitin and chitosan (a poorly acetylated or non-acetylated form of chitin; see Chapter 6) and polymers of uronic acids such as glucuronic acid instead of glucans. The oomycota have little chitin and, instead, they have a mixture of cellulose-like β-1,4-linked glucan and other glucans.

Table 2.1 Major polysaccharide components of fungal walls.

Phylum	Fibrillar components	Matrix components
Oomycota	Cellulose; β-(1,3)-β-(1,6)-glucans	Glucan
Chytridiomycota	Chitin; glucan	Glucan
Zygomycota	Chitin; chitosan	Polyglucuronic acid; glucuronomannoproteins
Ascomycota/ deuteromycota	Chitin; β-(1,3)-β-(1,6)-glucans	α-(1,3)-Glucan; galactomannoproteins
Basidiomycota	Chitin; β-(1,3)-β-(1,6)-glucans	α-(1,3)-Glucan; xylomannoproteins

Component (and corresponding polymer)	Yeast phase	Hyphae	Sporangiophores	Spores
N-acetylglucosamine (chitin)	8	9	16	2
Glucosamine (chitosan)	28	33	21	10
Mannose (mannans)	9	2	1	5
Glucuronic acid (glucuronans)	12	12	25	2
Glucose (glucans)	0	0	<1	43
Other sugars	4	5	3	5
Protein	10	6	9	16
Melanin	0	0	0	10

Table 2.2 Differences in wall composition of *Mucor rouxii* in different stages of growth or differentiation. Values shown are percentages of the wall dry weight. (Adapted from Bartnicki-Garcia, 1968.)

Despite these comments, it is important to recognize that the wall composition is not fixed. Indeed, it can change substantially at different stages of the life cycle, demonstrating that the wall structure and composition reflect the functional needs of a fungus. This is shown clearly in Table 2.2, for the fungus *Mucor rouxii* (zygomycota) which can grow as either hyphae or budding yeast-like cells. Compared with the hyphae, the yeast cells have considerably more mannan, probably in the form of mannoproteins. The asexual spores also differ, having high levels of glucans and melanin but lower levels of chitin, chitosan and polymers of glucuronic acid. The sporangiophores differ yet again. As another example, many yeasts such as *S. cerevisiae* have a high content of mannans but little chitin in their walls; the yeast forms of basidiomycota also have a high mannan content but correspondingly less glucan.

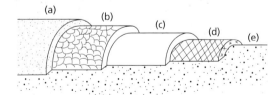

Fig. 2.8 Diagram to illustrate wall architecture in a 'mature' region of a hypha of *Neurospora crassa* (ascomycota) as evidenced by sequential enzymic digestion. (a) Outermost layer of amorphous β-1,3- and β-1,6-glucans; (b) glycoprotein reticulum embedded in protein; (c) more or less discrete protein layer; (d) chitin microfibrils embedded in protein; (e) plasmalemma. (Based on Hunsley & Burnett, 1970.)

Wall architecture

The major wall components as revealed by chemical analysis represent the bricks and mortar, but it was not until 1970 that we began really to understand the architecture of fungal walls. Hunsley and Burnett (1970) deserve much of the credit for this, because they developed an elegantly simple approach which they termed enzymic dissection. They mechanically disrupted fungal hyphae so that only the walls remained, and then treated the walls with different enzymes and examined them by electron microscopy. If, for example, the surface appearance of the wall changed after treatment with enzyme 'X' then the substrate of 'X' was likely to be the outermost wall component. So, by using various sequences and combinations of enzymes it was possible to strip away the major wall components and to see their relationships to one another.

Three fungi were used initially but we will take *Neurospora crassa* as an example to illustrate the essential features (Fig. 2.8). In mature regions of hyphae the wall of *Neurospora* was shown to have at least four concentric zones; they are shown as separate layers in Fig. 2.8, but in reality they grade into one another. The outermost zone consists of amorphous glucans with predominantly β-1,3- and β-1,6-linkages, which are degraded by the enzyme laminarinase. Beneath this is a network of glycoprotein embedded in a protein matrix. Then there is a more or less discrete layer of protein, and then an innermost region of chitin microfibrils embedded in protein. The total wall thickness in this case is about 125 nm. But, the wall at the growing tip is thinner (about 50 nm) and simpler, consisting of an inner zone of chitin embedded in protein and an outer layer of glucan. So, it is clear that the wall becomes stronger and more complex behind the extending hyphal tip, as further materials are added or as further bonding occurs between the components.

Neurospora seems to have an unusually complex wall architecture, because a glycoprotein network has not been seen in some other fungi. Nevertheless, the general pattern of wall architecture of hyphae is fairly consistent: the main, straight-chain microfibrillar components (chitin, or cellulose in the oomycota) are found predominantly in the inner region of the wall, and they are overlaid by non-fibrillar or 'matrix' components (e.g. other glucans, proteins and mannans) in the outer region. However, there is substantial bonding between the various components, serving to strengthen the wall behind the apex. In particular, some of the glucans are covalently bonded to chitin, and the glucans are bonded together by their side chains. We return to this topic in Chapter 3, when we consider the mechanisms of apical growth.

The extrahyphal matrix

In addition to the main structural components of the wall, some yeasts can have a discrete polysaccharide capsule, and both hyphae and yeasts can be surrounded by a more or less diffuse layer of polysaccharide or glycoprotein, easily removed by washing or mild chemical treatment. These extracellular matrix materials can have important roles in the interactions of fungi with other organisms. For example, the yeast *Cryptococcus neoformans* is an opportunistic pathogen of humans; its polysaccharide capsule masks the antigenic components of the cell wall so that the fungus is not engulfed by white blood cells and can proliferate in the tissues (Casadevall, 1995). In a different context, the fungus *Piptocephalis virginiana* (zygomycota; see Chapter 1, Fig. 1.7) parasitizes other zygomycota such as *Mucor* spp. on agar media, but does not parasitize them in liquid culture where the host fungi produce an extracellular polysaccharide (see Chapter 11). In fact, the production of an extracellular matrix often is influenced strongly by growth conditions. We noted in Chapter 1 that

pullulan is produced commercially from *Aureobasidium pullulans*; in this case its synthesis is favoured by an abundant sugar supply in nitrogen-limiting growth conditions (Seviour *et al.*, 1984). Some of the other gel-like materials around fungi also have potential commercial roles. For example, scleroglucan from *Sclerotium rolfsii* (basidiomycota) binds to the surface of some tumour cells and has been investigated for a possible therapeutic role.

Septa

The roles of septa have been debated for many years and still are not completely understood. They might help to provide structural support to hyphae, especially in relatively dry conditions, and in this respect it is notable that the septate fungi in general are much more tolerant of water-stress than are aseptate fungi (see Chapter 7). Septa also can help to defend against damage, because the septal pores become plugged by Woronin bodies or other materials if hyphae are physically disrupted (Markham, 1994) or if they are damaged by other organisms. An example of this is shown in Fig. 2.9, where hyphae of *Fusarium* have been damaged by contact with the mycoparasite *Gliocladium roseum*: only the contacted hyphal compartments have been disrupted because the septal pores were plugged immediately and *Fusarium* then regrew into the damaged compartments

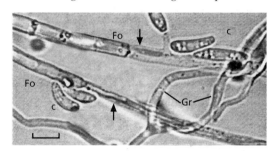

Fig. 2.9 Hyphae of *Fusarium oxysporum* (Fo) damaged 1–1.5 hours after being contacted by hyphae of *Gliocladium roseum* (Gr). Damage is confined to the contacted hyphal compartments, and *F. oxysporum* has regrown as narrow internal hyphae (arrows) from septa of viable hyphal compartments adjoining the damaged compartments. Some conidia (C) of *F. oxysporum* are seen. Print from a video copy processor, when hyphae were videotaped growing on a thin film of water agar. Bar = 10 μm.

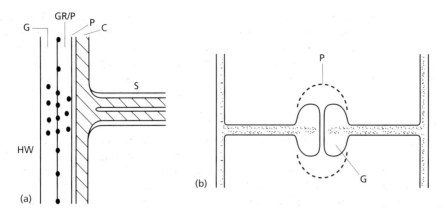

Fig. 2.10 Diagrammatic representation of two types of septum. (a) Part of the hyphal wall (HW) and simple septum (S) of *Neurospora crassa*, showing how the chitin-containing layers of the septal wall are closely related to the chitinous layer (C) of the hyphal wall. P = protein; G = glucan; GR/P = glycoprotein reticulum embedded in protein. Based on Trinci, A.P.J. (1978) *Science Progress*, **65**, 75–99. (b) Dolipore septum of basidiomycota, with a narrow central pore, flanges of predominantly glucan (G) around the pore, and membraneous parenthosomes (P).

by producing new hyphal tips from the plugged septal walls. However, even aseptate fungi can localize damage by developing 'barrier zones' of densely coagulated cytoplasm. So, the main significance of septa is thought to be in differentiation. If a septum is plugged then the compartments on either side of it can develop in different ways by differential gene expression or biochemical activity, whereas this would be difficult to achieve if the protoplasm were continuous. Consistent with this, the aseptate fungi produce complete cross-walls to isolate old parts of the hyphae or to produce sexual or asexual structures. In effect, the blocking of septal pores enables a fungus to transform from a continuous series of compartments to a number of independent cells or regions that can undergo separate development (see Chapter 4).

Several types of septum can be distinguished by electron microscopy, but not by normal light microscopy (Fig. 2.10). Typically, the ascomycota and deuteromycota have simple septa with a large central pore, 0.05–0.5 μm diameter, which allows the passage of cytoplasmic organelles and even nuclei. Variations on this are found in individual fungi; for example, *Geotrichum candidum* has septa with several smaller pores. The basidiomycota have a more complex **dolipore septum** (Fig. 2.10) with a narrow central channel (about 100–150 nm diameter) bounded by two flanges of amorphous (glucan) wall material. On either side of this type of septum there are bracket-shaped membranous

structures termed **parenthosomes**, which have pores that allow cytoplasmic continuity but preclude the movement of major organelles. This helps to ensure the regular distribution of nuclei in dikaryotic hyphae of the basidiomycota. We shall see in Chapter 4 that these septa are selectively degraded when the fungus begins to form a fruiting body, enabling the mass movement of materials to support the development of these large structures.

Septa provide interesting examples of localized wall growth behind the hyphal tip. They develop rapidly as centripetal ingrowths from the hyphal wall, involving a role of Golgi-derived vesicles which presumably deliver some of the enzymes required for development (see later). At maturity, a septum typically comprises an inner region of chitin, overlaid by protein or glucans. The existing hyphal wall also can undergo modification in the region of a septum; in *Neurospora crassa*, for example, there is a proliferation of the glycoprotein reticulum in this region (Fig. 2.10).

Nuclei and associated structures

Fungal nuclei are usually small (2–3 μm diameter) but exceptionally can be up to 20–25 μm diameter (e.g. *Basidiobolus ranarum*; see Chapter 3, Fig. 3.11). As in other eukaryotes, they are surrounded by a double nuclear membrane with pores, and they contain a ribonucleic acid (RNA)-rich nucleolus.

Most fungi are haploid, but the oomycota are diploid, and some other fungi can alternate between haploid and diploid generations. For example, *Saccharomyces cerevisiae* can grow as either a haploid or diploid yeast; *Allomyces* spp. (see Chapter 1, Fig. 1.6) have an alternation of haploid and diploid phases, and *Allomyces* and several *Phytophthora* spp. can exist as polyploids. We discuss these points further in Chapter 8, but here it is interesting to note that the peculiar organization of fungal hyphae, with several nuclei in a common cytoplasm, enables haploid fungi to exploit the advantages of both haploidy and diploidy. If a recessive mutation occurs in one nucleus in a hypha, then the fungus will still behave as wild-type, but it can also carry the mutant gene and thus store potential variability, which is the hallmark of diploid organisms. On the other hand, a mutation can be exposed to selection pressure periodically, when the fungus produces uninucleate spores or when a branch arises with only the mutant nuclear type in it. Diploidy predominates in the rest of the eukaryotic world (and also in oomycota) and even in yeasts such as *Candida* and *Saccharomyces* (see Chapter 8). But, it seems that mycelial fungi have remained haploid because they can exploit the advantages of a diploid lifestyle in any case.

Fungi are notable for several peculiar features of their nuclei and nuclear division (Heath, 1978). The nuclear membrane and the nucleolus usually remain intact during most stages of mitosis, and there is no clear metaphase plate; instead, the chromosomes seem to be dispersed randomly, and at anaphase the daughter chromatids pull apart along two tracks, on spindle fibres of different lengths. Also, most fungi do not have typical centrioles. Instead, they have a **spindle-pole body** (SPB) which functions like a centriole as a centre for microtubule assembly during nuclear division; in ascomycota it is a disc-shaped structure lying just outside the nuclear envelope, whereas in basidiomycota it is often composed of two globular ends connected by a bridge.

Cytoplasmic components

Organization of the hyphal apex

The hyphal apex is of special interest as the site of growth, and for many years it has been recognized as the key to understanding the behaviour of fungi. As we noted earlier, under a light microscope the tips of hyphae are seen to contain a phase-dark Spitzenkörper. It disappears when growth stops, reappears when growth restarts and it changes position in the apex when a hypha changes direction of growth. It also occurs to the exclusion of other major organelles in the growing tip, and so it was regarded as a possible 'apical growth organelle'. However, electron micrographs failed to identify a single organelle that corresponds to the Spitzenkörper. Instead, as shown in Fig. 2.2, the apex was seen to be filled with small, membrane-bound vesicles — the **apical vesicle cluster (AVC)**. These, also, were found to disappear when growth was stopped and to reappear when growth restarted. Moreover, the plasma membrane at the apex was found to have indentations of a similar size to the vesicles, suggesting that vesicles fuse with this membrane and this process was being captured during the fixation process.

The micrograph in Fig. 2.2 was prepared from a hyphal tip subjected to conventional chemical fixation, in which hyphae are killed by immersion in an aldehyde such as glutaraldehyde and then post-treated to obtain maximum contrast of the organelles. It shows that the AVC contains different classes of vesicles, some being larger (about 70–100 nm diameter) than others (about 30 nm), and that their contents vary from electron-lucent to electron-opaque. Since the early 1980s a new method of fixation has been developed (Howard & Aist, 1979). It is termed freeze-substitution, and it involves plunging a specimen into anhydrous acetone or an equivalent solvent at a very low temperature, and then progressively replacing the water in the specimen by post-fixation treatments as the temperature is raised. The value of this method is that it more instantaneously fixes the material, avoiding the short delay during which chemical fixatives must penetrate the cell, so that it more closely represents the living state of a hypha. Figure 2.4 shows a hyphal tip prepared in this way, and the image obtained is strikingly different (even disregarding the fact that Figs 2.2 and 2.4 are of different types of fungus). The AVC is still present, with vesicles of different sizes and contents, but now they are seen to be embedded in a

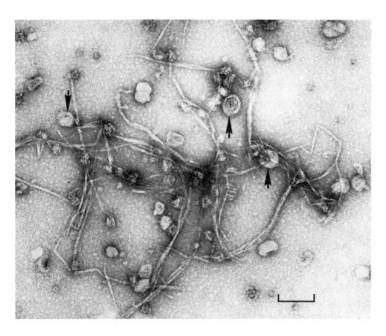

Fig. 2.11 Chitin microfibrils produced *in vitro* by chitosomes (arrows) isolated from *Mucor rouxii* and incubated with substrate (uridine diphosphate-*N*-acetylglucosamine) and proteolytic activator. Each ribbon-like microfibril is composed of several chitin chains. Bar = 0.1 μm. (Courtesy of C. E. Bracker; from Bartnicki-Garcia *et al.*, 1978.)

meshwork of actin microfilaments, and larger microtubules also are seen to run up to, and even through, the AVC, sometimes terminating at the plasma membrane. Microfilaments and micro-tubules are major components of the cytoskeleton (see later) and would be expected to be involved in growth. Some of the apical vesicles, especially on the shoulders of the apex and lining the lateral walls have a distinctly filamentous appearance, so they have been termed **filasomes**. Finally, the plas-malemma is seen to be smooth, suggesting that the indentations seen in chemically fixed material were artefacts, although it is important to note that the vesicles still are believed to fuse with the plasma membrane and to release their contents into the wall. The significance of all these features for apical growth is discussed in Chapter 3.

Most of the vesicles in hyphal tips have not been characterized chemically, but some of the smaller ones (microvesicles) resemble particles that have been purified *in vitro* and termed **chitosomes** (Fig. 2.11). These particles were discovered by homo-genizing hyphae, removing the wall material and the major membranous organelles by centrifuga-tion and then, after further purification, subjecting the supernatant to sucrose density gradient cen-trifugation so that its components separated out as bands. When one of the bands was examined by

electron microscopy it contained chitosomes — small spheroidal bodies, 40–70 nm diameter, each surrounded by a 'shell' about 7 nm thick. When this band was incubated in the presence of an acti-vator (a protease enzyme) and uridine diphos-phate (UDP)-*N*-acetylglucosamine (the sugar nucleotide from which chitin is made) each parti-cle was seen to produce a coil of chitin inside it and then to break open and release a chitin microfibril composed of several chitin chains. If the super-natant was first treated with digitonin (a saponin which solubilizes sterol-containing membranes) and then centrifuged in a density gradient, the chi-tosomes were no longer found; instead, the addi-tion of activator and UDP-*N*-acetylglucosamine resulted in the production of chitin in a different band corresponding to a lighter fraction of the homogenate. But, subsequent removal of the digi-tonin caused the chitosomes to reappear, indicat-ing that they are self-assembling aggregates of the enzyme **chitin synthase**, each particle containing sufficient enzyme molecules to produce, after pro-teolytic activation, a chitin microfibril in which the individual chitin chains coil round one another. So, this elegant collaborative study by biochemists and electron microscopists led to the identification of one of the types of vesicle in the hyphal apex. One of the questions still to be resolved is how these particles reach the hyphal tip after they have been assembled further back in the hyphae. One suggested mechanism is that they are packaged into membranes as the multivesicular bodies seen in electron micrographs (e.g. see Fig. 2.4). We consider this further in Chapter 3.

The plasma membrane

The plasma membrane of fungi is similar to that of other eukaryotes, consisting of a phospholipid bilayer and associated proteins and sterols. However, the fungal membrane is unique in one respect: the main membrane sterol is **ergosterol**, in contrast to cholesterol in animals and the cholesterol-like phytosterols in plants. This difference is important in practice, because some of the most effective antifungal antibiotics and fungicides act specifically on ergosterol or on its biosynthetic pathway (see Chapter 14). The oomycota differ from other fungi because they have plant-like sterols, and some oomycota such as *Pythium* and *Phytophthora* spp. do not synthesize their own sterols from non-sterol precursors but need to be supplied with sterols from a host plant.

The main role of the plasma membrane is to regulate the uptake and release of materials, discussed in Chapter 5. The membrane also can anchor some enzymes; in fact, the two main wall-synthetic enzymes, chitin synthase and glucan synthetase, are **integral membrane proteins**; they are anchored in the membrane in such a way that they produce polysaccharide chains from the outer membrane face (see Chapter 3). The plasma membrane has a third important role, in relaying signals from the external environment to the cell interior — the process termed **signal transduction** and discussed in Chapter 4.

Golgi, endoplasmic reticulum and vesicles

Fungi have a **secretory system**, consisting of the ER, Golgi apparatus and membrane-bound vesicles. Proteins that are destined for export from the cell are synthesized on ribosomes attached to the ER, enter the lumen of the ER as they are being synthesized and are then transported within the ER to the Golgi apparatus. Within the cisternae of the Golgi these proteins undergo various modifications, including partial cleavage and reassembly,

folding into a tertiary structure and the addition of sugar chains (glycosylation). Then they are packaged into vesicles which bud from the maturing face of the Golgi. The vesicles are transported to the plasma membrane and fuse with it, delivering the contents to the exterior or inserting them into the membrane. The system is like an intricate postal system in which the mail is sorted and delivered to specific destinations, but also processed during transit. Details of the secretory pathway are most fully developed for *S. cerevisiae* and discussed by Klis (1994).

The secretory system is involved not only in tip growth but also in the secretion of enzymes for degrading nutrients in the external environment. Additionally, it is important for the commercial production of foreign (heterologous) gene products (see Chapter 8). The ability of a protein to enter the ER and ultimately be exported is determined by a signal sequence at the N-terminus, which is subsequently removed. Without this sequence the protein will remain in the cell.

Vacuoles

The vacuolar system of fungi has several functions, including the storage of compounds and the recycling of cellular metabolites. For example, the vacuoles of several fungi, including mycorrhizal species (see Chapter 12), accumulate phosphates in the form of polyphosphate. Vacuoles also seem to be major sites for storage of calcium, which can be released into the cytoplasm as part of the intracellular signalling system (see Chapter 4). Vacuoles contain proteases for breaking down cellular proteins and recycling of the amino acids, and vacuoles also have a role in the regulation of cellular pH. All these important physiological roles are in addition to the potential role of vacuoles in cell expansion and in driving the protoplasm forwards or, at least, 'filling the space' from which cytosol and its components have been remobilized.

Vacuoles often are seen as conspicuous, rounded structures in the older regions of hyphae, but recent work has shown that there is also a tubular vacuolar system extending into the tip cells (see Fig. 2.4). It is an extremely dynamic system, consisting of narrow tubules which can dilate and contract, as inflated elements travel along them in a peristaltic manner (Fig. 2.12). This

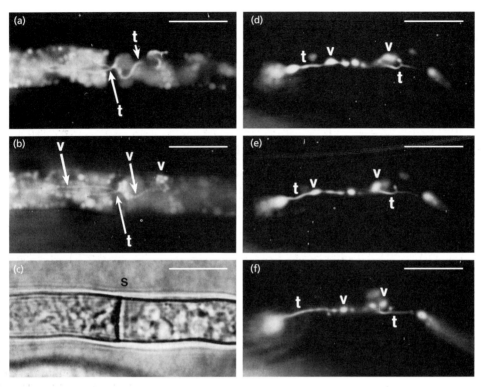

Fig. 2.12 Fluorescence micrographs of the dynamic tubular vacuole system in living fungal hyphae. Bars = 10 μm. The hyphae were treated with carboxyfluorescein-diacetate, which accumulated in the vacuole system where it was converted to the fluorescent compound, carboxyfluorescein (CF) which remains in the vacuoles. (a)–(c) Hypha of *Penicillium expansum* seen by bright field microscopy (c) and fluorescence microscopy (a, b) in the region of a septum (s). A vacuolar tubule (t) is seen to pass through the septal pore and branch on one side of it; tubular vacuoles (v) consisting of small dilations are also shown. (d)–(f) Hypha of *Aspergillus niger* photographed at 8-sec intervals, showing a succession of peristalsis-like movements of a tubule (t) such that the tubule dilations (v) are captured at various positions. (Photographs courtesy of Drs B. Rees, V. Shepherd and A. Ashford; from Rees *et al.*, 1994.)

was demonstrated quite recently by studying the movement of a fluorescent compound, carboxyfluorescein (CF), in living hyphae of a range of fungi. The hyphae were treated with a non-fluorescent precursor, CF diacetate (CFDA), which is lipid-soluble so it enters through the plasma membrane and is then rapidly absorbed from the cytosol into the vacuoles. There the acetate groups are cleaved by esterases to yield CF, which is membrane-impermeable so it remains in the vacuolar system. Rees *et al.* (1994) showed that the tubular vacuoles which contain CF form a more or less continuous system which passes through even the dolipore septa of basidiomycota, transporting materials backwards and forwards in the hyphae. This has important implications for the bidirectional translocation of materials within single hyphae (see Chapter 6), and perhaps especially for mycorrhizal systems (see Chapter 12).

The cytoskeleton

The cytoskeleton consists primarily of **microtubules** and **microfilaments**, with associated **motor proteins** (reviewed by Heath, 1994). Many of the components of this system remain poorly characterized in fungi, but their existence has been known for some time, based on circumstantial evidence. For example, the **benzimidazole** fungicides which are so important in agriculture (see Chapter 14) are known to bind to microtubules. They block nuclear division, presumably

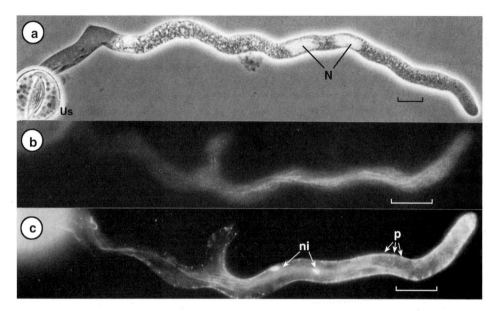

Fig. 2.13 Young (3 h) germling from a uredospore (Us) of the rust fungus *Uromyces phaseoli*, treated sequentially with different compounds to reveal internal components (Bars = 10 μm). (a) Treated with the fluorochrome DAPI which binds to A/T-rich regions of DNA, and observed by a combination of phase-contrast and fluorescence microscopy; the two nuclei (N) in the germling are revealed by DAPI fluorescence. (b) The same germling, fixed and stained with fluorochrome-labelled antibodies for tubulin and observed by fluorescence microscopy, showing the many microtubules that run longitudinally in the cytoplasm. (c) The same germling stained with rhodamine-conjugated phalloin which binds to actin, and observed by fluorescence microscopy. A conspicuous actin cap is seen in the hyphal tip. Actin is also seen as peripheral plaques (p), nuclear inclusions (ni) and in zones similar to those in which microtubules are found. (Photographs courtesy of H.C. Hoch; from Hoch & Staples, 1985.)

by binding to the spindle microtubules, and they disrupt the normal patterns of wall growth, presumably by binding to the cytoplasmic microtubules. The fact that microtubules can now be visualized at the extreme hyphal tip in freeze-substitution electron micrographs (see Fig. 2.4) adds weight to this; they are seen as long, straight structures, of about 25 nm diameter, occurring either singly or as parallel arrays along the axis of a hypha and extending up to the membrane at the tip. Similarly, cytoplasmic streaming in fungi and in other eukaryotic cells can be disrupted by applying **cytochalasins** ('cell relaxers'). These are fungal products and they bind to actin microfilaments. Again, actin has now been visualized in the tips of fungal hyphae, where it forms a meshwork in the AVC (see Fig. 2.4). The actin microfilaments are much narrower than microtubules, about 5–8 nm diameter. In the slime mould *Physarum polycephalum* the microfilaments are known to function in cytoplasmic contraction when actin associates with the motor protein **myosin**. This also seems to be true of fungi, where myosin-like proteins have recently been detected.

The full extent of the cytoskeleton has become clear by observing hyphae under a fluorescence microscope after treatment with fluorescently labelled compounds that bind to different components (Fig. 2.13). In Fig. 2.13(a) the young hypha emerging from a uredospore of a rust fungus has been treated with a fluorescent compound (DAPI) that binds to DNA and clearly shows the two nuclei in the young dikaryotic hypha. Figure 2.13(b) shows the same hypha when fixed and treated with a fluorescently labelled antibody to tubulin. It reveals that microtubules are common as longitudinal arrays in hyphae. Figure 2.13(c) shows the same hypha treated with phalloin conjugated to the fluorescent dye rhodamine. Phalloin is the deadly toxin in toadstools of the 'death cap',

Amanita phalloides, and it binds strongly to actin in both fungi and humans, with similar effects! Actin is seen as a meshwork in the hyphal apex and also as strands along much of the length of the hypha, as peripheral plaques near the cell surface and as localized plaques associated with the nuclei. The tubulin-rich regions in Fig. 2.13(b) correspond closely to the actin-rich regions in Fig. 2.13(c), perhaps because microfilaments and microtubules act together in the cell.

By using the same method of fluorescent labelling, the microtubules have been shown to be dynamic components of hyphae; they depolymerize in response to treatments such as cold shock, and conversely they can be stabilized by compounds such as **taxol**, the toxin from yew trees which, again, is deadly to humans. So, the microtubules are thought to be continuously degraded and reformed. Consistent with this, microtubules polymerize by self-assembly in cell-free systems: the two proteins α- and β-tubulin link together as a dimer, and then the dimers polymerize into chains. The microtubular system in fungal hyphae almost completely disappears if the hyphae are treated with benzimidazole fungicides which bind to the β-tubulin and prevent it from interacting with α-tubulin, thereby preventing the polymerization process. But, this system reappears when the inhibitors are washed out, and it is notable that the first signs of repolymerization are at the hyphal apex, suggesting that the apex is a **microtubule organizing centre** (MOC) (Hoch & Staples, 1985). This contrasts with the situation in the cells of other organisms, where repolymerization begins in the nuclear region, probably because the centrioles act as MOCs.

We shall return to the roles of the cytoskeleton in fungal growth (see Chapter 3), differentiation (see Chapter 4) and the behaviour of fungal zoospores (see Chapter 9). But, we end this account and this chapter by noting that fungal tubulins differ slightly from those of plants and animals. They are inhibited by the antibiotic griseofulvin and by the benzimidazole fungicides, whereas plant and animal tubulins are insensitive to these compounds. This is why griseofulvin and the benzimidazoles can be used to treat fungal infections of plants and of humans (see Chapter 14). Conversely, the fungal tubulins are unaffected by **colchicine** (the toxin from the autumn crocus),

which inhibits nuclear division in plant and animal cells. However, the oomycota are inhibited by colchicine and not by the antifungal agents, consistent with many other plant-like features of the oomycota (see Chapter 1, Table 1.1).

References

Bartnicki-Garcia, S. (1968) Cell wall chemistry, morphogenesis and taxonomy. *Annual Review of Microbiology*, **22**, 87–108.

Bartnicki-Garcia, S., Bracker, C.E, Reyes, E. & Ruiz-Herrera, J. (1978) Isolation of chitosomes from taxonomically diverse fungi and synthesis of chitin microfibrils *in vitro*. *Experimental Mycology*, **2**, 173–92.

Casadevall, A. (1995) Antibody immunity and *Cryptococcus neoformans*. *Canadian Journal of Botany*, **73**, S1180-6.

Grove, S.N. & Bracker, C.E. (1970) Protoplasmic organization of hyphal tips among fungi: vesicles and Spitzenkörper. *Journal of Bacteriology*, **104**, 989–1009.

Heath, I.B. (1978) *Nuclear Division in the Fungi*. Academic Press, New York.

Heath, I.B. (1994) The cytoskeleton in hyphal growth, organelle movements, and mitosis. In: *The Mycota*, Vol. 1 (eds J.G.H. Wessels & F. Meinhardt), pp. 43–65. Springer-Verlag, Berlin.

Hoch, H.C. & Staples, R.C. (1985) The microtubule cytoskeleton in hyphae of *Uromyces phaseoli* germlings: its relationship to the region of nucleation and to the F-actin cytoskeleton. *Protoplasma*, **124**, 112–22.

Howard, R.J. & Aist, J.R. (1979) Hyphal tip cell ultrastructure of the fungus *Fusarium*: improved preservation by freeze-substitution. *Journal of Ultrastructural Research*, **66**, 224–34.

Hunsley, D. & Burnett, J.H. (1970) The ultrastructural architecture of the walls of some hyphal fungi. *Journal of General Microbiology*, **62**, 203–18.

Klis, F.M. (1994) Protein secretion in yeast. In: *The Mycota*, Vol. 1 (eds J.G.H. Wessels & F. Meinhardt), pp. 25–41. Springer-Verlag, Berlin.

Markham, P. (1994) Occlusions of septal pores in filamentous fungi. *Mycological Research*, **98**, 1089–106.

Rees, B., Shepherd, V.A. & Ashford, A.E. (1994) Presence of a motile tubular vacuole system in different phyla of fungi. *Mycological Research*, **98**, 985–992.

Roberson, R.W. & Fuller, M.S. (1988) Ultrastructural aspects of the hyphal tip of *Sclerotium rolfsii* preserved by freeze substitution. *Protoplasma*, **146**, 143–9.

Roberson, R. W. & Fuller, M. S. (1990) Effects of the demethylase inhibitor, cyproconazole, on hyphal tip cells of *Sclerotium rolfsii*. II. An electron microscope study. *Experimental Mycology*, **14**, 124–35.

Seviour, R.J., Kristiansen, B. & Harvey, L. (1984) Morphology of *Aureobasidium pullulans* during polysaccharide elaboration. *Transactions of the British Mycological Society*, **82**, 350–6.

Chapter 3
Fungal growth

In this chapter we consider the mechanisms of fungal growth, both as hyphae and as yeasts. We also consider the kinetics of fungal growth in culture systems and we relate this to the exploitation of fungi in industrial processes.

Apical growth

Apical growth is the hallmark of fungi. It accounts for much of the importance of fungi in natural environments—as decomposers and as parasites. The apical growth mechanism has been a focus of attention for more than three decades and is still one of the most active fields of fungal research.

Microscopical observation

Apical growth can be observed with a compound microscope by placing a cover slip on the colony margin of a fast-growing fungus, such as *Neurospora crassa*, on agar. The tips are seen to extend across the field of view, and by focusing on a septum behind the apex the protoplasm is seen to move steadily towards the apex, passing through the septal pore from the older regions. Indeed, such a rapid rate of tip extension (up to 40 µm min^{-1} in *N. crassa*) is only possible if protoplasm is supplied from behind, because the tip cell cannot synthesize protoplasm fast enough to fill its volume. For this reason, what we term apical growth is actually apical extension; the true rate of growth, defined as increase in biomass per unit of time, is much slower. The length of hypha needed to support an extending apex can be estimated by making a diagonal cut across a colony margin with a scalpel, so that individual hyphae are severed at different distances from their tips. Hyphae that are cut too close to the tip die from physical damage.

Hyphae cut further back continue to extend but more slowly than usual, and eventually a point is reached at which the cut is so far back that it has no effect on the apical extension rate. This distance is termed the **peripheral growth zone** of a fungal colony, defined as the length of hypha needed to maintain the maximum extension rate of the **leading hyphae** at the colony margin; it varies between fungi, from below 200 µm up to several millimetres for the fastest extending fungi such as *N. crassa*.

By using tiny particles as reference points, it can be shown that extension of the hypha occurs only at the extreme tip—the tapered extension zone. A more sophisticated way of showing this is to supply a short pulse of radiolabelled wall precursor such as ^{3}H-acetylglucosamine or ^{3}H-glucose and then to autoradiograph a hypha. As shown in Fig. 3.1, the label is incorporated maximally in the wall at the tip and falls off sharply behind this. Thus, we have defined one of the central questions in fungal biology: how is extension growth localized to the extreme tip of the fungal hypha?

The mechanism of apical growth: early experiments

Robertson (1958, 1959) did the key early experiments on apical growth, using extremely simple methods but coupled with truly remarkable insight. He grew colonies of *Fusarium oxysporum* or *N. crassa* on nutrient-rich agar, flooded them with water and then observed the behaviour of the tips. As shown in Fig. 3.2, some of the tips stopped growing immediately but then resumed growth within 1 min, from a narrower apex than before. Others stopped growing for several minutes,

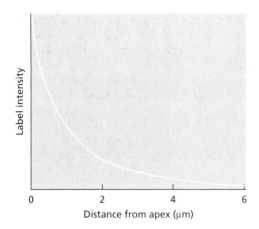

Fig. 3.1 Incorporation of label into the wall of a growing hypha during a brief (5 min) exposure to radiolabelled wall precursors.

swelled into a diamond shape during this time and eventually regrew by producing one or more narrow tips just behind the original apex. He then repeated the experiment, adding water but replacing it within 1 min by a solution of the same osmotic potential as the original agar (an isotonic solution). This caused all the tips to stop for several minutes and swell during this time, and eventually regrew from narrow subapical branches (Fig. 3.2).

To interpret these patterns, Robertson suggested that the normal process of apical growth involves two separate phenomena: (i) extension of a plastic, deformable tip; and (ii) rigidification of the wall behind the extending tip. He envisaged these two processes as occurring at the same rate, but with rigidification always slightly behind the tip, like two cars travelling along two lanes of a motorway at exactly the same speed but one always slightly behind the other. Then, if extension growth is halted by an osmotic shock the process of rigidification will continue and tend to 'overtake' the apex. If the tip readjusts to the new osmotic conditions in time it can grow on, but now from a thinner region of the apex where the wall has not yet rigidified. Roughly 50% of the tips in the original experiment seemed to be able to do this. However, if the tip cannot adjust in time then the apex will be sealed off by rigidification, and growth will only occur when new tips have been produced — in this case behind the original apex and by a process that takes several minutes. This would explain why all the tips stopped for several minutes in the second experiment, because they would need to make two osmotic adjustments and could not do so before the apex had rigidified.

We shall see later that this explanation was essentially correct, because now it is supported by many lines of evidence from wall enzymology and

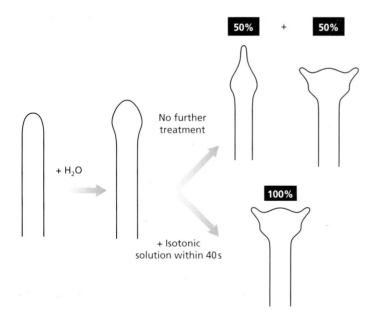

Fig. 3.2 The responses of hyphal tips of *Fusarium oxysporum* to osmotic shock caused by the addition of water or osmotica to a growing colony. (Redrawn from Robertson, 1959.) See text for explanation.

ultrastructural studies. But, at this stage we should note some further points. Sometimes the hyphal tips swell and burst in response to flooding with water, perhaps because the wall at the extreme apex is too fragile to withstand high turgor pressure or, as some workers argue, because the wall at the extreme apex is continuously being degraded by wall-lytic enzymes. Sometimes the tips grow on as usual in response to flooding, but a branch develops later from the position where the apex had reached at the time of flooding. In any case, growing hyphal tips are very sensitive to many types of disturbance and they tend to respond in the same way—by a 'stop–swell–branch' sequence as shown in Fig. 3.2. This response can be elicited by mild heat or cold shock, by a burst of intense light or even when hyphal tips encounter physical barriers. Some morphological mutants of *Neurospora* and *Aspergillus* even show this pattern regularly during growth, suggesting that there is a periodic endogenous 'shock' to the growing apex. In Chapter 4 we shall see that the 'stop–swell–branch' sequence occurs during the production of several differentiated structures, including the pre-penetration structures of fungal parasites of plant and animals.

Assembly of the wall at the apex

Electron micrographs always show an abundance of vesicles in the growing hyphal tip (see Chapter 2), suggesting that these vesicles are intimately involved in growth. They are thought to arise from Golgi bodies in the subapical regions, then are transported to the tip and fuse with the plasma membrane to deliver their contents for wall growth. Most of the vesicle contents have not been characterized, but from various lines of evidence we can construct a composite picture of wall growth at the apex. It is shown in Figs 3.3 and 3.4, and some of its main components are discussed below.

Chitin synthase

The enzyme chitin synthase catalyses the synthesis of chitin chains. These chains are known to be formed *in situ* at the apex; they do not arrive pre-formed in vesicles. When hyphal homogenates are tested for enzyme activity *in vitro*, chitin synthase is found in two forms: (i) as an inactive **zymogen** in chitosomes (see Chapter 2) and sometimes in membranes; and (ii) as an active form closely associated with the membranes. We saw in Chapter 2 that chitosomes resemble some of the microvesicles in the hyphal apex. However, the 'shell' around a chitosome is not a phospholipid membrane, so chitosomes might be packaged within phospholipid membranes for transport to the tip—perhaps in the multivesicular bodies that are sometimes seen in electron micrographs (see Chapter 2, Fig. 2.4). The zymogen form of chitin

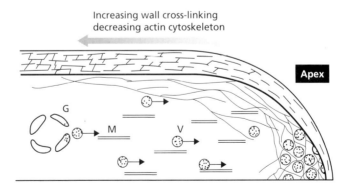

Fig. 3.3 Diagrammatic representation of the organization of wall growth at the hyphal tip. Only half of the tip is shown. Vesicles (V) derived from a Golgi body (G) are transported to the apex, perhaps by microtubule (M) mediated systems. The actin meshwork at the apex is thought to provide structural support where the wall is thinnest and where there is little or no cross-linking of wall polymers. Behind the extreme tip the wall is progressively rigidified by cross-linking of wall polymers.

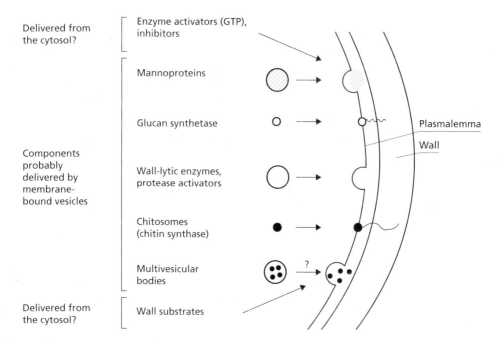

Delivered from the cytosol?
— Enzyme activators (GTP), inhibitors

Components probably delivered by membrane-bound vesicles
— Mannoproteins
— Glucan synthetase
— Wall-lytic enzymes, protease activators
— Chitosomes (chitin synthase)
— Multivesicular bodies

Delivered from the cytosol?
— Wall substrates

Plasmalemma
Wall

Fig. 3.4 Diagrammatic representation of the possible components of wall synthesis at the hyphal tip. Vesicles are thought to deliver some wall-synthetic enzymes, a few pre-formed wall components (e.g. mannoproteins) and perhaps some wall-lytic enzymes and enzyme activators when the vesicles fuse with the plasma membrane at the apex. Some of the wall-synthetic enzymes (chitin synthase, glucan synthetase) become integral membrane proteins. Some wall substrates, enzyme activators and possible enzyme inhibitors arrive in the cytosol. The roles of multivesicular bodies are unclear.

synthase, when inserted into the membrane, must be activated by a protease which probably arrives at the apex in other vesicles. Then the substrate is delivered from the cytosol to the part of the enzyme on the inner face of the plasma membrane, so that chitin chains are synthesized and extruded from the membrane outer face. The substrate is N-acetylglucosamine, but it is supplied as a sugar nucleotide, uridine diphosphate (UDP)-N-acetylglucosamine, to provide the energy for chitin synthesis as explained in Chapter 6.

Clearly there must be mechanisms for regulating the activity of chitin synthase during wall growth. This regulation could be achieved partly by controlling the delivery of proteases which activate the enzyme, partly by controlling the rate of delivery of the enzyme substrate, and partly by means of enzyme inhibitors, because the cytosol is known to contain a chitin synthase inhibitor. Further discussion of chitin synthesis can be found in Gooday (1995).

Glucan synthetase

This is another major enzyme involved in wall growth because it catalyses the synthesis of β-1,3-glucans. Like chitin synthase, it is thought to arrive in vesicles and is then inserted into the plasma membrane at the apex. The substrate is a sugar nucleotide (UDP-glucose) supplied from the cytosol. The activity of glucan synthetase is regulated in a different way from chitin synthase. The enzyme is composed of two subunits, one of which (on the membrane outer face) contains the catalytic site and the other is a guanosine triphosphate (GTP)-binding protein. So, the enzyme is thought to be activated when GTP arrives at the cytoplasmic face of the membrane; then glucan chains are synthesized and extruded into the wall. These glucan chains then seem to undergo further modification within the wall, where β-1,6-linkages are added to produce the typical branched glucans of fungi. The number of these branched linkages increases markedly behind the apex, but no enzyme has been found that performs this role, so it might occur spontaneously in the wall.

Mannoproteins

These wall polymers are characteristic of yeasts and yeast phases, and there is clear evidence that they are among the few wall components that are pre-formed in the Golgi apparatus and delivered in vesicles to the apex. The addition of mannan chains to the proteins is part of the normal glycosylation function of the Golgi, mentioned in Chapter 2.

Cross-linking

Various types of cross-linkage occur between the major wall polymers after these have been inserted in the wall, and this seems to occur progressively back from the hyphal tip. For example, essentially pure glucans can be extracted from newly formed fungal walls with water or hot alkali, but in the older wall regions an increasing proportion of the glucan is alkali-insoluble, apparently because it is complexed with chitin. In support of this view, the glucans can be extracted after treating walls with chitinase, which degrades chitin. The chitin and glucans are linked by covalent bonds; little is known about the process except that amino acids may be involved, because the amino acid lysine is associated with up to 50% of the glucan–chitin linkages in walls of *Schizophyllum commune* (basidiomycota). In addition to these intermolecular bonds, the individual chitin chains associate with one another by hydrogen bonding, to form microfibrils. The glucans also associate with one another. These additional bondings behind the growing apex could serve to convert the initially plastic wall into a progressively more cross-linked and rigidified structure.

Wall-lytic enzymes

There are opposing views on whether wall-lytic enzymes are necessary for apical growth. On the one hand, it has been suggested that the existing wall must be softened in order for new wall components to be inserted, so that wall growth involves a balance of wall lysis and wall synthesis.

Consistent with this, **chitinase**, **cellulase** (for oomycota) and **β-1,3-glucanase** activities can be found in hyphal wall fractions, although these enzymes might exist usually in latent form. Also, wall-lytic enzymes almost certainly are involved in hyphal branching, when new apices are created from an existing, mature hyphal wall (see later). The occasional bursting of hyphal tips when colonies are flooded with water (see earlier) is advanced as further evidence of the activity of lytic enzymes at the tip. The fact that hyphae can resist considerable turgor pressure (see Chapter 7) has been advanced as evidence that the wall at the apex must be fairly rigid and thus would need to be degraded continuously during growth. On the other hand, recent evidence has shown that growing hyphal tips have a well-developed cytoskeleton which could provide structural support (see Chapter 2), so that the wall at the apex could be truly plastic and not require the activities of lytic enzymes. If lytic enzymes are involved in wall growth then they would be delivered as vesicle components, like the other enzymes.

A steady state model of wall growth

Wessels (1990) proposed a steady state model of tip growth that makes the involvement of wall-lytic enzymes unnecessary. According to this model the newly formed wall at the extreme tip is viewed as being viscoelastic, so that it flows outwards and backwards as new components are added at the tip (Fig. 3.5), then the wall rigidifies progressively by the formation of extra bonds behind the tip. This is remarkably consistent with Robertson's original idea of a plastic, deformable tip wall that is rigidified behind the apex, but based now on much biochemical evidence of wall bonding. If an osmotic shock were to prevent new wall growth by, for example, disrupting the supply of vesicles to the apex, then the rigidification process would continue because it occurs within the wall and is, presumably, less sensitive to osmotic disturbance. But, it is still necessary to explain how hyphae with an essentially fluid (viscoelastic) wall could resist turgor pressure, and the answer to this could be that the meshwork of actin in the apex provides structural support. Jackson and Heath (1990) investigated this for *Saprolegnia ferax* (oomycota). They showed that

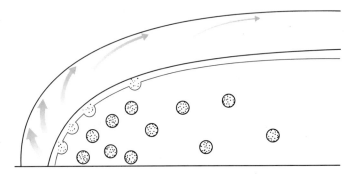

Fig. 3.5 Representation of the steady-state model of hyphal tip growth, in which the wall is envisaged as being viscoelastic. New wall polymers synthesized at the extreme tip are suggested to 'flow' outwards and backwards as more components are added at the extending tip. The decreasing thickness of arrows behind the tip signifies progressively reduced flow, as the wall components become cross-linked. (Based on a diagram in Wessels, 1990.)

treatment of hyphae with cytochalasin E caused disruption of the actin cap and led initially to an increase in the rate of tip extension, but then the tips swelled and burst. The weakest region of the tip, most susceptible to bursting, was not the extreme apex where the actin cap is densest but on the shoulders of the apex where the actin is less dense and where the wall, presumably, has not yet rigidified sufficiently to compensate for the weaker cytoskeleton. It will be recalled that the shoulder of the apex is where new tips were seen to originate when hyphae were flooded with water (see Fig. 3.2).

The driving force for apical growth

Having considered the dynamics of wall growth and wall rigidification, the remaining question concerns the driving force for apical extension. Over the last few years there has been much speculation that electrical or ionic fields could be involved in this, because growing hyphal tips, like many other tip-growing cells (pollen tubes, root hairs, etc.), generate an electric field around them. There could well be an involvement of individual ions, especially calcium, but recent evidence suggests that the electrical fields are intimately involved in nutrient uptake rather than in growth as such, so we will discuss them in Chapter 5. Instead, the cytoskeletal components have emerged as the strongest candidates for the driving force of apical growth. This is compatible with many studies on animal cells where protru-

sions such as pseudopodia are thought to be formed by the polymerization of actin.

Much remains to be learned about the cytoskeleton and its roles in fungal hyphae, but recent studies on *Saprolegnia* (oomycota) have shown that the apex can extend even when hyphae have negligible turgor pressure, presumably because actin polymerization drives this process (reviewed by Money, 1995). There is an abundance of actin in hyphal tips, and both tip extension and cytoplasmic streaming can be halted by treating fungi with the cytochalasins which bind to actin. In *Saccharomyces cerevisiae* there is strong evidence that F-actin is involved in bud formation and that it interacts with the motor protein myosin to transport vesicles to the bud site. Several myosin-like proteins have now been detected in fungi, although they have still to be characterized.

The role of microtubules in fungal growth is more problematical. Hyphal extension can be halted by the benzimidazole fungicides, the related azole drugs and griseofulvin (see Chapter 14), all of which interfere with microtubule function. Coinciding with this stoppage of growth, there is a progressive depletion of vesicles in the hyphal tip (Howard & Aist, 1980). Thus, microtubules must in some way be involved in growth, either by providing a framework for directing vesicles to the tip, or by actively moving the vesicles. In many studies on non-fungal cells the microtubules have been shown to have a principal role in moving nuclei and other major organelles, perhaps by interaction with the motor proteins dynein or kinesin. This also seems to be their role in *Saccharomyces cerevisiae*, where the antitubulin agents have no effect on localized growth of the bud but prevent nuclear division and other organellar movements. However, in other fungi the microtubules are thought to have a more direct role in vesicle movement and localized growth.

Calcium seems to be involved intimately in tip growth (Jackson & Heath, 1993) because the tips of several fungi including *Saprolegnia* (oomycota) and *Neurospora* (ascomycota) require external calcium for continued growth. Moreover, the plasmalemma at the extreme tip is reported to have a high concentration of stretch-activated calcium channels, which allow the ingress of calcium only when the membrane is stretched. The significance of this lies in the fact that the intracellular levels of free calcium are always tightly regulated in cells because calcium is sequestered into intracellular stores such as the endoplasmic reticulum, mitochondria and vacuoles. So, any localized ingress of calcium will cause a perturbation, including an interaction with the cytoskeleton: calcium is known to cause the contraction of F-actin.

The current evidence, therefore, suggests that the cytoskeleton, by interaction with motor proteins and calcium, has a central role in tip growth. The tip might be pushed forwards by actin polymerization (with a viscoelastic tip wall which would not resist this pressure), the protoplasm would stream towards the tip by motor proteins interacting with cytoskeletal components, and vesicles also might be transported to the tip by the cytoskeletal components. Further details of this topic can be found in Gow and Gadd (1994).

How are tips initiated?

Fungi are extremely responsive to environmental factors, and nowhere is this exhibited better than in their ability to produce hyphal tips or to change the orientation of tip growth in response to external signals. Below we consider several examples of this.

Studies on germinating spores

Some fungal spores, such as the uredospores of rusts (basidiomycota), have a fixed point of germination termed the **germ pore**, where the wall is conspicuously thinner than elsewhere. However, many spores seem to be able to germinate from any point, and the process follows a common pattern (Fig. 3.6). Initially, the spore swells by hydration, then it swells further by an active metabolic process and new wall materials are incorporated over most or all of the cell surface. Finally, a germ-tube (young hypha) emerges from a localized point on the cell surface, and all subsequent wall growth is localized to this region. The first sign that an apex will emerge is the localized development of an apical vesicle cluster, and in many cases this coincides with the development of a localized electrical field (see Chapter 5). In any case, we see that a hyphal apex develops *de novo* after an initial phase of non-polarized growth.

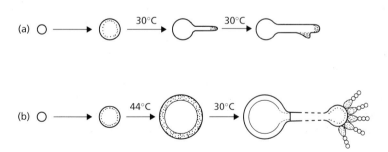

Fig. 3.6 Stages in the germination of spores of *Aspergillus niger*. (a) In normal conditions (e.g. 30°C) the spore swells and incorporates new wall materials uniformly over the cell surface, then a germ tube emerges and new wall materials are incorporated at the germ tube tip (shown by stippling). (b) At 44°C the spore continues to swell and incorporates wall material over its surface, producing a thick-walled giant cell. A germ tube emerges only if the temperature is lowered to about 30°C, and then it immediately forms a sporing head. (Based on Anderson & Smith, 1971.)

In *Aspergillus niger* the transition from non-polar to polar growth is temperature-dependent (Fig. 3.6). At a normal temperature of about 30°C, the spore initially incorporates new wall material over the whole surface and then an apex is formed. At 44°C, however, the spores continue to swell for 24–48 h, producing giant rounded cells of up to 20–25 μm diameter (a 175-fold increase in cell volume) with walls up to 2 μm thick. Then the cells stop growing. But, if these 'giant cells' are shifted down to 30°C before they stop growing they will form an apex, and this behaves in an unusual way: instead of forming a normal hypha it produces a small sporing head (Fig. 3.6). From these experiments we see two things. First, that some process required for normal apical growth of *A. niger* is blocked at high temperatures but, second, this fungus can still grow and 'mature' at the restrictive temperature, so that it reaches a stage of development at which it would, ordinarily, start to sporulate on an agar plate (in a zone aged about 18–24 h behind the growing colony margin). The giant cells become committed to sporulation, so that they produce a sporing head as soon as the temperature is lowered to enable polar outgrowth.

The production of spores from germinating spores with a minimum of intervening growth is termed **microcycle sporulation**. It occurs naturally in some fungi, especially if they grow in water films in nutrient-limited conditions. For example, microcycle sporulation has been reported for some saprotrophs on leaf surfaces (e.g. *Cladosporium, Alternaria* spp.; see Chapter 10), some leaf-infecting pathogens (e.g. *Septoria nodorum*), several vascular wilt pathogens that colonize xylem vessels (e.g. *Fusarium oxysporum*; see Chapter 11) and the rhizosphere fungus *Idriella bolleyi* which is a biological control agent of root pathogens (see Chapter 10). All these fungi will germinate to form normal hyphae in nutrient-rich conditions, so microcycling might be a means of spreading to new environments in nutrient-poor conditions.

Spore germination tropisms

A **tropism** is defined as a directional growth response of an organism to an external stimulus. The spores of some fungi show this very markedly, a classic example being *Geotrichum candidum* (deuteromycota), which is a common spoilage fungus of dairy products. As shown in Fig. 3.7, the cylindrical spores of this fungus germinate typically from one or other pole, but the site of germ-tube emergence is influenced strongly by the presence of neighbouring spores when the spores are seeded densely on agar and covered with a cover-slip. In these conditions the germ-tubes always emerge furthest away from a touching spore—a phenomenon termed **negative autotropism**. The causes of this are still unclear. On the one hand, it has been suggested to involve the release of auto-inhibitors, so that these would accumulate maximally in the zone of contact of two spores but

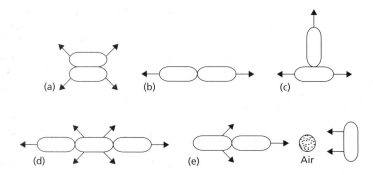

Fig. 3.7 Behaviour of germinating spores of *Geotrichum candidum* when incubated in a thin liquid film beneath a cover-slip. The arrows indicate positions of germ tube outgrowth. (a)–(d) Negative autotropism of spores touching in pairs or in groups: the spores always germinate near their poles, but furthest from a touching spore. (e) The presence of oxygen (a perforation in the cover-slip) overcomes the negative autotropism, and the spores germinate from a point nearest to the oxygen source. (Redrawn from Robinson, 1973.)

could diffuse away from the 'free' ends, leading to germination there. On the other hand, oxygen depletion in the zone of spore contact could be a critical factor for *G. candidum* because the spores always germinated towards an oxygen source (a small hole in a plastic cover-slip placed over the spores) and this positive tropism to oxygen could

overcome the negative autotropism of touching spore pairs (Fig. 3.7).

The spores of *Idriella bolleyi* (deuteromycota) show negative autotropism, but they show an even more spectacular response when placed in contact with cereal root hairs (Fig. 3.8). They germinate away from living root hairs but consistently towards dead root hairs, leading rapidly to penetration of the dead cells. This seems to be ecologically relevant because *I. bolleyi* is a weak parasite of cereal and grass roots. It exploits the root cortical cells as they start to senesce naturally behind the growing root tip, and in doing so it competes with aggressive root pathogens that otherwise would use the dead cells as a food base for infection. Thus, the spore germination tropisms of *I. bolleyi* help to explain its role as a biological control agent of cereal root pathogens (see Chapter 11). The tropic signals for *I. bolleyi* spores seem to be quite specific, because *G. candidum* and some other fungi tested in the same conditions showed quite different responses; for example, *G. candidum* germinated towards both living and dead root hairs (Allan *et al.*, 1992).

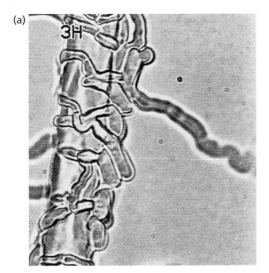

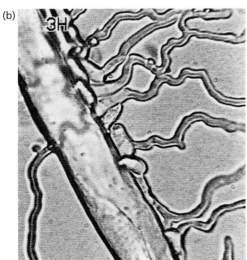

Fig. 3.8 Behaviour of spores of *Idriella bolleyi* on wheat root hairs in aseptic conditions. (a) Spores germinate towards a dead root hair and the germ tubes envelop and penetrate the root hair. (b) Spores germinate from a point away from a living root hair and the germ tubes continue to grow away. (From Allan *et al.*, 1992.)

Fungal spores can respond to electrical fields of sufficiently high strength ($5–20\,V\,cm^{-1}$) in electrophoresis cells. For example, the spores of *Neurospora crassa* and *Mucor mucedo* were found to germinate towards the anode, whereas spores of *Aspergillus nidulans* showed no significant response. The somatic hyphae of these and other fungi showed an array of orientation responses: *Neurospora* hyphae grew towards the anode and formed branches towards the anode; but *Aspergillus* and *Mucor* hyphae grew and branched towards the cathode, and *Trichoderma* hyphae grew slightly towards the cathode but branched towards the anode. Nevertheless, a more recent study (Lever *et al.*, 1994) showed that the galvanotropic responses of somatic hyphae were pH- and calcium-dependent. *Neurospora* hyphae even changed from strongly cathodotropic at pH 4.0 to strongly anodotropic at pH 7. In almost all of these studies the field strengths were considerably higher than those measured around plant roots and probably higher than fungi encounter elsewhere in nature. But, they might help to identify some of the mechanisms of tropism if they can be shown to alter the distribution or charge of plasma membrane proteins (including ion channels) that direct the orientation of growth.

Hyphal tropisms

Ironically, fungi require organic nutrients for growth and yet the hyphae of most fungi have never been shown to exhibit tropism to nutrients. Only the oomycota show this behaviour, and even in these fungi the response can be strain-specific because some strains of a particular species of *Saprolegnia* or *Achlya* orientate towards mixtures of amino acids, whereas other strains show no response. Where it occurs, the response can be very striking: on a nutrient-poor medium the hyphae will turn through 180° to an agar disc containing casein hydrolysate or other amino acid mixtures (Fig. 3.9) and the hyphae also branch from the side closest to the amino acid source. Manavathu and Thomas (1985) investigated this for a strain of *Achlya ambisexualis* and found that, of all the single amino acids tested, only the sulphur-containing amino acid methionine could elicit hyphal tropism. However, gradients of many other single amino acids would elicit tropism if the medium contained a uniform background of cysteine. In explanation of this it was suggested that cysteine, when taken up by cells, could donate one of its sulphydryl (SH) groups to other amino acids and thereby generate methionine. In bacteria the attraction (chemotaxis) of cells to several types of compound is mediated

by chemoreceptor complexes, and this involves a role for methionyl derivatives which donate methyl groups to the interior domains of the receptor complexes (Armitage & Lackie, 1990). A similar system might be involved in chemotropism by the oomycota, because Manavathu and Thomas found that several methyl-donor compounds could elicit a tropic response. In work with a strain of *Achlya bisexualis*, Schreurs *et al.* (1989) found that the hyphal tips orientated towards the tips of micropipettes containing either methionine or phenylalanine, whereas glutamate or tyrosine caused only weak responses, and arginine was a repellent. Also, when the micropipettes containing attractant amino acids were placed behind the hyphal tips, then branches emerged from the hyphae and grew towards the attractants. Thus, it seems that the initiation of branching and the tropism of hyphal tips are closely related responses to environmental signals, and in hyphae of some oomycota these responses might be mediated by receptors for specific amino acids in the plasmalemma. In some way, as yet unknown, the binding of attractants to these receptors causes the localized accumulation of apical growth vesicles.

The hyphae of many fungi show tropic responses to non-nutrient factors of potential ecological relevance. For example, the germ-tubes arising from spores of the arbuscular mycorrhizal fungi (zygomycota; see Chapter 12) can grow towards volatile metabolites (perhaps aldehydes) from roots; some wood-rotting fungi (e.g. *Chaetomium globosum*; see Chapter 10) orientate towards volatile compounds from freshly cut wood blocks, and hyphae of the seedling pathogen *Sclerotium rolfsii* (basidiomycota) orientate towards methanol and other short-chain alcohols from freshly decomposing organic matter (see Chapter 12). Sexual pheromones also elicit orientation responses, as discussed in Chapter 4.

The yeast cell cycle

The process of yeast growth is not fundamentally different from hyphal growth. It is localized to the bud site, probably by the same mechanisms as in apical growth. Consistent with this, electron micrographs show the presence of small vesicles and of cytoskeletal components at the growing tip of bud sites or when septa form to separate a bud

Fig. 3.9 Orientation of hyphae and hyphal branches of an *Achlya* or *Saprolegnia* species when an agar disk rich in amino acids is placed on a colony growing on nutrient-deficient agar.

from the parent cell. The swollen, rounded form of yeasts could also be explained by the mechanisms we have discussed already: non-polar wall growth leads to the production of rounded cells, as in the early stages of spore germination (see Fig. 3.6), and a delay in rigidification of the newly formed cell wall would cause a cell to swell due to turgor pressure. We return to these points in Chapter 4 when we consider mould–yeast dimorphism.

The yeast cell cycle has been studied intensively as a model of the regulation of cell growth and division. As shown in Fig. 3.10, the cycle is divided into four phases, termed G1 (first gap), S (deoxyribonucleic acid (DNA) synthesis), G2 (second gap) and M (mitosis). In each turn of the cycle a bud emerges, grows to nearly full size, receives one of the daughter nuclei from nuclear division and then separates from the parent cell. At its fastest the cycle takes about 1.5 h in *Saccharomyces cerevisiae*, but the time can vary within wide limits, depending on the availability of nutrients. Most of this variation occurs in G1, because together S, G2 and M occupy a more or less constant time.

One of the most important points in this cycle is termed **start**. It occurs during G1 in *S. cerevisiae* and is the stage at which the cell integrates all the information from intracellular and environmental signals to determine whether it will continue the cycle or enter a stationary phase or undergo sexual reproduction. Several cell division cycle (CDC) mutants have been obtained that are blocked at this point when yeast is grown at high temperature (36°C) but not at 30°C; being **temperature-sensitive** mutants, their gene products can be identified by comparing gene expression (e.g. messenger ribonucleic acid (RNA) production) at the different temperatures. Over 50 such genes have been identified, coding for products such as DNA ligase (for splitting DNA), adenylate cyclase (for producing cyclic adenosine monophosphate (AMP)) and enzymes involved in DNA synthesis. Perhaps the most interesting finding, although it should not be surprising, is that these genes are highly conserved in evolutionary terms. One of the CDC genes (termed *cdc2*) of the fission yeast *Schizosaccharomyces pombe* resembles the *CDC28* gene of *S. cerevisiae*; both code for a protein kinase—an enzyme that phosphorylates proteins, using phosphate derived from adenosine triphosphate (ATP), and the phosphorylated proteins then act as regulators of gene expression. A human gene has been found to have 63% base sequence homology with *cdc2*, and it can be transformed into a *cdc2* mutant, overcoming the mutation in the yeast gene so that the cell cycle can continue. Thus, the yeast cell cycle can be used to understand cel-

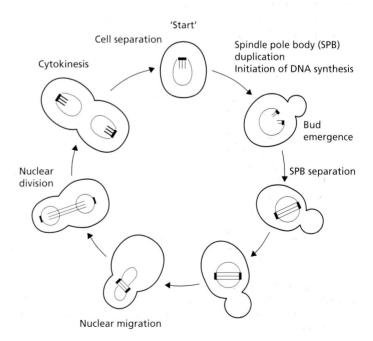

Fig. 3.10 Cell cycle of the yeast *Saccharomyces cerevisiae*. (Based on a drawing by Hartwell, 1974.)

lular events of wider significance, including some virus-induced cancers of humans (Wheals, 1987).

Do mycelial fungi have a cell cycle?

Despite its biochemical complexity, the yeast cell cycle is easy to understand in principle. During G1 the cell accumulates nutrients and synthesizes cellular components, then a decision is made (at 'start') whether to continue the cycle or abort. If the cycle continues then the cell number is doubled and the cells enter G1 again. This raises the question of whether an equivalent cycle occurs in mycelial fungi. The first clear evidence that there is such a cycle was obtained with *Basidiobolus ranarum* (zygomycota), an unusual fungus which can grow as single large cells in the hind gut of frogs but as hyphae with complete, unperforated septa on agar plates. *B. ranarum* also has extremely large nuclei which are easy to see by light microscopy and are arranged regularly, one per cell. As shown in Fig. 3.11, the hyphae extend by tip growth on agar, synthesizing protoplasm and drawing it forwards as the tip grows. When a critical volume of protoplasm has been synthesized, the large central nucleus divides and a septum is laid down at the point of nuclear division, creating two cells. The new apical cell then grows on, and repeats the whole process when enough proto-

plasm has been synthesized. The penultimate cell, which has been isolated by the complete septum, produces a new branch apex and its protoplasm flows into this. In effect, therefore, two hyphal tips (each with a given volume of cytoplasm) are formed from the original one, and there is a clear relationship between cytoplasmic volume, nuclear division and branching, exactly like the cell cycle of yeasts. However, it is termed a **duplication cycle** rather than a cell cycle.

A similar duplication cycle has now been shown to occur in many fungi, even those with normal, perforated septa (Trinci, 1984). To demonstrate this, spores were allowed to germinate on agar and the young colonies were photographed at intervals. From the photographs, the number of hyphal tips and the total hyphal length were recorded at different times and used to calculate a **hyphal growth unit** (G) where:

$$G = \frac{\text{total length of mycelium}}{\text{number of hyphal tips}}$$

After initial fluctuations in the very early stages of growth, the value of G became constant and characteristic of each fungal species or strain. For example, a G value of 48 µm (equivalent to a hyphal volume of 217 µm³) was calculated for *Candida albicans*, and values of 32 µm (629 µm³),

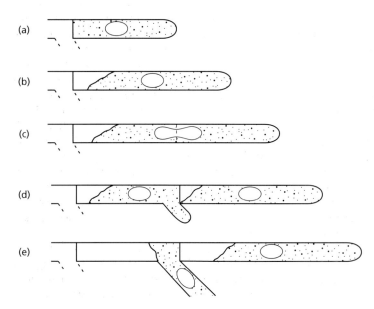

(a)

(b)

(c)

(d)

(e)

Fig. 3.11 The duplication cycle of *Basidiobiolus ranarum*, a fungus that has complete, unperforated septa. (a)–(b) An apical cell extends and synthesizes protoplasm, drawing it forwards during growth. (c) When the protoplasmic volume attains a critical size the nucleus divides and a septum is formed. (d)–(e) The new apical cell grows on to repeat the process; the sub-apical cell forms a branch and the protoplasm migrates into this, forming another apical cell.

130 µm (4504 µm³) and 402 µm (11 986 µm³) were found for a wild-type strain and two 'spreading' mutants of *Neurospora crassa*.

The constancy of these values for individual strains demonstrates that, as a colony grows, the number of hyphal tips is directly related to the cytoplasmic volume. For example, when a colony of *C. albicans* had produced an additional 48 µm length of hypha (or a hyphal volume of 217 µm³) it had synthesized enough protoplasm to produce a new tip. We can thus consider the colony to be composed of a number of 'units' (the hyphal growth units), each of which represents a hyphal tip plus an average length of hypha (or volume of cytoplasm) associated with it. They are not seen as separate units because they are joined together, but they are equivalent to the separate cells produced in the yeast cell cycle. In fact, the duplication cycle of *Aspergillus nidulans* has been shown to be associated closely with a nuclear division cycle. The apical compartment grows to about twice its original length, then the several nuclei divide more or less synchronously and a septum is laid down near the middle of the apical compartment. After this, a series of septa are formed in the new subapical compartment to divide it into smaller compartments, each with just one or two nuclei, while the multinucleate tip grows on and will repeat the process in due course.

Earlier in this chapter we mentioned the peripheral growth zone of a fungal colony—the length of hypha needed to support the normal extension rate of tips at a colony margin; it can be estimated by cutting the hyphae at different distances behind the tips, and it can be as large as 5–7 mm. Clearly, this is quite different from the hyphal growth unit which ranges from about 30 to 400 µm. The difference is explained by the fact that the hyphal growth unit is measured for young colonies in nutrient-rich conditions and is a true reflection of **growth** (increase in biomass, or numbers of tips), whereas the peripheral growth zone is a reflection of the rate of **extension** of a colony margin, and it applies to older colonies, where the hyphae have differentiated. As a colony matures, some of the hyphae become **leading hyphae**, which are much wider and have much faster extension rates than the rest; but, they depend heavily on a supply of protoplasm from behind in order to maintain their high extension rates.

Colony branching and branch behaviour

We have seen that branches arise by the development of new apices as a colony grows and synthesizes new protoplasm. The new apices can arise from almost any point along a hypha, although they seldom develop near the hyphal tip unless this has been disturbed (e.g. see Fig. 3.2). Most often they arise immediately behind the septa, presumably because septa interrupt the flow of cytoplasm to some degree so that vesicles might accumulate there. In any case, new apices develop from the previously mature hyphal wall, so this must be preceded by softening of the wall. It might involve the localized delivery of wall-lytic enzymes, but another possibility is that lytic enzymes might already be present in the wall, awaiting activation.

These points are of more than academic interest, because they relate directly to the behaviour of fungi in natural situations. For example, the mycoparasite *Pythium oligandrum* (oomycota) can penetrate and destroy the hyphae of other fungi within 5 min of hyphal contact, by forming a branch at the point where the contact occurred (Fig. 3.12). Enzymes almost certainly are involved in this penetration process, and they have been assumed to be produced by the mycoparasite. However, the production of wall-lytic enzymes by mycoparasites is an inducible phenomenon — induced by the presence of host wall components (see Chapter 5) — and these enzymes are typically assayed from culture filtrates of the mycoparasites about 24 h after the inducing substrates were supplied. It is debatable whether these inducible extracellular enzymes could be produced (and have time to act) by a mycoparasite within 5 min of contacting a host hypha. The alternative possibility, although unproven, is that *P. oligandrum* might locally activate the wall-lytic enzymes of the host hyphae (Laing & Deacon, 1991). In any case, this example demonstrates the rapidity and precisely localized control of branching and wall dissolution during hyphal interactions (see Chapter 11).

More generally, the pattern of colony branching by fungi is ecologically relevant. When fungi are inoculated onto nutrient-poor agar they form sparsely branched colonies, whereas they form

(a)

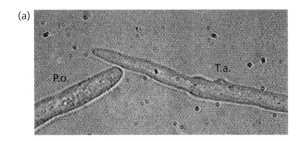

(b)

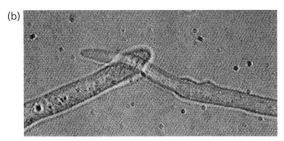

(c)

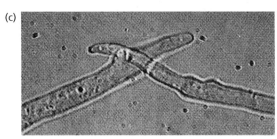

(d)

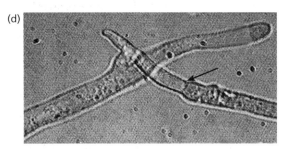

Fig. 3.12 Interaction of the mycoparasite *Pythium oligandrum* (P.o.) with a fungal host *Trichoderma aureoviride* (T.a.), videotaped through a thin film of water agar. (a) A hyphal tip of P.o. contacts the lateral wall of a hypha of T.a. (b) 90 sec later, the hyphal tip of P.o. has grown over the T.a. hypha. (c) 4 min after the initial contact, the P.o. hypha has branched at the contact point and penetrated the T.a. hypha. (d) 7 min after contact, the P.o. branch tip (arrow) is now growing within the host hypha. (See Laing & Deacon, 1991.)

densely branched colonies on nutrient-rich media. Yet, over a wide range of nutrient levels, and even on water agar, the colony margin extends across the agar at much the same rate. In other words, fungi respond to nutrient availability by producing fewer or more branches, but the hyphal tips at the colony margin will continue to extend, by drawing protoplasm forwards from the older regions if necessary. This also occurs in soil, as shown in Fig. 3.13. In this case, a mycorrhizal fungus has grown from the roots of a tree seedling into a non-nutritive peat medium. It grew as an extensive but sparsely branched network in the peat (a **nutrient-searching** strategy) but it formed densely branched patches at localized points

Fig. 3.13 Young larch (*Larix*) seedling, inoculated with the mycorrhizal fungus *Boletinus cavipes*, growing in peat against the face of a transparent root-observation chamber. The mycorrhizal fungus has formed a fan-like system of aggregated hyphae (mycelial cords) in the peat but has also produced densely branched patches of hyphae where it encountered partly decomposed pieces of leaf litter (arrows). (Courtesy of D. J. Read; from Read, 1991.)

where there were small fragments of leaf-litter that provided nutrients (a **nutrient-exploiting** strategy).

Kinetics of growth

Growth in a strict sense can be defined as an orderly, balanced increase in cell numbers or biomass with time. All components of an organism increase in a coordinated way during growth—the cell number, dry weight, protein content, nucleic acid content, and so on.

Figure 3.14 shows a typical growth curve of a yeast in shaken liquid culture, when the logarithm of cell number or dry weight is plotted against time. An initial **lag phase** is followed by a phase of **exponential** or **logarithmic growth**, then a **deceleration** phase, **stationary** phase and phase of **autolysis** or cell death. During exponential growth one cell produces two cells in unit time, two produce four, four produce eight, and so on. This continues until an essential nutrient or oxygen becomes limiting, or until metabolic products accumulate to inhibitory levels. This type of curve is typical of a **batch culture**, i.e. an essentially closed culture system such as a flask in which all the nutrients are present initially. The rate of growth during exponential phase is termed the specific growth rate (μ) of the organism, and if all conditions are optimal then the maximum specific growth rate, μ_{max}, is obtained. This is characteristic of a particular organism or strain. The value of μ is calculated by measuring $\log_{10}$ of the number of cells (N_o) or the biomass at any one time (t_o) and $\log_{10}$ of cell number (N_t) or biomass at some time later (t), according to the equation:

$$\log_{10} N_t - \log_{10} N_o = \frac{\mu}{2.303}(t - t_o)$$

where 2.303 is the base of natural logarithms. Rewritten, this becomes:

$$\mu = \frac{\left(\log_{10} N_t - \log_{10} N_o\right)}{(t - t_0)} 2.303$$

If, for example, $N_o = 10^3$ cells ml^{-1} and $N_t = 10^5$ cells ml^{-1}, 4 h later, then:

$$\mu = \frac{(5-3)2.303}{4} = \frac{2.303}{2} = 1.15 h^{-1}$$

From this we can compute the **mean doubling time**, or generation time (g) of the organism, as the time needed for a doubling of the natural logarithm, according to the equation:

$$g = \frac{\log_e 2}{\mu} = \frac{0.693}{1.15}$$

In our example, $g = 0.60$ h. For *S. cerevisiae* at 30°C, near-maximum values of μ and g are 0.45 h^{-1} and 1.54 h, respectively. For the yeast *Candida utilis* at 30°C, $\mu = 0.40$ h^{-1} and $g = 1.73$ h.

Mycelial fungi also grow exponentially because they have a duplication cycle: averaged for a colony as a whole, they grow as hypothetical 'units', one producing two in a given time interval, two producing four, and so on. Representative values of μ_{max} and g for mycelial fungi are: 0.35 h^{-1} and 1.98 h for *Neurospora crassa* at 30°C; 0.28 h^{-1} and 2.48 h for *Fusarium graminearum* at 30°C, and 0.80 h^{-1} and 0.87 h for *Achlya bisexualis* at 24°C. These values compare quite favourably with those of yeasts. However, it is difficult to maintain exponential growth of mycelial fungi, because the hyphae do not separate from one another. The colonies in a shaken culture tend to form spherical pellets, with growth restricted to the margins of the pellets because nutrients and oxygen diffuse slowly into the pellets and metabolic waste products diffuse slowly outwards. This problem can be overcome to some degree by adding compounds

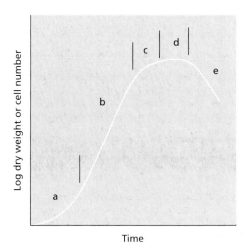

Fig. 3.14 Typical growth curve of a batch culture. (a) Lag phase; (b) exponential or logarithmic growth phase; (c) deceleration phase; (d) stationary phase; (e) phase of autolysis.

that alter the hyphal branching pattern. These **paramorphogens** include sodium alginate, carboxymethylcellulose and other anionic polymers. They cause fungi to grow as more dispersed, loosely branched mycelia, perhaps by binding to hyphae and causing ionic repulsion. This can be desirable in some industrial processes but not in others. For example, dispersed filamentous growth favours the production of fumaric acid by *Rhizopus arrhizus* and pectic enzymes by *Aspergillus niger*, but pelleted growth is preferred for production of itaconic acid and citric acid by *A. niger* (Morrin & Ward, 1989).

Batch culture versus continuous culture

Batch culture systems approximate to the conditions for growth in some natural environments—a 'feast and famine' existence. For example, many *Aspergillus* and *Penicillium* spp. grow rapidly when nutrients are available and they produce many spores for dispersal to new environments, but they stop growing when the nutrients are depleted. Batch cultures also are used commonly in industry because useful primary products such as organic acids and secondary metabolites such as antibiotics (see Chapter 6) are produced in the deceleration and early stationary phases. They are used also in brewing and wine-making, where the culture broth is the marketable product.

Continuous culture systems (Fig. 3.15) are an alternative to batch culture. In these systems, fresh culture medium is added at a continuous slow

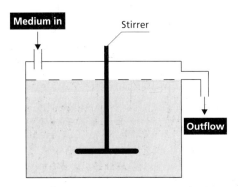

Fig. 3.15 Diagrammatic representation of a chemostat for continuous production of mycelial biomass.

rate, and a corresponding volume of the old culture medium together with some of the fungal biomass is removed continuously by an overflow device. Such culture systems are monitored automatically so that the pH, temperature and dissolved oxygen concentration are maintained at the desired levels. They are stirred vigorously to keep the organism in suspension and to facilitate diffusion of nutrients and metabolic by-products. However, for mycelial fungi the stirrer blades need to be sharp, to fragment the mycelium.

There are various types of continuous culture system but the most common is the **chemostat**, in which the growth medium is designed so that one essential nutrient is at a low concentration while all other components are present in excess. Once growth has started, therefore, one of the nutrients becomes the growth-limiting nutrient, and the rate of growth of the culture can be controlled precisely by the rate at which fresh culture medium is supplied; this is termed the **dilution rate**. However, it is important to note that the fungus is always growing **exponentially**—only the **rate** of exponential growth is governed by the dilution rate. In theory, by adjusting the dilution rate the culture growth rate can be adjusted to any desired level, up to μ_{max} (any further increase in dilution rate would cause 'wash-out' because the cells would be removed by overflow faster than they can grow). In practice, however, these cultures become unstable as they approach μ_{max} because then even a minor, temporary fluctuation in growth rate can cause wash-out.

Chemostats are useful for many experimental purposes, because the physiology of fungi can change at different growth rates or in response to different growth-limiting nutrients. These changes can be studied in detail in chemostats, whereas they occur transiently in batch cultures. For example, when *Saccharomyces cerevisiae* is grown at low dilution rates (slow growth) in glucose-limited culture it uses a substantial proportion of the substrate for production of biomass, but it produces ethanol at the expense of biomass at higher dilution rates. It switches from cell production to alcohol production in conditions favouring rapid metabolism, even though glucose is the growth-limiting nutrient in both cases. Chemostats also are useful for industrial processes, because cells or cell products (e.g. antibiotics) can be retrieved continuously from the overflow medium. In practice, however, most of the traditional industrial processes rely on batch cultures, either because the cost of converting to continuous culture systems does not justify the increased efficiency or because it is difficult to design and operate full production-scale chemostat systems. However, the batch cultures used industrially often are 'fed-batch' systems, in which nutrients or other substrates are added periodically to sustain the production of metabolites, and the cultures are monitored closely to maintain the pH, oxygen level, etc., within strict limits.

Commercial production of fungal biomass (Quorn)

Continuous culture systems have found a recent, major application in the production of a novel foodstuff, termed Quorn mycoprotein. The scientific and technical background to this was described by Trinci (1992). Quorn is produced commercially from chemostat cultures of the fungus *Fusarium graminearum* (deuteromycota), grown at 30°C in a medium of glucose, ammonium and other mineral salts. The mycelium is retrieved continuously from the culture outflow, aligned to retain the fibrous texture and vacuum dried on a

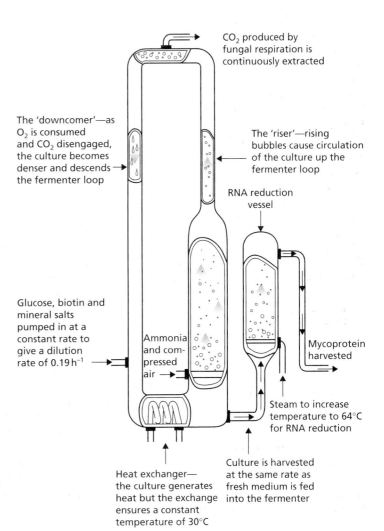

The 'downcomer'—as O_2 is consumed and CO_2 disengaged, the culture becomes denser and descends the fermenter loop

CO_2 produced by fungal respiration is continuously extracted

The 'riser'—rising bubbles cause circulation of the culture up the fermenter loop

RNA reduction vessel

Glucose, biotin and mineral salts pumped in at a constant rate to give a dilution rate of $0.19\,h^{-1}$

Ammonia and compressed air

Mycoprotein harvested

Steam to increase temperature to 64°C for RNA reduction

Heat exchanger—the culture generates heat but the exchange ensures a constant temperature of 30°C

Culture is harvested at the same rate as fresh medium is fed into the fermenter

Fig. 3.16 Diagrammatic representation of the air-lift fermenter used by Marlow Foods for the production of mycoprotein in continuous flow culture. (From Trinci, 1992.)

filter bed before being constituted into meat-like chunks. In this case a continuous culture system was deemed necessary in order to ensure a high degree of reproducibility of the product, and also for economic reasons because the yield of mycelium over a period of time was about five times higher than if a series of batch cultures were used. Glucose is used as the growth-limiting nutrient in the production system, and the dilution rate is set to give a doubling time (μ) of 0.17–0.20 h^{-1}, which is below the μ_{max} of 0.28 h^{-1}. The rate of substrate conversion to protein is extremely high, about 136 g protein being produced from every 1000 g sugar. For comparison with this, the equivalent production of protein by chickens, pigs and cattle would be 49, 41 and 14 g, respectively.

Trinci (1992) has described the many stages in the development of this technology. Initially, a large number of potential fungi was screened to find a suitable organism for commercial use. Then, a suitable large-scale fermenter system had to be found because fungal mycelia have viscous properties in solutions, so the cultures are difficult to mix to achieve adequate oxygenation—the fungus uses 0.78 g oxygen for every 1 g biomass produced. The system used commercially is a 40 m^3 air-lift fermenter (Fig. 3.16) where compressed air is used to aerate and circulate the culture. This avoids the heating that would be caused by a mechanical stirrer, saving on cooling costs. There is a potential problem with the high nucleic acid content of any type of microbial biomass as a human food source, because RNA is metabolized to uric acid which can cause gout-like symptoms. So, the nucleic acid content of the harvested mycelium needed to be reduced while retaining as much of the protein as possible. This was achieved by exploiting the higher heat tolerance of RNAases than of proteases. The culture outflow containing the biomass is collected in a vessel and its temperature is raised rapidly to 65°C and maintained at this for 20–30 min. The growth of the fungus is stopped and the proteases are destroyed so that relatively little protein is lost, but the ribosomes break down and the fungal RNAases degrade much of the RNA to nucleotides which are released into the culture filtrate. Then, the mycelium can be harvested for drying.

A final problem has still not been overcome completely and it limits the efficiency of commercial production. During prolonged culture in fermenter vessels the fungus is subjected to selection pressure and it mutates to 'colonial forms' with a high branching density but relatively low extension rate of the main, leading hyphae. These forms predominate over the wild-type after about 500–1000 h of continuous culture. Their hyphal growth unit lengths range from 14 to 174 μm, compared with 232 μm for the wild-type, so they give a significantly less fibrous biomass, which is undesirable in the end-product. The production runs have to be terminated prematurely to avoid this problem. Nevertheless, Quorn mycoprotein (to comply with legislation it cannot be called 'single-cell protein') is significant as the only human food product to have emerged from the much-heralded 'single-cell protein' revolution. This is a fitting end to our discussion of fungal growth.

References

Allan, R.H., Thorpe, C.J. & Deacon, J.W. (1992) Differential tropism to living and dead cereal root hairs by the biocontrol fungus *Idriella bolleyi*. *Physiological and Molecular Plant Pathology*, **41**, 217–26.

Anderson, J.G. & Smith, J.E. (1971) The production of conidiophores and conidia by newly germinated conidia of *Aspergillus niger* (microcycle conidiation). *Journal of General Microbiology*, **69**, 185–97.

Armitage, J.P. & Lackie, J.M. (eds) (1990) *Biology of the Chemotactic Response*. Society for General Microbiology Symposium 46. Cambridge University Press, Cambridge.

Gooday, G.W. (1995) The dynamics of hyphal growth. *Mycological Research*, **99**, 385–94.

Gow, N.A.R. & Gadd, G.M. (1994) *The Growing Fungus*. Chapman & Hall, London.

Hartwell, L.L. (1974) *Saccharomyces cerevisiae* cell cycle. *Bacteriological Reviews*, **38**, 164–98.

Howard, R.J. & Aist, J.R. (1980) Cytoplasmic microtubules and fungal morphogenesis: ultrastructural effects of methyl benzimidazole-2-yl-carbamate determined by freeze-substitution of hyphal tip cells. *Journal of Cell Biology*, **87**, 55–64.

Jackson, S.L. & Heath, I.B. (1990) Evidence that actin reinforces the extensible hyphal apex of the oomycete *Saprolegnia ferax*. *Protoplasma*, **157**, 144–53.

Jackson, S.L. & Heath, I.B. (1993) Roles of calcium ions in hyphal tip growth. *Microbiological Reviews*, **57**, 367–82.

Laing, S.A.K. & Deacon, J.W. (1991) Video microscopical comparison of mycoparasitism by *Pythium oligandrum*, *P. nunn* and an unnamed *Pythium* species. *Mycological Research*, **95**, 469–79.

Lever, M.C., Robertson, B.E.M., Buchan, A.D.B., Miller,

P.F.P., Gooday, G.W. & Gow, N.A.R. (1994) pH and Ca^{2+} dependent galvanotropism of filamentous fungi: implications and mechanisms. *Mycological Research*, **98**, 301–6.

Manavathu, E.K. & Thomas D. des S. (1985) Chemotropism of *Achlya ambisexualis* to methionine and methionyl compounds. *Journal of General Microbiology*, **131**, 751–6.

Money, N.P. (1995) Turgor pressure and the mechanics of fungal penetration. *Canadian Journal of Botany*, **73**, S96–102.

Morrin, M. & Ward, O.P. (1989) Studies on interaction of Carbopol-934 with hyphae of *Rhizopus arrhizus*. *Mycological Research*, **92**, 265–72.

Read, D.J. (1991) Mycorrhizas in ecosystems — nature's response to the 'Law of the Minimum'. In: *Frontiers in Mycology* (ed. Hawksworth, D.L.), pp. 101–30. CAB International, Wallingford.

Robertson, N.F. (1958) Observations of the effect of water on the hyphal apices of *Fusarium oxysporum*. *Annals of Botany*, **22**, 159–73.

Robertson, N.F. (1959) Experimental control of hyphal branching forms in hyphomycetous fungi. *Journal of the Linnaean Society, London*, **56**, 207–11.

Robinson, P.M. (1973) Oxygen — positive chemotropic factor for fungi? *New Phytologist*, **72**, 1349–56.

Schreurs, W.J.A., Harold, R.L. & Harold, F.M. (1989) Chemotropism and branching as alternative responses of *Achlya bisexualis* to amino acids. *Journal of General Microbiology*, **135**, 2519–28.

Trinci, A.P.J. (1984) Regulation of hyphal branching and hyphal orientation. In: *The Ecology and Physiology of the Fungal Mycelium* (eds D.H. Jennings & A.D.M. Rayner), pp. 23–52. Cambridge University Press, Cambridge.

Trinci, A.P.J. (1992) Myco-protein: a twenty-year overnight success story. *Mycological Research*, **96**, 1–13.

Wessels, J.G.H. (1990) Role of cell wall architecture in fungal tip growth generation. In: *Tip Growth in Plant and Fungal Cells* (ed. I.B. Heath), pp. 1–29. Academic Press, New York.

Wheals, A.E. (1987) Biology of the cell cycle in yeasts. In: *The Yeasts*, Vol. 1 (eds A.H. Rose & J.S. Harrison), pp. 282–390. Academic Press, London.

Chapter 4
Differentiation

Differentiation can be defined as the regulated change of an organism from one state to another. These states can be physiological, morphological or both. So, the germination of fungal spores (see Chapters 3 and 8) or the switch from primary to secondary metabolism (see Chapter 6) are examples of differentiation. But, here we will focus on the changes in fungal morphology (morphogenesis) that lead to a wide range of differentiated structures which serve particular functions. In doing so, we consider both the underlying control mechanisms and the functions of the differentiated structures.

Mould–yeast dimorphism

In general, yeasts and yeast phases are found in environments with high levels of soluble sugars that can diffuse towards the cells, or where the cells can be dispersed in liquid films or circulating fluids. Yeasts have little or no ability to degrade insoluble polymers such as cellulose, and they also have no penetrating power, unlike the mycelial fungi which commonly have these abilities. So, the yeast form and the hyphal form represent two different growth strategies, suited to particular environments, and the behaviour of the relatively few **dimorphic** fungi which can change between a mycelial and a yeast phase illustrates this point. For example, *Candida albicans* grows as a yeast on the moist mucosal membranes of humans, but converts to hyphae for invasion of host tissues (see Chapter 13). The insect pathogens, *Metarhizium* and *Beauveria*, penetrate the insect cuticle by hyphae but then form single, dispersed cells in the circulating fluids of the host (see Chapter 13). The vascular wilt pathogens of plants (e.g. *Fusarium oxysporum*, *Ophiostoma novo-ulmi*) penetrate plants by hyphae but then spread as yeast-like forms in the xylem vessels (see Chapter 12).

The switch between mycelial (M) and yeast-like (Y) phases occurs in response to environmental signals, and it can be reproduced experimentally, as shown in Table 4.1. Some of the opportunistic pathogens of humans grow in the M phase as saprotrophs in plant and animal remains (their normal habitat; see Chapter 13) and grow as mycelia in laboratory culture at 20–25°C, but they convert to budding yeasts or swollen cells in the body fluids or at 37°C in laboratory culture. Other opportunistic pathogens of humans require more specific conditions for the M–Y switch, such as a combination of high temperature and high levels of carbon dioxide or particular nutrients. In contrast, the dimorphic saprotrophic *Mucor* spp. (*M. rouxii*, *M. racemosus*, etc.) do not respond to temperature changes but respond specifically to oxygen levels; they grow as budding yeasts in anaerobic conditions but as mycelia in the presence of even micromolar concentrations of oxygen. *Ustilago maydis* and other plant-pathogenic smut fungi (basidiomycota) are Y in their monokaryotic phase but hyphal in the dikaryotic form (see Chapter 1), so their transition is governed genetically. The vascular wilt pathogens of plants are Y in submerged liquid culture, and often change from M to Y form when colonies are flooding with water. Several insect-pathogenic fungi also are Y in liquid culture but grow as hyphae on solid media.

This range of responses shows that there is no common environmental cue that governs the M–Y transition, and instead we need to consider the underlying control mechanisms. The literature on this topic is vast and is not easy to summarize (see Gow, 1995; Orlowski, 1995) so we deal with only selected aspects to identify some major themes.

Table 4.1 Environmental factors that cause the mycelial (M)–yeast (Y) transition in dimorphic fungi.

	M form	Y form or similar swollen, single-celled or budding phase
HUMAN PATHOGENS		
Histoplasma capsulatum	20–25°	37°
Blastomyces dermatitidis	20–25°	37°
Paracoccidioides brasiliensis	20–25°	37°
Sporothrix schenckii	20–25°	37°
Coccidioides immitis	20–25°	37°
Candida albicans	Serum	High sugars levels
SAPROTROPHS		
Mucor rouxii and other zygomycota	Aeration	Anaerobiosis
PLANT PATHOGENS		
Ophiostoma ulmi	Calcium Some nitrogen sources	Low calcium
Gaeumannomyces graminis		Flooding with water
Phialophora asteris		Flooding with water
Ustilago maydis	Dikaryon	Monokaryon
INSECT PATHOGENS		
Metarhizium anisopliae	Solid media	Submerged liquid culture
Beauveria bassiana	Solid media	Submerged liquid culture

Control of the dimorphic switch

The usual approach to identifying the underlying basis of dimorphism is to grow a fungus in conditions that are, as near as possible, identical except for one factor that changes the growth from M to Y. Then, synchronous populations of M and Y forms can be compared for differences in biochemistry, physiology or gene expression. Even so, it is difficult to establish an obligatory, causal relationship between these differences and a change of cell shape, because the altered environmental factor might cause coincidental changes in biochemistry and gene expression. In fact, most of the differences that have been found are quantitative rather than qualitative. Some examples are given below.

Differences of wall composition

The wall components of M and Y forms sometimes differ: in *M. rouxii* the Y form has more mannose than the M form; in *Paracoccidioides brasiliensis* the Y form has α-1,3-glucan, whereas the M form has β-1,3-glucan; in *C. albicans* the M form has more chitin than the Y form; in *Histoplasma capsulatum* and *Blastomyces dermatitidis* the M form has less chitin than the Y form. Perhaps these differences are not surprising, given that wall composition and wall bonding are intimately linked with cell shape (see Chapter 3). However, the lack of consistency between the fungi suggests that wall composition cannot provide a common basis for understanding dimorphism.

Differences in cellular signalling and regulatory factors

Environmental signals often affect cellular behaviour through a signal transduction pathway, leading to altered metabolism or gene expression. The intracellular factors involved in this include calcium and calcium-binding proteins, pH, cyclic adenosine monophosphate (cAMP) and protein phosphorylation mediated by protein kinases.

The complexing of calcium with the calcium-binding protein calmodulin was found to be essential for mycelial growth of *Ophiostoma ulmi* (Dutch elm disease); otherwise, the fungus grew in a Y form. Consistent with this, the levels of calmodulin typically are low (0.02–0.89 µg g^{-1} protein, for 14 species) in fungi that always grow as yeasts, but seem to be higher (2.0–6.5 µg g^{-1} protein) in mycelial fungi (Muthukumar *et al.*, 1987). An external supply of calcium is required for apical growth of *Neurospora crassa* and many other fungi, and calcium is needed in larger amounts for initiation of the M form than for budding in *C. albicans*. High intracellular levels of cyclic AMP are associated with yeast growth in *Mucor* spp., *C. albicans*, *H. capsulatum*, *B. dermatitidus* and *P. brasiliensis*, whereas low cyclic AMP levels are always associated with hyphal growth. The supply of cyclic AMP externally can also cause a dimorphic switch. Changes in cytosolic pH have been associated with the M–Y transition of *C. albicans*, and with polar outgrowths in some other cell types. Thus, it seems clear that intracellular signalling components are associated with phase transitions, but they are the messengers and mediators not the direct cause of changes in cell morphology.

Differences in gene expression

Differences in gene expression in the M and Y phases can be detected by extracting messenger ribonucleic acid (mRNA) and comparing the mRNA banding patterns by gel electrophoresis. Alternatively, this can be done by comparing the extracted protein profiles, but this does not distinguish between gene expression and the differential translation of existing mRNA. In several cases it has been shown that a few polypeptides are constantly associated with only the M or the Y phase. But, for dimorphic fungi there seems to be no case in which a gene or gene product is obligatorily involved in the generation of cell shape. Harold (1990), in a review of the control of cell shape in general, wrote: '. . . form does not appear to be hard-wired into the genome in some explicit, recognizable fashion. It seems to arise epigenetically . . . from the chemical and physical processes of cellular physiology.'

A potentially unifying theme: the vesicle supply centre (VSC)

Bartnicki-Garcia *et al.* (1995) have used computer simulations as a basis for understanding the dimorphic switch and fungal morphogenesis in general. The central feature of the simulations is a postulated VSC such as a Spitzenkörper, envisaged as releasing vesicles in all directions to 'bombard' the cell membrane and synthesize the wall. As shown in Fig. 4.1, it is then possible to simulate almost any aspect of fungal growth merely by changing the spatial location of the VSC: if it remained fixed then the cell would expand uniformly as a sphere; if it moved rapidly and continuously it would produce an elongated structure, like a hypha; if it moved more slowly it would produce an ellipsoidal cell; if it moved to one side of the hypha it would cause bending; and so on. Thus, the key to morphogenesis could be the rate of displacement of the VSC, and other variations in shape and size could occur if the rate of vesicle release from the VSC were altered relative to its rate of movement. We saw in Chapters 2 and 3 that the apical vesicle cluster is intimately associated with cytoskeletal components that are implicated in cellular movements, so it is not unreasonable to assume that the rate of displacement of a VSC could be altered by factors that affect the cytoskeleton. Recent video-enhanced microscopy of hyphal tips has shown that the Spitzenkörper exhibits random oscillations in growing hyphal tips, associated with oscillations in the direction of growth, and the Spitzenkörper also can divide to leave a 'daughter' VSC at a future branch point.

Infection structures of plant pathogens

Fungal parasites often penetrate directly through an intact host surface, but sometimes they use natural openings such as stomata which they locate by recognizing host surface signals. In any case, infection is preceded by production of a specialized pre-penetration structure (Fig. 4.2). In this section we focus on plant pathogenic fungi, but equivalent structures are produced by insect pathogens, discussed in Chapter 13. The simplest pre-penetration structures are terminal swellings

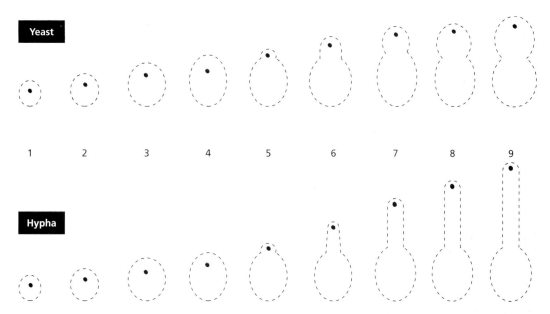

Fig. 4.1 Computer simulation of dimorphism in *Candida albicans*, based on the assumption that wall growth occurs by bombardment of the wall by apical vesicles generated from a vesicle-supply centre (VSC, black dot). The yeast (top series) and hyphal (lower series) shapes were 'grown' simultaneously at the same rate (10 000 vesicles/ frame). Each frame indicates a unit of time. The VSC was moved at different rates to generate the shapes. Between frames four and five the speed of VSC movement was increased four-fold to produce a cell outgrowth; then it was returned to its previous rate to produce the yeast bud but continued at a four-fold rate to generate the hypha. (Based on Bartnicki-Garcia & Gierz, 1993.)

called **appressoria** (singular appressorium) if they occur on germ-tubes, or **hyphopodia** if they develop on short lateral branches of hyphae. More complex **infection cushions** can be formed if a fungus infects from a saprotrophic food base and needs to overcome substantial host resistance; in these cases the fungus penetrates from several points beneath the infection cushion, to overwhelm the host defences.

All these pre-penetration structures serve to anchor the fungus to the host surface, usually by secretion of a mucilaginous matrix. Enzymes such as cutinase sometimes are secreted into this matrix. The penetration process is achieved by a narrow hypha, termed an **infection peg**, which develops beneath the pre-penetration structure. There has been much debate about the relative roles of enzymes and mechanical forces in this process. Enzymes probably are involved locally, to aid the passage of the penetration peg through the host cell wall, but they do not cause generalized dissolution of host walls. Mechanical forces almost

certainly are involved, as demonstrated for *Magnaporthe grisea*, the pathogen that causes rice blast. The penetration pegs of this fungus can penetrate inert materials such as Mylar and even Kevlar, the polymer used to manufacture bullet-proof vests! A turgor pressure of about 8 MPa is generated in the appressorium of this fungus by the conversion of stored glycogen into osmotically active products before the penetration peg develops (Howard *et al.*, 1991). This force is channelled into the narrow peg because the wall of the upper surface of the appressorium is heavily melanized and resists deformation. In contrast, the underside of the appressorium has a very thin wall, or perhaps none at all, but the adhesive released by the appressorium is extremely strong and forms an O-ring seal on the host surface.

Melanization is not found in all appressoria or infection cushions, so its role must not be overstated. However, some of the fungi that depend on melanized infection structures can be controlled by chemicals that specifically block the pathway of

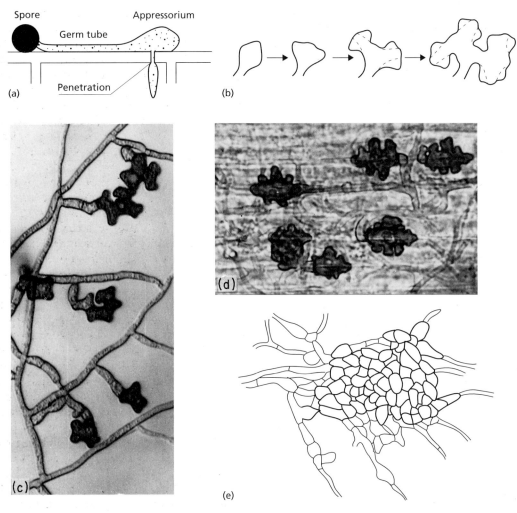

Fig. 4.2 Pre-penetration structures of plant pathogens. (a) Appressorium, produced as a terminal swelling on a germ tube; an equivalent structure formed on a short hyphal branch is termed a hyphopodium. (b)–(d) Lobed, melanized hyphopodia of one of the take-all fungi, *Gaeumannomyces graminis* var. *graminis*; each consists of a single cell. (b) Suggested mode of development of lobed structures by repeated stoppage, swelling and branching of a hyphal tip (see Fig. 3.2). (c) Lobed hyphopodia produced against the base of a Petri dish in agar culture. (d) The same structures formed on the surface of a wheat stem base as a prelude to infection. (e) Infection cushion of the eyespot fungus *Pseudocercosporella herpotrichoides* formed on the base of a Petri dish or on a cereal stem; they consist of plaques of swollen cells produced by repeated apical branching and septation.

melanin biosynthesis. Both *M. grisea* and *Rhizoctonia oryzae* (which causes sheath blight on rice) are examples of this, although control in practice has been disappointing because pathogens can develop resistance by mutations that alter the melanin biosynthesis enzymes (see Chapter 14). Melanized appressoria also are characteristic of *Colletotrichum* spp. that cause leaf and fruit spots, including *C. musae* which causes the small brown flecks on the skins of ripe bananas. These fungi are weak parasites that infect fruit tissues after ripening, when the host tissue resistance has declined. Their splash-dispersed spores can land on the host surface at any time and, being thin-walled and hyaline (colourless), they cannot survive desiccation or exposure to ultraviolet (UV) irradiation. So, the spores often germinate immediately and the germ-tube produces a melanized appressorium which survives on the host surface until the onset of host senescence.

The morphogenetic triggers

The simple sequence in Fig. 4.2(a), where a germ-tube tip swells into an appressorium and a narrow penetration peg develops beneath it, is reminiscent of the 'stop–swell–branch' sequence of hyphal tips that are subjected to osmotic or other shocks (see Chapter 3, Fig. 3.2). The lobed hyphopodia shown in Fig. 4.2(b) could be formed by repetition of the same process. Likewise, infection cushions develop by multiple localized branching. So, all of these structures could represent variations on a simple, basic developmental pattern—the stoppage of hyphal tips, followed by swelling and branch proliferation. Numerous physical and chemical factors have been reported to influence the development of appressoria but the only consistent and absolute requirement is contact with a surface of sufficient physical hardness. *In vivo* this would be a leaf cuticle or insect cuticle, etc. *In vitro* it can be simulated by various artificial surfaces. Thus, **contact-sensing** seems to be a key morphogenetic factor.

Contact-sensing

Contact-sensing can be either **topographical** or **non-topographical**. In the latter case, a fungus responds merely to the presence of a hard surface, and this is true of the air-borne conidia of *Erysiphe graminis* (powdery mildew of cereals). These spores have minute warts on their surface, and within a few minutes of landing on a leaf or of being placed on glass, the warts in contact with the surface secrete an adhesive which contains some wall-degrading enzymes. The warts on the rest of the spore surface do not do this. Topographical sensing is more specific and precise, because several fungi will respond to ridges or grooves of particular heights (or depths) or spacing, and they use this as a means of locating their preferred infection sites. For example (Fig. 4.3) the germ-tubes emerging from spores of *E. graminis* or from

the uredospores of cereal rust fungi (*Puccinia graminis*, *P. recondita*, etc.) initially grow at random on cereal leaves, but when they encounter the first groove on the leaf surface (the junction of two leaf epidermal cells) they orientate and then grow across the leaf surface, perpendicular to the parallel lines of the epidermal cells. This is thought to maximize the chances of locating a stomatal opening, because the stomata are arranged in staggered rows along the leaf (Fig. 4.3). These fungi show precisely the same behaviour on inert grooved surfaces such as leaf replicas or polystyrene, confirming that the response is to topographical signals and not chemical factors.

Evidence is beginning to accumulate for the mechanism of topographical sensing. The germ-tube tip grows along a surface 'nose-down', with the Spitzenkörper displaced towards the surface, and parallel arrays of microtubules are seen to occur just beneath the plasma membrane of the lower surface of the germ-tube, where they could be involved in transducing the contact signals. All of this depends on close adhesion to the surface, mediated by the extracellular matrix, and in the rust *Uromyces appendiculatus* the digestion of this matrix by applying a protease, Pronase E, prevents the topographical signalling. By preparing protoplasts of this fungus and using the patch-clamp technique it has been shown that the plasma membrane at the germ-tube tip contains stretch-activated ion channels. So, it is believed that stretching of the membrane when the germ-tube tip encounters a ridge or groove leads to an ion flux (possibly calcium (Ca^{2+})) through these channels and initiates the cellular response.

The alignment of a germ-tube is the first stage in an intricate developmental sequence for rust fungi, because infection from uredospores always occurs through a stoma. As shown in Fig. 4.4 for the bean rust fungus *U. appendiculatus*, the germ-tube locates a stomatal ridge, stops growing after about 4 min and the apex swells to form an appressorium. The original two nuclei in the germ-tube migrate into the appressorium, then divide and a septum develops to isolate the appressorium from the germ-tube. About 120 min after contacting the stoma, a penetration peg grows into the substomatal cavity, produces a substomatal vesicle, and an infection hypha develops from this to produce a haustorium mother cell on one of the leaf

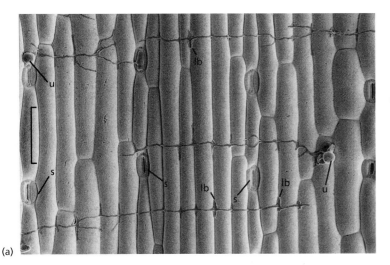

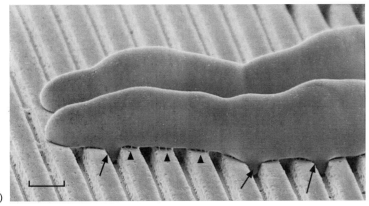

(a)

(b)

Fig. 4.3 Directional growth of rust germ tubes in response to topography. (a) Hyphae growing from uredospores (u) of *Puccinia graminis*, with growth orientated perpendicularly across a replica of the lower surface of a wheat leaf. Bar = 100 μm. Periodically, the hyphae produce short lateral branches (lb) where they encounter grooves between the parallel rows of leaf epidermal cells. The positions of stomata are shown (s). (b) Scanning electron micrograph of two germ tubes growing perpendicularly over the ridges and grooves of a polystyrene replica of a microfabricated silicon wafer. Bar = 4 μm. The germ tube tips have a 'nose-down' orientation which might be involved in topographical sensing. Dried remains of the surface mucilage that adhered the germ tubes to the surface are shown by arrowheads; the hyphae also form projections into the grooves (arrows) resembling the lateral branches seen in Fig. 4.3a. (Photographs courtesy of N. D. Read; from Read *et al.*, 1992.)

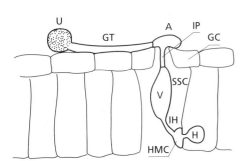

Fig. 4.4 Infection structures of the bean rust fungus, *Uromyces appendiculatus*, penetrating a stomatal opening from a germinating uredospore. U, uredospore; GT, germ tube; A, appressorium; GC, stomatal guard cell; IP, infection peg; V, sub-stomatal vesicle; SSC, sub-stomatal cavity; IH, infection hypha; HMC, haustorial mother cell; H, haustorium. (Based on a drawing by H. C. Hoch & R. C. Staples; see Hoch *et al.*, 1987.)

parenchyma cells. The infection process is completed when the fungus penetrates the host cell to form a **haustorium**, a specialized nutrient-absorbing structure (see Chapter 12).

Initial studies of this sequence for *U. appendiculatus* showed that all the events up to, and including, the development of the infection hypha could be induced in response to 'stomata' on nail varnish replicas of leaf surfaces. They could also be induced by scratches on other artificial surfaces. But, the formation of the haustorial mother cell usually depends on chemical recognition of a leaf cell wall. Thus, most of this developmental process is pre-programmed and it requires only an initial topographical signal. To investigate this further, Hoch *et al.* (1987) exploited the techniques of microelectronics to make silicon wafers with precisely etched ridges and grooves of different heights and spacings. The wafers were then used as templates to produce transparent polystyrene replicas on which rust spores would germinate and the responses to surface topography could be studied. *U. appendiculatus* produced appressoria in response to single ridges or grooves of precise height (or depth), about 0.5 µm, but little or no differentiation occurred in response to ridges lower or higher than this (Hoch *et al.*, 1987). The inductive height corresponds to the height of the lip on the guard cells of bean stomata, which probably are the inductive signal *in vivo*. In a further study, a total of 27 rust species were tested on ridges of different heights (Allen *et al.*, 1991). The species were categorized in four groups.

1 Group 1 included *U. appendiculatus* and seven other species, which produced appressoria in response to a single ridge or groove of narrowly defined height; ridges or grooves higher or lower than this had little effect (Fig. 4.5a).

2 Group 2 included *Puccinia menthae* (mint rust) and three other species, which needed a minimum ridge height (0.4 µm for *P. menthae*) for production of appressoria but also responded to all ridge heights above this, up to at least 2.25 µm (Fig. 4.5b).

3 Group 3 was represented by a single species, *Phakopsora pachyrhizi* (soybean rust) which could form appressoria even on flat surfaces.

4 Group 4 included many cereal rusts (*Puccinia*

graminis, P. recondita, etc.) that did not respond to single ridges of any dimension. However, the cereal rusts have since been shown to differentiate in response to multiple, closely spaced ridges of optimal 2.0-µm height and 1.5-µm spacing.

The finding that either ridges or grooves elicit the same response indicates a minimum requirement for two consecutive right angles as the topographical signal. It is suggested that the different requirements for ridge heights or spacings by different rust species could reflect adaptation to the stomatal topography of the host, consistent with the high degree of host specificity of these biotrophic parasites (see Chapter 12). Further details of contact-sensing and the possible underlying mechanisms can be found in Read *et al.* (1992).

Sclerotia

Sclerotia (singular sclerotium) are specialized hyphal bodies involved in dormant survival. At their most complex they can be dense, three-dimensional structures up to 1-cm diameter, with clearly defined internal zonation (Fig. 4.6). Sclerotia of this type are produced by the omnivorous plant pathogens *Sclerotium rolfsii* and *Sclerotinia sclerotiorum*, the ergot fungus *Claviceps purpurea* and some mycorrhizal fungi such as *Cenococcum geophilum* and *Paxillus involutus* (Fig. 4.6). At the

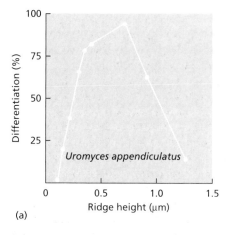
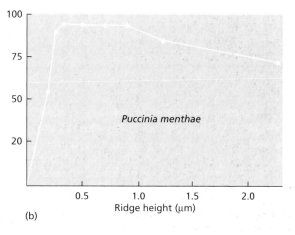

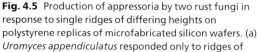

Fig. 4.5 Production of appressoria by two rust fungi in response to single ridges of differing heights on polystyrene replicas of microfabricated silicon wafers. (a) *Uromyces appendiculatus* responded only to ridges of defined height. (b) *Puccinia menthae* responded to all ridges above a minimum height. (Based on Allen *et al.*, 1991.)

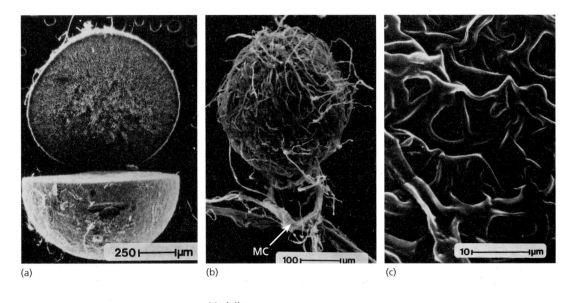

(a)   (b)   (c)

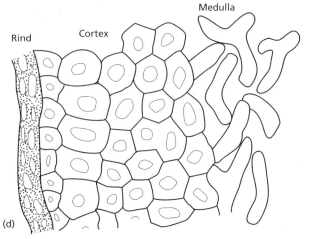

(d)

Fig. 4.6 (a) Scanning electron micrograph of a cut sclerotium of *Cenococcum geophilum*, showing a compact fungal tissue beneath a thin outer rind. (b) Scanning electron micrograph of a young sclerotium of *Paxillus involutus* which developed on a mycelial cord (MC). (c) Outer surface of a mature sclerotium of *P. involutus*, showing a network of modified, flattened and heavily melanized hyphae of the sclerotial rind. (d) Diagrammatic representation of a cross section of the sclerotium of *Sclerotium rolfsii*, showing zonation of the tissues. ((a)–(c) Courtesy of F.M. Fox; see Fox, 1986.)

other extreme, the microsclerotia of the plant pathogen *Verticillium dahliae* are less than 100 μm diameter and are merely clusters of melanized chlamydospore-like cells. Between these extremes lie a range of types, such as the spherical or crust-like sclerotia of *Rhizoctonia solani*, commonly seen as brown patches on potato tubers (black scurf disease). Details of the developmental patterns of sclerotia are given by Willetts and Bullock (1992). Essentially, all sclerotia develop initially by repeated, localized hyphal branching, followed by adhesion of the hyphae and anastomosis of the

branches. As the sclerotial initials expand and mature, the outer hyphae can be crushed and form a rind, while the interior of the sclerotium differentiates into a tissue-like **cortex** of thick-walled, melanized cells, and a central **medulla** consisting of hyphae with substantial nutrient-storage reserves of glycogen, lipids or trehalose (see Chapter 6). These later developmental stages are poorly understood because they cannot be followed in intact sclerotia. Sclerotia can survive for considerable periods, sometimes years, in soil. They germinate in suitable conditions, either by producing hyphae (the **myceliogenic** sclerotia of *S. rolfsii*, *Cenococcum*, *R. solani*, etc.) or by producing a sexual fruiting body (the **carpogenic** sclerotia of *Claviceps* and *Sclerotinia sclerotiorum*). We discuss this further in Chapter 12.

Several experimental treatments can induce the production of sclerotia, but nutrient stress is probably the major environmental trigger. It causes sclerotia to develop rapidly from pre-existing mycelia or from sclerotial initials laid down at an earlier time, and a large proportion of the mycelial reserves are conserved in the sclerotia. Christias and Lockwood (1973) demonstrated this by growing four sclerotium-forming fungi in potato–dextrose broth, collecting the mycelial mats before they had initiated sclerotia, then washing the mats and placing them either on the surface of normal, unsterile soil or on sterile glass beads through which water was percolated continuously to impose a nutrient-stress equivalent to that in soil (see Chapter 9). In all cases the mycelia responded by initiating sclerotia within 24 h, and the sclerotia had matured by 4 days. When these sclerotia were harvested and analysed for nutrient content, they contained up to 58% of the original carbohydrate in the mycelia and up to 78% of the original nitrogen. An example is shown in Fig. 4.7, for *S. rolfsii* on the leached glass beads. Such high levels of carbohydrate conservation could only be explained if some of the original hyphal wall polymers had been broken down and the products were mobilized into developing sclerotia. As we shall see later, wall polymers can be degraded by controlled lysis and used as nutrient reserves to support differentiation.

Translocating organs

All fungi translocate nutrients in their hyphae, but some fungi produce conspicuous differentiated organs for bulk transport of nutrients across nutrient-free environments. Depending on their structure and mode of development, these structures are termed **mycelial cords** or **rhizomorphs**. They are quite common in wood-rotting fungi, including the wood-rotting pathogens of tree roots which spread from one root system to another, translocating nutrients for establishment at a new site. They are formed also by the ectomycorrhizal fungi of tree roots (see Chapter 12), translocating carbohydrates from the roots to the mycelium in soil and translocating mineral nutrients and water back towards the roots. Additionally, mycelial cords are found at the bases of the larger mushrooms and toadstools, serving to channel nutrients for fruitbody development from the mycelium, and some types of sclerotia also are formed on mycelial cords (e.g. see Fig. 4.6b).

Mycelial cords

Mycelial cords have been studied most intensively in *Serpula lacrymans* (basidiomycota) which causes dry rot of timbers in buildings. Once established in timbers, this fungus can spread several metres beneath plaster or brickwork to initiate new sites of decay. It spreads across non-nutritive surfaces as fan-like mycelia which draw nutrients forwards from an established site of decay and which differentiate into mycelial cords behind the colony margin (see Fig. 6.6). The early stage of differentia-

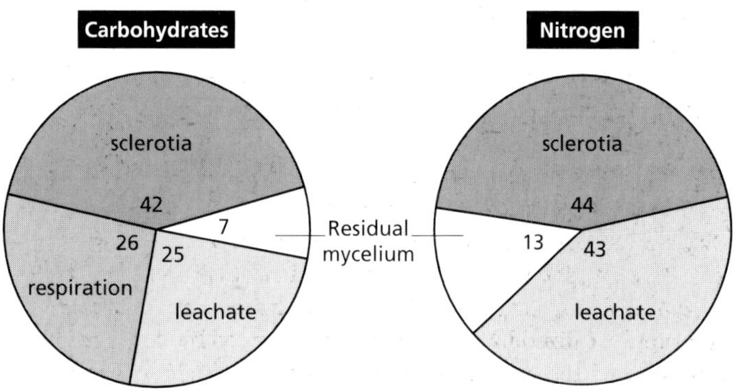

Fig. 4.7 Conservation of mycelial nitrogen (glycine equivalents) and carbon (glucose equivalents) into newly-formed sclerotia of *Sclerotium rolfsii*, four days after mycelial mats were transferred to starvation conditions. The numbers shown are percentages of the original carbohydrate or nitrogen in the mycelia that were recovered in newly-formed sclerotia or in the small amounts of residual mycelium, or that had been 'lost' through respiration or leakage into the glass beads (see text). (From Christias & Lockwood, 1973.)

tion occurs when branches emerge from the main hyphae and, instead of radiating, they branch to form a T-shape and these branches grow backwards and forwards alongside the parent hypha. The branches produce further branches that repeat the process, so that the cord becomes progressively thicker, with many parallel hyphae. Consolidation occurs by intertwining and anastomosis of the branch hyphae and by secretion of an extracellular matrix which cements them together. Some of the main hyphae then develop into wide, thick-walled **vessel hyphae** with no living cytoplasm, while some of the narrower hyphae develop into **fibre hyphae** with thick walls and almost no lumen. Interspersed with these hyphae are normal, living hyphae rich in cytoplasmic contents. The cords of other fungi, such as the mycorrhizal species *Leccinum scabrum* (Fig. 4.8), do not have fibre hyphae but otherwise show a similar pattern of development. In mature hyphal cords there is evidence of a large degree of degeneration of hyphal contents and of the deposition of large amounts of cementing material between the hyphae (Fig. 4.8). So, this seems to be another example of controlled hyphal

lysis contributing to the development of a differentiated structure.

The factors that control the development of mycelial cords are poorly understood, but some of the older studies on *S. lacrymans* suggest that the availability of nitrogen is a key factor. Cords were found to develop on media containing inorganic nitrogen (e.g. nitrate) but not on media containing amino acids. Also, cords growing from a mineral nutrient medium onto an organic nitrogen medium gave rise to normal, diverging hyphal branches. So, it was suggested that cords develop when the parent hyphae leak organic nitrogen in nitrogen-poor conditions, causing branch hyphae to grow close to the parent hyphae in the nitrogen-rich zone. Regulatory control by nitrogen seems logical for wood-decay fungi, because wood has a very low nitrogen content and these fungi could have evolved special mechanisms for conserving and remobilizing their organic nitrogen (see Chapter 10). This could apply also to the cords of ectomycorrhizal fungi, because these fungi have a significant role in degrading organic nitrogen in otherwise nitrogen-limiting soils (see Chapter 12).

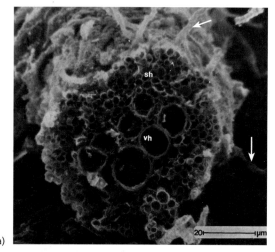

(a)

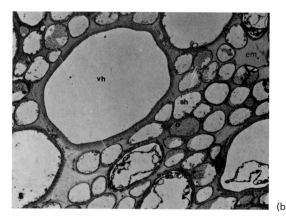

(b)

Fig. 4.8 Mycelial cords of the mycorrhizal fungus, *Leccinum scabrum*. (a) Scanning electron micrograph of a cord, broken to reveal the internal hyphae. Central vessel hyphae (vh) are surrounded by sheathing hyphae (sh). The surface hyphae of the cord are covered with extracellular matrix; the arrows show single hyphae radiating from the cord. Bar = 20 μm. (b) Transmission

electron micrograph of a cross-section of the cord, showing wide, thick-walled vessel hyphae (vh) which are usually empty but a developing vessel hypha (bottom right) contains cytoplasmic remains. Some of the sheathing hyphae (sh) are empty but others contain viable protoplasm (arrows); extracellular matrix is also seen (em). (Courtesy of F. M. Fox; from Fox, 1987.)

In terms of function, mycelial cords have been shown to translocate carbohydrates, organic nitrogen and water over considerable distances between sources and sinks of these materials. The vessel hyphae seem to act like xylem vessels of plants, transporting water by osmotically driven mass flow (see Chapter 6). The combination of their thick walls, the extensive extrahyphal matrix and reinforcement by fibre hyphae could enable the vessel hyphae to withstand considerable hydrostatic pressure.

Rhizomorphs

Rhizomorphs serve similar functions to mycelial cords but have a more clearly defined organization. The notable example is the rhizomorph of *Armillaria mellea*, a major root-rot pathogen of broad-leaved trees. It spreads from tree to tree by growing as rhizomorphs through soil, and it also spreads extensively up the trunks of dead trees by forming thick, black rhizomorphs beneath the bark. These rhizomorphs resemble boot laces, giving *A. mellea* one of its common names, the boot-lace fungus. As shown in Fig. 4.9, the rhizomorph has a specially organized apex or growing point similar to a root tip, with a tightly packed sheath of hyphae over the apex, like a root cap. Behind the apex is a fringe of short hyphal branches. The main part of the rhizomorph has a fairly uniform thickness and is differentiated into zones: an outer cortex of thick-walled melanized cells in an extracellular matrix; a medulla of thinner-walled, parallel hyphae; and a central channel where the medulla has broken down, serving a role in gaseous diffusion. Rhizomorphs branch by producing new multicellular apices,

either behind the tip or by bifurcation of the tip.

Rhizomorphs extend much more rapidly than the undifferentiated hyphae of *A. mellea*, and they can grow for large distances through soil. However, they need to be attached to a food base because their growth depends on translocated nutrients, so one of the traditional ways of preventing spread from tree to tree was by trenching of the soil to sever the rhizomorphs. Almost nothing is known about the developmental triggers of rhizomorphs, except that ethanol and other small alcohols can induce them; similarly, almost nothing is known about their mode of development because they originate deep within an established colony in laboratory culture. However, the behaviour of rhizomorphs is of considerable interest, as shown by the work of Smith and Griffin (1971) on *Armillariella elegans* (related to *A. mellea*). The rhizomorph apex will only grow if it remains hyaline, and this means that the partial pressure of oxygen at the surface of the apex must be 0.03 or less (compared with about 0.21 in air). Above this level, the apex rapidly melanizes, stopping its growth. Yet, growth of the apex is strongly oxygen-dependent, and the fungus seems to resolve this dilemma by a combination of factors. A high respiration rate is maintained at the apex, supported partly by diffusion of oxygen along the central channel, while the surface of the apex is covered by a water film which limits the rate of oxygen diffusion; at 20°C, oxygen diffuses about 10 000 times more slowly through water than through air. The dependence on a water film ensures that rhizomorphs grow naturally at a specific depth in soil, depending on the soil type and the climate. If a tip grows too close to the soil surface then the width of the water film is reduced

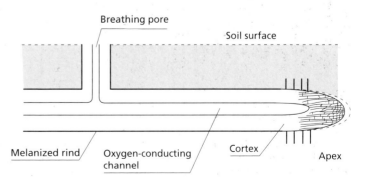

Fig. 4.9 Diagrammatic representation of a rhizomorph of *Armillaria mellea*.

and oxygen diffuses to the tip more rapidly, causing melanization. These tips near to the soil surface then break down to produce 'breathing pores' connected to the central channel. Conversely, if the apex grows too deep into moist soil then the water film increases and the rate of growth becomes oxygen-limited. Thus, the peculiar organization of a rhizomorph helps to regulate growth to specific zones in the soil, and these zones are where tree roots occur, maximizing the opportunities for infection.

Asexual reproduction

The diversity and roles of asexual spores are discussed in Chapter 9. Here, we are concerned only with their development, and in this respect there are two fundamentally different patterns.

1 Sporangiospores are formed by cleavage of the protoplasm within a multinucleate sporangium (chytridiomycota, oomycota, zygomycota).

2 Conidia develop directly from hyphae or special hyphal cells (ascomycota, deuteromycota, basidiomycota).

Sporangiospores

During the development of sporangiospores a hyphal cell undergoes repeated mitotic divisions to produce a sporangium. Then the cytoplasm is cleaved around the individual nuclei by the alignment and fusion of membranes. This is followed by production of a wall around each of the spores in the zygomycota, or by the development of flagella in the motile wall-less spores of chytridiomycota and oomycota. Finally, the spores are released by controlled lysis of all or part of the sporangium wall.

The process of cytoplasmic cleavage seems to occur in several ways. In *Gilbertella persicaria* (zygomycota; Fig. 4.10) and *Phytophthora cinnamomi* (oomycota) a large number of cleavage vesicles are produced by the Golgi body and migrate around the nuclei so that they are aligned; then, they fuse with one another so that their membranes become the plasma membranes of the spores. However, in *Saprolegnia* and *Achlya* (oomycota; see Chapter 1, Fig. 1.5) a large central vacuole develops in the sporangium and forms radiating arms between the nuclei, then the arms fuse with the plasma membrane of the sporangium to delimit the spores. The cytoskeleton is involved intimately in these processes, as Heath and Harold (1992) showed by the use of an actin-specific stain (phalloidin conjugated to rhodamine). The cleavage planes in sporangia of *Saprolegnia* and *Achlya*

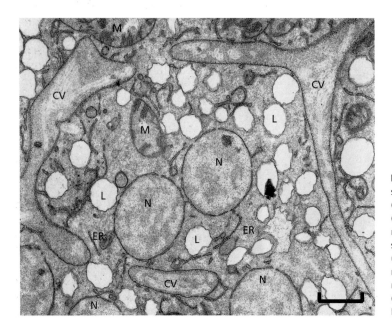

Fig. 4.10 Mid-cleavage stage in the sporangium of *Gilbertella persicaria* (zygomycota). The cleavage vesicles (CV) fuse and extend between the nuclei (N) to separate the sporangium contents into uninucleate spores. M, mitochondrion; ER, endoplasmic reticulum; L, lipid. Bar = 1 μm. (Courtesy of C. E. Bracker; from Bracker, 1968.)

(the expanding vacuolar membranes) were associated with sheet-like arrays of actin which only appeared at the beginning of cleavage and disappeared when cleavage was complete.

In a study of cleavage in *P. cinnamomi* (Hyde *et al.*, 1991) the flagellar axonemes (microtubular shafts) were found to develop in a different class of vesicle: the flagellar vacuoles near each nucleus. These vacuoles then fused with the fusing cleavage vesicles so that the flagella were located on the outside of the spores and surrounded by flagellar membranes. This difference in origin of the flagellar membrane and the plasmalemma of the zoospore body could be significant for zoospore function because there is evidence that different putative chemoreceptors occur on the flagella and on the body of the zoospore (see Chapter 9).

In experimental studies, the cleavage and release of zoospores is strongly influenced by environmental factors; typically, the sporangia must be washed to remove nutrients and then flooded with water to induce cleavage and zoospore release. Several oomycota are sensitive to the antibacterial agent streptomycin; at high concentrations it is toxic, but at sublethal concentrations it interferes with zoospore cleavage, causing the sporangium contents to be released as a multinucleate mass with several flagella and uncoordinated swimming. For this reason, streptomycin was used at one time to control *Pseudoperonospora humuli* (hop downy mildew), although now it has been replaced by conventional fungicides (see Chapter 14). The mode of action of streptomycin in these cases is unclear but might not involve the inhibition of protein synthesis on 70S ribosomes, as occurs in bacteria. Griffin and Coley-Smith (1975) investigated this by adding radiolabelled streptomycin to *P. humuli*. Up to 95% of the label remained on or near the cell surface and could not be removed with water, but it was readily displaced by Ca^{2+}. Consistent with this, streptomycin acts like a divalent cation in solution, so it might interfere with Ca^{2+}-mediated processes by competing for Ca^{2+}-binding sites on or near the cell surface.

There is an interesting evolutionary development in the oomycota. Most of its members have remained essentially aquatic, producing zoospores from sporangia attached to the hyphae; but some of the plant-pathogenic *Phytophthora* spp. and the related downy mildew fungi have

detachable, wind-dispersed sporangia. In the case of *Phytophthora infestans* (potato blight) and *P. erythroseptica* (pink rot of potatoes) these sporangia will undergo cleavage and release zoospores when incubated in water at 12°C or lower, but at higher temperature (around 20°C) the sporangia germinate to produce a hypha. In Britain and other cool regions, zoospores are thought to be the main infective agents of *P. infestans* because the sporangia are produced and land on potato leaves early in the growing season when the cool, wet conditions would favour zoospore production. But, 'direct' germination of sporangia (by hyphal outgrowth) might be more important for infection of the tubers later in the season. Some of the downy mildew pathogens such as *P. humuli* on hops and *Plasmopara viticola* on grapevine typically produce zoospores for infection through the host stomata. However, some other downy mildew pathogens (*Bremia lactucae* on lettuce, *Peronospora parasitica* on cruciferous hosts) have sporangia that usually or always germinate by hyphae which then invade through the host epidermal walls. Sporangia that function in this way are, to all intents and purposes, conidia.

Conidia

Conidia are formed in many ways (see Chapter 1, Fig. 1.9) such as ballooning of a hyphal tip, sequential budding, hyphal fragmentation or extrusion from flask-shaped cells (phialides). These developmental (ontogenic) patterns have been studied intensively, at least partly in an attempt to find natural relationships and thus a natural approach to the classification of the deuteromycota. Details can be found in the books by Kendrick (1971) and Cole and Samson (1979). Although these patterns are diverse, a basic distinction can be made between **blastic** conidia which are formed by a budding or swelling process and then become separated from the parent cell, and **thallic** conidia which are formed essentially by a fragmentation process but might then undergo a swelling phase. At least superficially, these two forms of development parallel the budding process of yeasts and the septation process of fungal hyphae.

Neurospora crassa can be used as an example of blastic conidial development (Fig. 4.11). This fungus sporulates by producing aerial hyphae

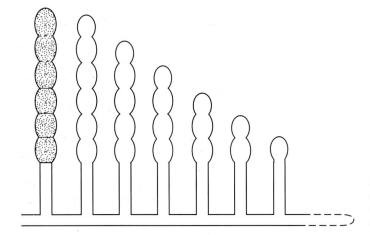

Fig. 4.11 Diagrammatic representation of stages in the development of chains of conidia on aerial hyphae of *Neurospora crassa*.

(conidiophores) that grow for some distance away from an agar surface then swell at their tips. The tip of the swelling produces a broad, bud-like outgrowth which swells and repeats this process, so that a chain of **proconidia** is formed. The proconidia become conidia when septa develop to separate them, starting at the base of the chain. At this stage they develop their characteristic pink colour. Each conidium contains several nuclei because it has developed by the budding of a hyphal (conidiophore) tip. However, in some other types of blastic development the conidia are characteristically uninucleate. For example, in *Penicillium*, *Aspergillus* and *Trichoderma* the conidiophore produces flask-shaped phialides (see Chapter 1, Fig. 1.9), each of which is uninucleate. The phialide extrudes a spore from its tip, and the nucleus divides so that a daughter nucleus enters the developing spore. Then the spore is cut off by a septum and the process is repeated from the phialide. Essentially the same process occurs in *Fusarium* (see Chapter 8, Fig. 8.4), but in this case the single nucleus of the spore undergoes repeated division while the spore elongates and produces septa, so that it becomes a multicellular, multinucleate spore.

Geotrichum candidum (see Chapter 1, Fig. 1.9) is a good example of thallic conidial development. In this case a hyphal branch grows to some length, then stops and develops multiple septa which separate it into short compartments. The septal pores are then plugged and the middle zone of each septum is enzymatically degraded to separate the spores.

Regulation and control of conidiation

Asexual sporulation occurs during normal colony growth, but in zones behind the extending colony margin or in the aerial environment rather than on the substrate. The controlling factors are difficult to study in these conditions because development is not synchronized over the whole colony. Ng *et al.* (1973) overcame this problem by growing *Aspergillus niger* in a chemostat so that the hyphae were in synchronous growth, then the culture conditions could be adjusted to trigger developmental changes. Conidiophores were found to be produced only in nitrogen-limited growth conditions but when the medium was carbon-rich. Apparently, this is the trigger that switches the fungus from vegetative growth to sporulation. However, the conidiophores did not develop further unless the medium was changed to contain nitrogen and a tricarboxylic acid cycle intermediate such as citrate (see Chapter 6). Then, the tips of the conidiophores swelled into vesicles (the large swollen heads) which produced phialides. The production of the conidia from phialides required yet another change — to a medium that contained both nitrogen and glucose. On agar plates this whole developmental sequence occurs in a zone of about 1–2 mm diameter a few millimetres behind the colony margin, possibly because there is a succession of physiological changes in the hyphae, but only by growing the fungus in chemostat culture was it possible to show that each stage is differently regulated. An even more comprehensive study of sporulation has been made with

Aspergillus nidulans, where the sequential activation of many sporulation-related genes has been demonstrated (Adams, 1995). The details are complex, but essentially the genes are suggested to fall into three categories:

1 those involved in the switch from somatic growth to sporulation;
2 those that regulate the developmental stages of sporulation;
3 those that govern secondary aspects such as spore colour.

Although many fungi such as *Aspergillus* and *Penicillium* can sporulate in darkness, some require a light trigger. The most common response is to near-UV irradiation (NUV; 330–380 nm wavelength) which can induce sporulation after a short (1 h) exposure if the colony is then kept in darkness. However, a subsequent exposure to blue light can reverse the process because the photoreceptor exists in alternating forms, one responsive to NUV and one responsive to blue light. *Botrytis cinerea*, which causes grey mould of strawberries and other soft fruits, behaves in this way. It never becomes wholly committed to sporulation, as Suzuki *et al.* (1977) demonstrated by subjecting the colonies to 1-h exposures of NUV and blue light in different sequences. As shown in Fig. 4.12, at almost any stage in the sporulation sequence an exposure to blue light caused the fungus to form hyphal outgrowths instead of continuing the developmental pathway.

Role of hydrophobins in differentiation

Hydrophobins are a recently discovered class of small proteins secreted by fungi (Wessels, 1996). They are the products of genes expressed abundantly during differentiation, suggesting that they might have important roles in the differentiation process. All the hydrophobins studied to date have similar structure and properties. They consist of about 100 amino acids, including eight cysteine residues that occur in a specific pattern so that the protein can fold to produce a highly hydrophobic domain. The hydrophobins are soluble in water, but at an interface with air they self-assemble into a film with a hydrophobic and a hydrophilic face. As shown in Fig. 4.13, this is thought to happen when a hypha forms an aerial branch so that the aerial portion is coated by a hydrophobic film. The potential significance for differentiation is that hyphae with hydrophobic surfaces might interact with one another in specific ways, as discussed later. For conidial fungi the most obvious significance is that hydrophobins could affect the surface properties and thus the functions of the spores (see Chapter 9). The conidia of *A. nidulans*, *A. fumigatus* and *N. crassa* are seen to have a pattern of rodlets on their surface, but the rodlets are absent when the hydrophobin genes are inactivated by targeted gene disruption. Identical rodlets are formed when solutions of pure hydrophobins are allowed to dry.

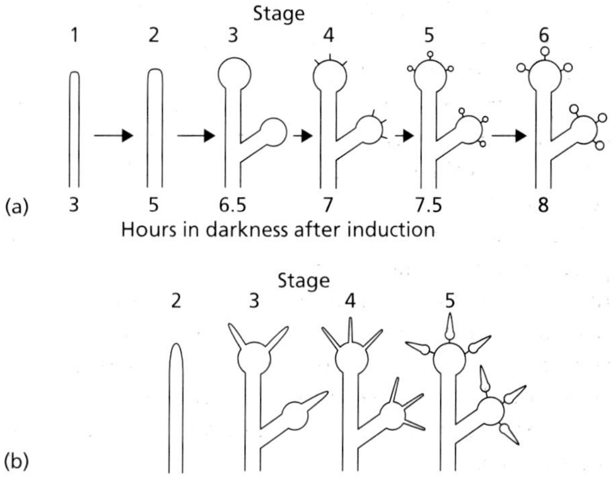

Fig. 4.12 Diagrammatic representation of stages in the development of conidia of *Botrytis cinerea*. (a) After exposure to near-ultraviolet (NUV) irradiation, the aerial hyphae are transformed into branched conidiophores, which swell at their tips to form small vesicles; the conidia are formed on minute projections from the vesicles, so that the mature clusters of conidia resemble bunches of grapes. (b) The effect of a short exposure to blue light at different stages after triggering of development by NUV. Differentiation is arrested at whatever stage had been reached and the fungus grows on as hyphal projections. (Based on Suzuki *et al.*, 1977.)

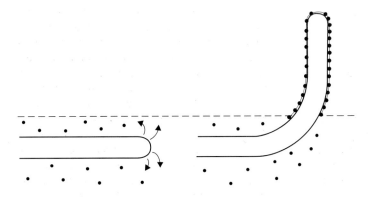

Fig. 4.13 Diagram to illustrate the release of hydrophobin proteins from hyphae of *Schizophyllum commune* that are submerged in a culture medium, and polymerization of the hydrophobins on the hyphal surface when hyphae grow into an aerial environment. (Based on a diagram in Wessels, 1996.)

Sexual development

Sexual reproduction in all organisms involves three fundamental events: (i) the fusion of two haploid cells (**plasmogamy**) so that their nuclei are in a common cytoplasm; (ii) nuclear fusion (**karyogamy**) to form a diploid; and (iii) **meiosis** to produce recombinant haploid nuclei. Depending on the fungus, these events can occur in close succession or separated in time; also, they occur at different stages of the life cycle according to whether the fungus is normally haploid or diploid. These points were outlined in Chapter 1 (see Figs 1.5–1.10).

Two further points must be made. First, some fungi are **homothallic** (self-fertile) but many are **heterothallic** (out-crossing) in which case sexual reproduction is governed by **mating-type** (compatibility) genes. Often, there are two mating types, governed by a single gene locus (**bipolar compatibility**); in these cases the alternative genes are not an allelic pair but usually are quite different from one another so they are called **idiomorphs**. Some basidiomycota have two mating-type loci (**tetrapolar compatibility**) with multiple idiomorphs at each locus. In these cases a successful mating occurs between two fungal strains that differ from one another at each gene locus. In most fungi the mating-type genes are regulatory genes, producing protein products that bind to deoxyribonucleic acid (DNA) and control the expression of several other genes.

The second point is that the sexual spores of many fungi function as dormant spores (e.g. zygospores, oospores, ascospores and some basidiospores), so sexual reproduction serves an important role in survival and the sexual spores are produced at the onset of unfavourable conditions for growth. Fungi that do not undergo regular sexual reproduction sometimes produce alternative survival structures such as chlamydospores, sclerotia or melanized hyphae. Other fungi adopt different strategies. For example, *Pythium oligandrum* produces sexual spores by parthenogenesis, and *Saccharomyces cerevisiae* and a few other ascomycota undergo regular **mating-type switching** to ensure that there will always be a mixture of the two mating types (termed **a** or α) in a population. In this case, every time that an **a** cell buds, the parent cell switches to α while the daughter cell remains **a**. In fact, wild populations of *S. cerevisiae* are always diploid because the **a** and α cells fuse and undergo karyogamy, then the diploid cell buds to form further diploids which can respond rapidly to unfavourable conditions by undergoing meiosis and producing ascospores. For this reason, all the laboratory strains used by yeast geneticists have been mutated so that they do not undergo mating-type switching, and this enables the sexual crossings to be controlled experimentally.

Against this background, the rest of this chapter will focus on two topics: (i) the roles of the mating-type genes, especially in hormonal regulation (reviewed by Gooday & Adams, 1993); and (ii) the development of fruiting bodies of basidiomycota because these are the most advanced differentiated structures in the fungal kingdom.

Mating and hormonal control

Chytridiomycota

Most of the information on the control of sexual reproduction in chytridiomycota has come from studies on *Allomyces* spp. (see Chapter 1, Fig. 1.6 for the life cycle). These fungi are homothallic, but they produce motile male and female gametes (sex cells) from different gametangia. The female gametes are larger, hyaline and they release a pheromone, **sirenin** (Fig. 4.14), to attract the male gametes. The male gametes are small, and orange coloured due to the presence of a carotenoid pigment. Sirenin is a powerful attractant, active at concentrations as low as 10^{-10} M, but optimally at 10^{-6} M. Compounds that attract cells at these concentrations can be assumed to alter the swimming pattern by binding to a surface-located receptor (see Chapter 9). Although there is no information on this possible receptor, the male gametes are known to rapidly inactivate sirenin, which could aid their movement up a concentration gradient. It is interesting that both sirenin (from the female gametes) and the carotenoid pigment of the male gametes are produced from the same precursor isoprene units $[CH_2=C(CH_3)-CH=CH_2]$ (see Chapter 6). So, this is an example where cells of the same genetic make-up show different biochemical properties because they have been produced in different gametangia, separated from one another by a complete cross-wall.

Oomycota

The oomycota can be homothallic (e.g. most *Pythium* spp.) or heterothallic with two mating types. But, in all cases a single colony produces both the 'male' and 'female' sex organs (antheridia and oogonia; see Chapter 1, Fig. 1.5), so the mating-type genes govern compatibility, not the development of the sex organs themselves. In a few cases, notably in *Achlya*, a single colony can show 'relative sexuality' — it will behave as a male (fertilizing the oogonium from its antheridium) or 'female' depending on the strain with which it is paired.

The hormonal control of mating has been studied intensively in *Achlya*. The hormones are steroids (Fig. 4.14) derived from the isoprenoid pathway, like sirenin discussed above. They were discovered by growing 'strong female' and 'strong male' strains in separate dishes of water, then passing the water over colonies of the opposite strain. The hyphae of the female produce the hormone **antheridiol**, which causes the male strain to increase its rate of cellulase enzyme production and to form many hyphal branches (the antheridial branches; see Chapter 1, see Fig. 1.5), perhaps because cellulase helps to soften the wall. Once it has been triggered by antheridiol, the male strain produces other steroid hormones, termed **oogoniols**, which trigger the development of oogonia (sex organs) in the female strain. In normal conditions the antheridial hyphae then grow towards the oogonia, clasp onto them and produce fertilization tubes to transfer the 'male' nuclei into the oogonium (see Chapter 1, Fig. 1.5). Similar hormonal systems might occur in other oomycota such as *Pythium* and *Phytophthora* spp., but they have not been characterized. A notable feature of *Pythium* and *Phytophthora* is that they cannot synthesize sterols from non-sterol precursors, and often do not need sterols for somatic growth, but they always require trace amounts of sterols for both sexual reproduction and asexual reproduction. They would be able to obtain them from a plant host or other sources in nature.

Zygomycota

The zygomycota can be homothallic, or heterothallic with two mating types (termed 'plus' and 'minus'). One of the roles of the mating-type genes is to regulate the production of hormone precursors (**prohormones**) from β-carotene, another product of the isoprenoid pathway already mentioned. This hormonal system is shared by many members of the subgroup Mucorales because, for example, a plus strain of *Mucor* can elicit a sexual response from a minus strain of *Rhizopus*, even though the development of a hybrid is blocked at a later stage due to incompatibility. As shown in Fig. 4.14, the prohormones of the plus and minus strains are similar molecules but they differ slightly, owing to the different enzymes of the two strains. The prohormones are volatile compounds which diffuse towards the opposite mating type;

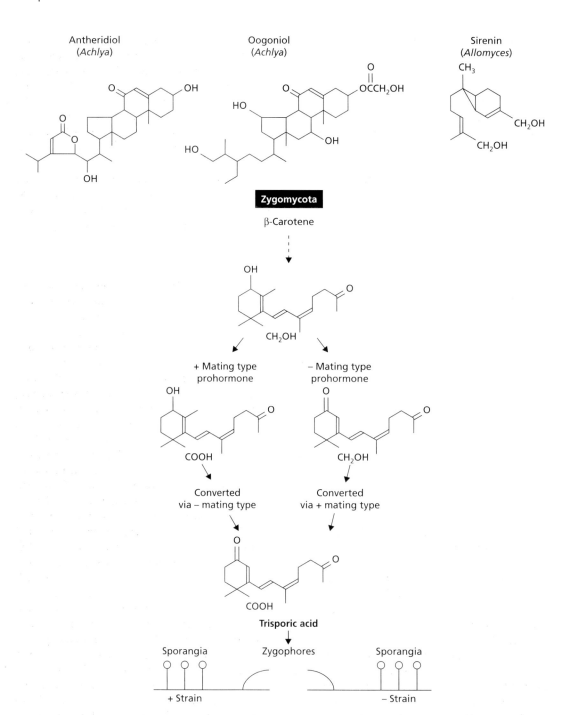

Fig. 4.14 Some pheromones that regulate sexual reproduction of fungi. In zygomycota the two mating types (+ and −) produce different volatile hormone precursors (prohormones) from β-carotene. The prohormones diffuse towards the strain of opposite mating type and are converted to active hormones, the trisporic acids, by the complementary enzyme systems of the strains. The active compounds induce the production of aerial sexual hyphae (zygophores) which grow towards one another. Further development then occurs as shown in Fig. 1.7.

then they are absorbed and converted to active hormones, the **trisporic acids**, by the complementary enzymes of the opposite strain. Initially, only low levels of the prohormones are produced by each strain, but trisporic acids cause gene derepression so that increasing levels of prohormones are produced. The trisporic acids produced by this mutual escalation cause the production of **zygophores** (aerial sexual branches) which then grow towards one another and lead to sexual fusion (see Chapter 1, Fig. 1.7).

Ascomycota

The ascomycota usually have two mating types, termed **A** and **a**, but **a** and α in *Saccharomyces*. These mating-type genes are regulatory elements, controlling the activities of other genes, but only the system in yeasts has been well characterized. One of the chromosomes (III) of the yeast *S. cerevisiae* has a *MAT* gene locus, which is flanked by two other loci termed *MATa* and *MATα*. Haploid cells will normally behave as α mating type, but can change to **a** mating type when the information at the *MATa* locus is copied and transferred into the *MAT* locus (the expression of the *MATα* locus is then shut down). This is a natural transposition event, which causes the switching of mating type every time a cell produces a bud. The *MAT* locus is a regulatory locus, governing the expression of several genes. These include the following.

1 The genes responsible for producing hormones termed **a-factor** and **α-factor**, consisting of 13 and 12 amino acids, respectively.

> **a**-Factor: NH_2–Trp–His–Trp–Leu–Gln–Leu–Lys–Pro–Gly–Gln–Pro–Met–Tyr–COOH.
> **α**-Factor: NH_2–Tyr–Ile–Ile–Lys–Gly–Val–Phe–Trp–Asp–Pro–Ala–Cys(S-farnesyl)–$COOCH_3$.

2 The genes that code for hormone receptors on the cell surface.

3 Genes that code for cell-surface agglutinins.

The α strains produce α-factor constitutively and it diffuses to **a** cells where it is recognized by a specific receptor. This receptor binding causes the growth of **a** cells to be arrested at 'start' during G1 of the cell cycle—the only stage at which they are competent to mate (see Chapter 3). The **a** strain then produces **a**-factor which diffuses to the α strain where it is, again, recognized by a specific receptor. Receptor binding in both cases leads to

other changes in the cells: they produce short outgrowths which function as conjugation tubes, and the surface of these is covered by a strain-specific glycoprotein, so that when an **a** cell and an α cell make contact they adhere tightly by their complementary agglutinins. The conjugation tubes then fuse, and the two nuclei fuse to form a diploid cell. At this stage the cell can go on to produce a diploid budding colony or it can undergo meiosis, depending on whether the environmental conditions are suitable for growth. An essentially similar hormone and receptor system to this is found in other yeasts such as *Pichia* and *Hansenula* spp.

Basidiomycota

Most of the basidiomycota are heterothallic, with mating-type genes at either one locus or two loci, termed A and B. There are multiple idiomorphs at each locus, and a successful mating will occur between any two strains that differ from one another at each locus. This greatly increases the chances of finding a mate, compared with fungi that have only two idiomorphs. The basic mating behaviour of basidiomycota was described in Chapter 1 (see Fig. 1.10). The strains derived from haploid basidiospores are monokaryons and they can fuse with other compatible monokaryons to form a dikaryon. All subsequent growth involves the synchronous division of the two nuclei in each compartment and their regular distribution as nuclear pairs throughout the mycelium. In several basidiomycota this regular arrangement is aided by the production of clamp connections (Fig. 4.15), whereby one of the daughter nuclei enters a clamp branch and migrates through this to the cell behind. However, the functional significance of this is unclear, because other basidiomycota do not form clamp connections and yet still have a regular nuclear distribution. Eventually, the dikaryotic colony will produce a fruitbody, and nuclear fusion and meiosis occur in the basidia.

By making experimental pairings between strains with the same A or the same B allele, it has been possible to deduce the regulatory roles of the A and B loci (see Casselton *et al.*, 1995). As summarized in Table 4.2, pairings of fully compatible strains (with different A and B idiomorphs) lead to reciprocal exchange of nuclei, because the dolipore septa which normally prevent nuclear migration

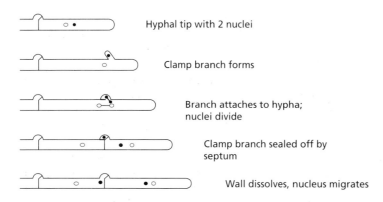

Hyphal tip with 2 nuclei

Clamp branch forms

Branch attaches to hypha;
nuclei divide

Clamp branch sealed off by
septum

Wall dissolves, nucleus migrates

Fig. 4.15 Development of clamp
connections, characteristic of some
members of the basidiomycota.

are digested, and clamp branches are formed at each septum to fuse with the compartment behind the septum. In pairings of strains with the same A and B idiomorphs (common A, common B) there is no septal breakdown, no nuclear migration, no dikaryotization and no clamp connections. Pairings of strains with different A idiomorphs but common B idiomorphs show nuclear pairing, synchronous division of the nuclei and formation of clamp branches; but, the dolipore septa do not break down, and the clamp branches do not fuse with the parent hypha. Pairings of strains with different B but common A idiomorphs lead to septal dissolution but none of the other events. Thus, it can be concluded that the **A locus** (common B pairings) controls the pairing and synchronous division of nuclei and also the formation of clamp branches, whereas the **B locus** controls septal dissolution and the fusion of clamp branches. The dissolution of septa coincides with a marked

increase in the activity of a β-**glucanase** in the hyphae, indicating that the B locus controls the derepression of glucanase genes. In some basidiomycota there is evidence of pheromonal control of mating, again regulated by the mating-type loci.

Development of fruitbodies

The toadstools, brackets and other fruitbodies of basidiomycota are the largest and most complex differentiated structures in the fungal kingdom. Their development is correspondingly complex and still only poorly understood. Here, we consider one example where a start has been made to dissect this process at the biochemical and molecular level, and we end with a discussion of the commercial mushroom because of its economic importance. We shall return to this topic in Chapter 7, where we consider the effects of environmental factors on fruitbody production.

Pairing of strains with:	Events observed
Different A, different B idiomorphs (**dikaryon**)	**1** Septal dissolution **2** Nuclear migration **3** Clamp branches arise and fuse with hypha
Common A, different B idiomorphs	**1** Septal dissolution **2** Nuclear migration
Common B, different A idiomorphs	**1** Septa remain intact **2** No nuclear migration **3** Clamp branches arise but do not fuse
Common A, common B idiomorphs	**1** No septal dissolution **2** No nuclear migration **3** No clamp connections

Table 4.2 Roles of mating-type genes in basidiomycota.

Schizophyllum commune

S. commune is specially suited for laboratory studies because it grows readily in agar culture and produces its small (about 1 cm) fan-shaped fruitbodies in response to light. Actually, this trigger leads only to the development of fruitbody primordia—compact clusters of hyphae which are overarched by other hyphae. Further development from the primordia occurs when carbon nutrients are depleted from the medium, and is then fuelled by carbon reserves within the mycelium. Early in this process the mycelial storage compounds such as glycogen are converted to sugars, which are translocated to the developing primordia. Then, as the sugar levels in the hyphae decline, the hyphal walls begin to break down and the breakdown products are translocated to the primordia. The wall glucans seem to provide the major source of sugars, because fruitbody development is associated with a marked rise in glucanase activity in the mycelia. We have already seen that synthesis of this enzyme is derepressed by the B mating-type locus, but it is still subject to catabolite repression by sugars; so, its generalized activity in the hyphae, as opposed to its localized activity in degrading septa, depends on depletion of the mycelial sugar reserves. The breakdown of hyphal walls to recycle nutrients for differentiation is, in fact, quite common in fungi. An example was seen earlier in the production of sclerotia (see Fig. 4.7). The breakdown of wall glucans also fuels the developing ascocarps of *Emericella nidulans* (sexual stage of *Aspergillus nidulans*).

Wessels and his colleagues (see Wessels, 1992) identified several differentiation-associated genes in *S. commune*. In order to do this, they crossed and repeatedly back-crossed strains to generate monokaryons that were essentially isogenic except for the mating-type genes. Then, the monokaryons, and dikaryons synthesized from them, were compared for their production of polypeptides and mRNAs when grown in different conditions. Any differentiation-specific mRNAs in the dikaryon were confirmed by making complementary DNA (cDNA) and testing this for lack of hybridization to the mRNA from the monokaryons. All these comparisons were made in two sets of conditions.

1 For 2-day-old colonies, when the monokaryons and dikaryons were growing as mycelia with similar colony morphology.

2 For 4-day-old colonies grown in light, when the monokaryon had produced copious aerial hyphae but the dikaryon had produced numerous small fruitbodies.

The following principal findings emerged from this work.

1 For the 2-day colonies, 20 or so proteins were found only in the monokaryon and 20 or so only in the dikaryon. Yet, there was no detectable difference in the bands of mRNA, suggesting that the same mRNAs are transcribed but their products undergo different post-translational modification in monokaryons and dikaryons.

2 For the 4-day colonies, eight proteins were found only in the monokaryon, and 37 only in the dikaryon. Some of these 37 occurred only in the fruitbodies; others were found in both the fruitbodies and the mycelium of the dikaryon. These proteins included some that were secreted into the growth medium.

3 For the 4-day colonies, about 30 unique mRNAs were found only in the dikaryon, whereas no unique mRNA was found in the monokaryon. cDNA was used as a probe to assess the levels of the 'fruiting-associated mRNAs' during development. They were scarce in young vegetative colonies of both strains, and they remained scarce in the monokaryon, but they increased in the dikaryon when this began to fruit.

4 Some of the excreted proteins were the cysteine-rich **hydrophobins**, mentioned earlier; in fact, the hydrophobins were first discovered in this work on *S. commune*. The gene (*SC3*) for one of these hydrophobins was expressed by both the monokaryon and dikaryon during the emergence of aerial hyphae; this hydrophobin is now known to cover the aerial hyphae and the hydrophobic hyphae on the fruitbody surface, but the gene is not expressed by hyphae that make up the main fruitbody tissue. Three other hydrophobin genes (*SC1*, *SC4* and *SC6*) were highly expressed only in the dikaryon and especially in the developing fruitbody tissues (Wessels, 1996). It is suggested that the specific properties of different hydrophobins might affect the surface interactions of hyphae, contributing to the development of fruitbody tissues.

Commercial mushrooms: the exploitation of differentiation

Mushroom production is a substantial industry. The major cultivated mushroom *Agaricus bisporus* (or *A. brunnescens*) had an annual global farm-gate value of approximately £1 billion in 1991, and this represented only 37% of the total production of cultivated mushrooms because other species grown commercially include the oyster mushroom *Pleurotus ostreatus* (22%), the Shiitake mushroom *Lentinula edodes* (12%) which is grown on logs, and *Volvariella volvacea* (6%) grown on rice straw in south-east Asia (see Chapter 10).

The commercial production process for *A. bisporus* is described by Flegg (1985). A mixture of composted straw and animal dung is pasteurized and placed in wooden trays, then inoculated with a commercially supplied 'spawn' consisting of sterilized cereal grains permeated with hyphae of *A. bisporus*. The spawn is allowed to 'run' for 10–14 days so that the fungus thoroughly colonizes the compost. Then, a thin casing layer of pasteurized, moist peat and chalk is added to the compost surface. Over the next 18–21 days the fungus colonizes this casing layer by means of mycelial cords, and fruitbodies are produced on these cords. The cropping of fruitbodies is done over a 30–35-day period, because the fungus produces 'flushes' of fruitbodies at 7–10-day intervals.

In terms of differentiation there are several interesting features of this system. The casing layer is essential for a high fruitbody yield, and part of its role involves the activities of pseudomonads which are stimulated to grow in the casing layer by volatile metabolites, including ethanol, of *A. bisporus*. In experimental conditions the role of the casing layer can be replaced by using activated charcoal, suggesting that the fungus produces auto-inhibitors of fruiting which are removed by pseudomonads in the normal production process. The casing layer also provides a non-nutritive environment in which the fungus produces mycelial cords. As we saw earlier, these translocating organs develop in nutrient-poor conditions and they would be necessary for channelling large amounts of nutrients to the developing fruitbodies. The regular periodicity of fruiting is also of interest. In commercial conditions the crop must be harvested regularly at the 'button' stage to achieve this, and any delay in harvesting until the fruitbodies have opened will cause a corresponding delay in the next flush. Yet, most of the fruitbody primordia are already present at the time of the first flush, and mushrooms at the button stage have already received all or nearly all of their nutrients from the mycelium. So, the effect of delayed picking must be to delay the release of other primordia for further development. The mechanism of this is not fully known, but the cellulase activity of the mycelium in the compost increases markedly as each flush of fruitbodies develops. It seems that the expression of cellulase genes is closely linked to fruiting, presumably when the mycelial sugar reserves are depleted (see Chapter 5). Removal of the existing fruitbodies might act as a signal for a further round of mycelial activity, providing extra nutrients for the next batch of fruitbodies.

A final point of interest concerns the mechanism of fruitbody expansion from the button stage to the fully expanded T-shaped mushroom. This must involve the differential expansion of tissues, and studies on this have been carried out mainly with a much simpler experimental system — the expansion of stipes (stalks) of *Coprinus* spp. In field conditions these 'ink-cap' toadstools elongate very rapidly from the button stage to the mature stage. In laboratory conditions the initially short stipes can be severed and incubated in humid conditions, when they will elongate to more than seven times their length within 24 h. Most of this increase occurs by cell expansion rather than cell division, and it correlates with an increase in chitinase activity for loosening of the existing hyphal walls, and an increase in chitin synthase activity for new wall synthesis. But, unlike the apical growth of normal hyphae, new wall material is inserted along the length of the existing hyphae in the stipes. So, the wall extension is mainly by **intercalary growth**. This form of growth is also found in other rapidly extending structures, such as the sporangiophores of *Phycomyces* (zygomycota) on dung (see Chapter 10). It might play a role during the formation of fruitbody tissues, although this is difficult to investigate. In any case, the rapid expansion of mushrooms from the button stage depends on the intake of water from the mycelium. The driving force for this would be the conversion of storage reserves into osmotically active compounds (see

Chapter 6). Consistent with this, mushroom fruit-bodies contain high levels of mannitol which accounts for 25% or more of the dry weight, compared with only 1.5–4.5% in the mycelia. Reviews of this and other aspects of the developmental biology of basidiomycota can be found in Moore *et al.* (1985).

References

Adams, T.H. (1995) Asexual sporulation in higher fungi. In: *The Growing Fungus* (eds N.A.R. Gow & G.M. Gadd), pp. 367–82. Chapman & Hall, London.

Allen, E.A., Hazen, B.A., Hoch, H.C. *et al.* (1991) Appressorium formation in response to topographical signals in 27 rust species. *Phytopathology*, **81**, 323–31.

Bartnicki-Garcia, S. & Gierz, G. (1993) Mathematical analysis of the cellular basis of fungal dimorphism. In: *Dimorphic Fungi in Biology and Medicine* (eds H. Van den Bossche, F.C. Odds & D. Kerridge), pp. 133–44. Plenum Press, New York.

Bartnicki-Garcia, S., Bartnicki, D.D. & Gierz, G. (1995) Determinants of fungal cell wall morphology: the vesicle supply center. *Canadian Journal of Botany*, **73**, S372–8.

Bracker, C.E. (1968) The ultrastructure and development of sporangia in *Gilbertella persicaria*. *Mycologia*, **60**, 1016–67.

Casselton, L.A., Asante-Owusu, R.N., Banham, A.H. *et al.* (1995) Mating type control of sexual development in *Coprinus cinereus*. *Canadian Journal of Botany*, **73**, S266-72.

Christias, C. & Lockwood, J.L. (1973) Conservation of mycelial constituents in four sclerotium-forming fungi in nutrient-deprived conditions. *Phytopathology*, **63**, 602–5.

Cole, G.T. & Samson, R.A. (1979) *Patterns of Development in Conidial Fungi*. Pitman, London.

Flegg, P.B. (1985) Biological and technological aspects of commercial mushroom growing. In: *Developmental Biology of Higher Fungi* (eds D. Moore, L.A. Casselton, D.A. Wood & J.C. Frankland), pp. 529–39. Cambridge University Press, Cambridge.

Fox, F. M. (1986) Ultrastructure and infectivity of sclerotia of the ectomycorrhizal fungus *Paxillus involutus* on birch (*Betula* spp.). *Transactions of the British Mycological Society*, **87**, 627–31.

Gooday, G.W. & Adams, D.J. (1993) Sex hormones and fungi. *Advances in Microbial Physiology*, **34**, 69–145.

Gow, N.A.R. (1995) Yeast-hyphal dimorphism. In: *The Growing Fungus* (eds N.A.R. Gow & G.M. Gadd), pp. 403–22. Chapman & Hall, London.

Griffin, M.J. & Coley-Smith, J.R. (1975) Uptake of streptomycin by sporangia of *Pseudoperonospora humuli* and the inhibition of uptake by divalent metal cations. *Transactions of the British Mycological Society*, **65**, 265–78.

Harold, F.M. (1990) To shape a cell: an inquiry into the causes of morphogenesis of microorganisms. *Microbiological Reviews*, **54**, 381–431.

Heath, I.B. & Harold, R.L. (1992) Actin has multiple roles in the formation and architecture of zoospores of the oomycetes, *Saprolegnia ferax* and *Achlya bisexualis*. *Journal of Cell Science*, **102**, 611–27.

Hoch, H.C., Staples, R.C., Whitehead, B., Comeau, J. & Wolfe, E.D. (1987) Signalling for growth orientation and differentiation by surface topography in *Uromyces*. *Science*, **235**, 1659–62.

Howard, R.J., Ferrari, M.A., Roach, D.H. & Money, N.P. (1991) Penetration of hard surfaces by a fungus employing enormous turgor pressures. *Proceedings of the National Academy of Science, U.S.A.*, **88**, 11281–4.

Hyde, G.J., Lancelle, S., Hepler, P.K. & Hardham, A.R. (1991) Freeze substitution reveals a new model for sporangial cleavage in *Phytophthora*, a result with implications for cytokinesis in other eukaryotes. *Journal of Cell Science*, **100**, 735–46.

Kendrick, B. (1971) *Taxonomy of Fungi Imperfecti*. University of Toronto Press, Toronto.

Moore, D., Casselton, L.A., Wood, D.A. & Frankland, J.C. (1985) *Developmental Biology of Higher Fungi*. Cambridge University Press, Cambridge.

Muthukumar, G., Nickerson, A.W. & Nickerson, K.W. (1987) Calmodulin levels in yeasts and mycelial fungi. *FEMS Microbiology Letters*, **41**, 253–5.

Ng, A.M.L., Smith, J.E. & McIntosh, A.F. (1973) Conidiation of *Aspergillus niger* in continuous culture. *Archives of Microbiology*, **88**, 119–26.

Orlowski, M. (1995) Gene expression in *Mucor* dimorphism. *Canadian Journal of Botany*, **73**, S326–34.

Read, N.D., Kellock, L.J., Knight, H. & Trewavas, A.J. (1992) Contact sensing during infection by fungal pathogens. In: *Perspectives in Plant Cell Recognition* (eds J.A. Callow & J.R. Green), pp. 137–72. Cambridge University Press, Cambridge.

Smith, A.M. & Griffin, D.M. (1971) Oxygen and the ecology of *Armillariella elegans* Heim. *Australian Journal of Biological Sciences*, **24**, 231–62.

Suzuki, Y., Kumagai, T. & Oka, Y. (1977) Locus of blue and near ultraviolet reversible photoreaction in the stages of conidial development in *Botrytis cinerea*. *Journal of General Microbiology*, **98**, 199–204.

Wessels, J.G.H. (1992) Gene expression during fruiting in *Schizophyllum commune*. *Mycological Research*, **96**, 609–20.

Wessels, J.G.H. (1996) Fungal hydrophobins: proteins that function at an interface. *Trends in Plant Science*, **1**, 9–15.

Willetts, H.J. & Bullock, S. (1992) Developmental biology of sclerotia. *Mycological Research*, **96**, 801–16.

Chapter 5
Nutrition

Fungi need organic nutrients as both an energy source and to provide carbon skeletons for biosynthesis, so they are classed as **chemoheterotrophs** like animals and most bacteria. However, they have a characteristic mode of nutrition because they absorb soluble nutrients through the wall and plasma membrane, and in many cases these soluble nutrients are derived from complex polymers (polysaccharides, proteins, lipids) which are broken down by the release of extracellular enzymes (**depolymerases**). Fungi are the major degraders of organic matter in natural environments, and almost every naturally occurring organic compound can be degraded by one fungus or another. These decomposer activities of fungi are discussed in Chapter 10, but here we focus on the basic aspects of fungal nutrition and how this is related to fungal physiology.

Fungal adaptations for nutrient capture

Apical growth

The need to obtain nutrients has literally shaped the way that fungi grow. The rate of diffusion of soluble nutrients to a stationary cell would always tend to be growth-limiting, but fungi overcome this problem by apical growth. As we saw in Chapter 3, the apex extends continuously into new zones of substrate, supplied by protoplasm from behind, so that, in effect, fungi continuously move their protoplasm into new zones of substrate and evacuate the older zones. They behave like an amoeba in a continuously extending tube. Apical growth also confers a remarkable penetrating power, so that fungal pathogens can penetrate directly into living tissues (see

Chapter 4), or into other solid materials such as wood, leaf-litter, etc.

Electrical fields

The growing tips of fungi generate an electrical field around them. It can be mapped by placing a tiny vibrating electrode in the liquid film around a hypha, giving a trace like that shown in Fig. 5.1. Typically, the exterior of the hypha is more electronegative at the apex than further back, showing that current (which is positive by convention) enters at the tip and exits in the subapical region. In most fungi that have been examined in sufficient detail (e.g. *Achlya*, *Neurospora*) the current seems to be carried by hydrogen ions (H^+), because it corresponds to a gradient of pH along the hyphal surface and the current still occurs when other candidate ions are reduced or eliminated from the growth medium. In marine fungi, however, the current might be carried by potassium ions (K^+). For several years it was speculated that the electrical field might be the driving force for apical growth, perhaps by affecting the activities of membrane proteins or the cytoskeleton. However, this now seems unlikely because the direction of current flow can be reversed without affecting tip growth (Cho *et al.*, 1991). Instead, the electrical field seems to be involved intimately in nutrient uptake.

The uptake of organic nutrients is an energy-dependent process, which can be explained in general by the **chemiosmotic** theory. According to this, ions (e.g. H^+) are pumped to the outside of the cell by membrane-located ion pumps, using energy derived from the dissociation of adenosine triphosphate (ATP). This leads to an electrical potential difference across the membrane (mem-

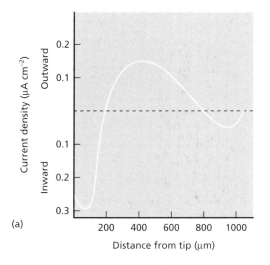

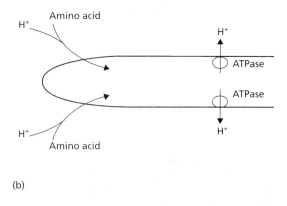

(a)

(b)

Fig. 5.1 (a) Typical current profile around an individual hypha of a growing fungus (*Achlya* sp). (From Gow, 1984.) (b) Interpretation of the current, involving proton export through ATPase-driven proton pumps behind the hyphal tip and proton symport with amino acids near the tip. (Based on Kropf *et al.*, 1984).

brane potential) so that ions re-enter the cell down an energy gradient. This re-entry of ions occurs through specific membrane proteins termed **symport** (co-uptake) proteins that also enable the transport of an organic molecule, so that the uptake of the molecule is energized by ion transport. This seems to apply generally in fungi, because their plasma membranes have been shown to have H$^+$-coupled transport systems for sugars, amino acids, nitrate, ammonium, phosphate and sulphate, as well as other ion channels such as stretch-activated channels for calcium and non-selective channels for both cations and anions (Garrill, 1995). However, fungi are thought to differ from most other cells in one significant respect: in most cell types the ion pumps (which extrude ions) and the symport proteins probably occur in close proximity, whereas in tip-growing fungi the ion pumps are thought to be most active behind the apex, whereas the symport proteins may be most active close to the tip. This spatial separation would explain why hyphae generate an external electrical field, as shown in Fig. 5.1(b). It has been argued that this makes sense intuitively, because nutrient uptake would occur at the tip which extends continuously into fresh zones of

substrate, while the ion pumps would be most active behind the tip where mitochondria are abundant to supply the ATP needed for pump activity (see Chapter 2).

Enzyme secretion

In general, fungi can only absorb small soluble nutrients such as monosaccharides and amino acids, or peptides composed of two or three amino acids. Even disaccharides such as sucrose may need to be degraded to monosaccharides before they are taken up by most fungi. So, the nutrition of most fungi is strongly dependent on the release of degradative enzymes. However, this imposes constraints because enzymes are large molecules, about 20 000–60 000 Da in the case of fungal cellulases, so they do not diffuse far from the hyphal surface. As a result, fungi create localized zones of erosion of insoluble substrates such as cellulose, and the hyphae must extend continuously into fresh zones—even more so than when growing on soluble substrates (Fig. 5.2). This is evidenced by the fact that yeasts never produce depolymerase enzymes (for starch, cellulose, etc.) because they would have no way of escaping from the erosion zones that they had created. Instead, yeasts occur in environments rich in simple soluble nutrients— on leaf, fruit and root surfaces, or on mucosal membranes in the case of *Candida albicans*. They depend either on a continuous supply of nutrients from the underlying substratum or on water currents that bring fresh supplies of nutrients.

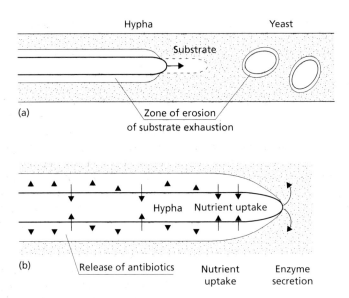

Fig. 5.2 Fungal strategies for growth on insoluble polymers. (a) Hyphae extend continuously at the apex, drawing protoplasm forwards to evacuate the zones of enzymic erosion of the substrate; yeasts do not utilize insoluble (non-diffusible) polymers because they would become trapped in the substrate erosion zones. (b) Suggested defence of substrate by a hypha of a polymer-degrading fungus: enzymes are secreted at the apex to degrade the polymer, and soluble nutrients are absorbed sub-apically; antibiotics or other inhibitors (arrowheads) may be released sub-apically into the substrate erosion zone to prevent competitors from exploiting the enzymic digestion products.

The large size of enzymes also creates a problem in terms of their release through the hyphal wall, because this would require the presence of continuous pores of significant size. Estimates of wall porosity can be obtained by studying the permeability of walls to molecules of known molecular mass such as polyethylene glycols and dextrans, and these studies suggest a cut-off at about 700–5000 Da, much lower than the size of most enzymes. These findings apply mainly to yeasts, so they might not reflect the wall porosity of fungi in general, and it is possible that enzymes might be folded so as to enable their passage through the wall. However, recent studies indicate that enzymes are released mainly and perhaps exclusively in the regions of new wall growth. This was first shown when a cellulase gene was cloned into yeast (*Saccharomyces cerevisiae*) and the enzyme was found to be released from the growth sites of the bud. More recently a glucoamylase was shown to be secreted exclusively at the tips of *Aspergillus niger*. So, it seems that enzymes destined for release through the wall are transported in vesicles from the Golgi body to the hyphal tip, where they are released by exocytosis and might flow outwards with the wall components themselves according to the steady state model of wall growth described in Chapter 3 (Wessels, 1990). Some of these enzymes would reach the cell surface and be released into the external environment, whereas others might become locked in the wall to serve as wall-bound enzymes.

Defence of territory

The release of enzymes to degrade a polymer represents a substantial investment of resources, so we can expect it to be coupled with some mechanism for defence of territory. Otherwise, any organism could grow in the zone of substrate erosion, sharing the breakdown products. There are, in fact, a few examples of non-cellulolytic fungi that grow in close association with cellulose degraders (see Chapter 11), but they might grow as symbionts, benefiting the cellulolytic partner. In general, however, three factors might help in the defence of territory.

1 The synthesis of depolymerase enzymes is tightly regulated by feedback mechanisms, discussed later, so that the rate of enzyme production is matched to the rate at which the breakdown products are utilized.

2 The final stages of polymer breakdown are achieved by wall-bound enzymes, so that the most readily utilizable monomers are not made available to other organisms; we will see this later in the case of cellulose.

3 A polymer-degrading fungus might produce antibiotics or other growth-suppressing metabolites. This is difficult to demonstrate because the inhibitors would be needed in only small, localized amounts, and they may be transitory. Burton and Coley-Smith (1993) reported antibacterial compounds of this type in the hyphae of *Rhizoctonia* spp. It may be significant that antibiotics are produced mainly by polymer-degrading fungi (see Chapter 11) and their production *in vitro* is associated with nutrient-limiting growth conditions (see Chapter 6). These are the conditions that polymer-degrading fungi will experience most of the time, because the production of depolymerase enzymes only occurs when readily available nutrients are in short supply. Thus, contrary to popular belief, antibiosis might not have evolved as an aggressive strategy but for the defence of territory.

The nutrient requirements of fungi

The nutrient requirements of fungi are reviewed comprehensively in the books by Garraway and Evans (1984) and Griffin (1994). For routine laboratory culture, most fungi are grown on chemically undefined media such as **potato–dextrose agar** (from boiled potatoes and glucose), **malt extract agar** (from malted barley and a protein digest) or **cornmeal agar**. These standard culture media are acidic (pH 5–6) and carbohydrate-rich, in contrast to the standard bacterial media such as 'nutrient agar' which are neutral or alkaline (pH 7–8) and rich in organic nitrogen. This difference reflects, in part, the difference in habitats of the more common fungi and bacteria, but it also reflects the generally higher requirement for organic nitrogen on the part of bacteria. Many other types of agar

are used to culture specific fungi, or to promote specific developmental stages, or for selective isolation of particular species from mixed communities (Booth, 1971).

In order to determine the **minimum** nutrient requirements, a fungus must be grown on chemically defined culture media like that shown in Table 5.1. This basal medium contains only mineral salts and glucose, and yet many common fungi will grow on it, demonstrating that fungi have essentially simple requirements. Given a sugar source, they can synthesize all their cellular components from this and the mineral elements. The exceptions tend to fall into a few basic categories.

1 Some fungi need to be supplied with one or more vitamins, the most common requirements being for thiamine (vitamin B_1), biotin, or both. These vitamins play essential roles in basic cellular metabolism (see Chapter 6), so the fungi that do not synthesize them will almost certainly obtain them from the natural habitat. However, multiple vitamin requirements are rare in fungi.

2 Some fungi cannot utilize nitrate or ammonium, but instead require a source of organic nitrogen. In these cases the requirement usually can be met by supplying a single amino acid such as asparagine. Several basidiomycota fall in this category.

3 Some fungi have more specific individual requirements, again related to their habitats where these requirements will be met. For example, some oomycota (e.g. *Phytophthora infestans*) need to be supplied with sterols for growth; the order Leptomitales within the oomycota (e.g. *Leptomitus lacteus*) need sulphur-containing amino acids such as cysteine because they cannot use inorganic sulphur; some of the fungi from animal dung need the iron-containing haem group, a component of cytochromes in the respiratory pathway.

It should be emphasized that all these comments refer to the minimum requirements for growth, and not necessarily to the optimal growth requirements. Also, they refer only to somatic growth and not necessarily to the complete life cycle. For example, several *Pythium* and *Phytophthora* spp. do not require sterols for hyphal growth but they need them for both sexual and asexual reproduction. Finally, many fungi will have extra nutrient requirements for growth in suboptimal environmental conditions. As an example of this, *S. cere-*

Table 5.1 Common laboratory media for the growth of fungi.

'*Natural media*'	
Potato–dextrose agar. Prepared from boiled potatoes, with added glucose	
Malt extract agar. Prepared from malt extract and peptone (a beef protein digest)	
Cornmeal or oatmeal agar. Prepared from boiled cornmeal or oatmeal	
Chemically defined liquid medium (similar components to commercial Czapek–Dox agar)	
$NaNO_3$ or NH_4NO_3 or asparagine at equivalent N content	2 g
KH_2PO_4 (or a buffered mixture of KH_2PO_4 and K_2HPO_4)	1 g
$MgSO_4$	0.5 g
KCl	0.5 g
$CaCl_2$	0.5 g
$FeSO_4$, $ZnSO_4$, $CuSO_4$	0.005–0.01 g each
Sucrose or glucose	20 g
Distilled water	1 l

Common supplements include: biotin (10 μg), thiamine (100 μg) or yeast extract (0.5 g) to supply unknown vitamin or trace element requirements.

visiae has relatively simple requirements for growth in aerobic culture but needs a wide range of vitamins and other compounds for growth in anaerobic conditions, when some of the basic metabolic pathways are inoperative. We return to this in Chapter 6; it illustrates that the nutrition of fungi cannot be divorced from an understanding of cellular metabolism in general.

Carbon and energy sources

An enormous range of organic compounds can be utilized by fungi, but this does not imply that all fungi can use all of the compounds. This is illustrated by the simple bell-shaped curve in Fig. 5.3. At one extreme of the spectrum, the simplest organic compound, methane (CH_4), is utilized by only a few yeasts. If their metabolism is similar to that of the methane-utilizing bacteria then they will convert methane via methanol to formaldehyde before it is incorporated into cellular components. Several more yeasts (e.g. *Candida* spp.) and a few mycelial fungi can grow on the longer-chain hydrocarbons (C_9 or larger) that are common in petroleum products. The main limitation with these compounds, as with methane, is that they are not miscible with water and so growth is restricted to a water–hydrocarbon interface. Two mycelial fungi are notorious for their ability to utilize the longer-chain hydrocarbons in aviation kerosene: *Amorphotheca resinae* (ascomycota) and *Paecilo-*

myces varioti (deuteromycota) grow commonly in fuel-storage tanks, and *A. resinae* has the dubious distinction of having caused at least one major air-crash in this way (see Chapter 14).

Moving further along the spectrum in Fig. 5.3, a larger number of fungi can utilize the common alcohols such as methanol and ethanol. In fact, ethanol is an excellent carbon source for *Candida utilis*, *Aspergillus nidulans* and *Armillaria mellea*, and it can even be the preferred substrate. Glycerol and fatty acids will support the growth of several fungi, and can be the preferred substrates for a few fungi such as *Leptomitus lacteus*. Amino acids also can be utilized, but a limitation in this respect is that amino acids contain excess nitrogen in relation to their carbon content for fungi, so that ammonium is released during their metabolism and can raise the pH to growth-inhibitory levels.

The vast majority of fungi can utilize glucose and several other monosaccharides and disaccharides. Some also can use sugar derivatives such as aminosugars (e.g. *N*-acetylglucosamine), sugar acids (e.g. galacturonic acid) or sugar alcohols (e.g. mannitol). However, the utilization of sugars depends on the membrane carrier proteins, which have at least a degree of substrate-specificity. Fungi usually have a constitutive carrier for glucose. It transports glucose in preference to other sugars in a mixture, but it has a relatively low binding specificity and so, in the absence of glucose, it will transport some other sugars. Then,

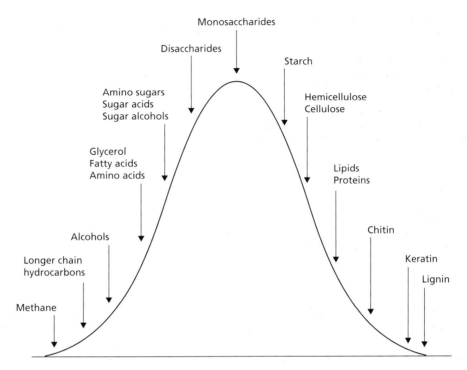

Fig. 5.3 Some representative carbon substrates of fungi, arranged approximately (left to right) in terms of chemical or structural complexity and (vertically) according to the proportion of fungi that utilize them.

the uptake of these sugars leads to the induced synthesis of their specific carrier proteins. A classic demonstration of this is the **diauxic** growth curve (Fig. 5.4) when *S. cerevisiae* is grown in liquid culture containing a disaccharide such as lactose. The fungus degrades lactose to glucose and galactose by means of a wall-bound enzyme, **β-galactosidase**, then grows rapidly at the expense of glucose while galactose remains in the medium. When glucose is depleted the growth rate slows for about 30 min while a galactose carrier protein is synthesized and inserted in the membrane, then the growth rate increases again. Sucrose is used in a similar way: it is cleaved to glucose and fructose by the wall-bound enzyme **invertase**, then the glucose is taken up before fructose. The relatively few fungi that cannot use sucrose (e.g. *Rhizopus nigricans*, zygomycota; *Sordaria fimicola*, ascomycota) cannot produce invertase. However, perhaps the most ironic finding in relation to sugar utilization is that many fungi cannot grow on **mannitol**

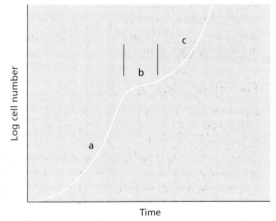

Fig. 5.4 Diauxic growth curve of a fungus growing on a disaccharide such as lactose, composed of glucose and galactose. (a) Phase of exponential growth as the fungus degrades the disaccharide externally and absorbs the glucose. (b) Temporary slowing of growth when the available glucose is depleted and the fungus is induced to synthesize a galactose uptake protein for insertion in the membrane. (c) Second phase of growth using galactose as substrate.

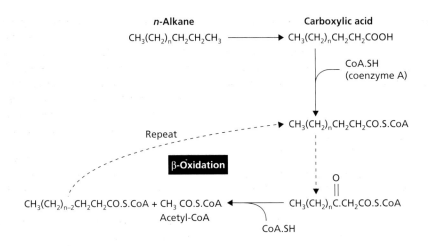

Fig. 5.5 Utilization of fatty acids and long-chain *n*-alkanes as growth substrates by the β-oxidation sequence, in which successive 2-carbon (acetyl) units are removed from the chain to enter the energy-yielding pathways (see Fig. 6.2).

or other **sugar alcohols** because they do not synthesize the necessary carrier proteins, and yet these sugar alcohols are abundant as internal components of fungi (see Chapter 6), where they are produced and metabolized easily.

The utilization of polymers requires the appropriate enzymes. Many fungi can use starch (by producing amylase), some can grow on lipids (using lipases), some on proteins (using proteases), and some on cellulose, hemicellulose, pectic compounds (see Chapter 12) or chitin, etc. We will consider this further below, using the utilization of cellulose to illustrate the roles of extracellular enzymes. At the extreme are a few highly specialized fungi that degrade the most structurally complex, cross-linked polymers. For example, lignin is degraded by a few of the wood-rotting basidiomycota termed 'white-rot fungi' (see Chapter 10). Similarly, 'hard keratin' is degraded by a few **keratinophilic** fungi, including the dermatophytic pathogens of humans (see Chapter 12) and related saprotrophic species in soil. It is a major component of nails and hooves, and is unavailable to most fungi that can degrade other proteins because it contains numerous disulphide bonds between the cysteine residues.

The breakdown of xenobiotics

The term **xenobiotic** refers to non-natural, synthetic compounds such as pesticides and plastics. Fungi do not play a major role in their breakdown; instead, the substituted benzene rings of the major pesticides are degraded by bacteria, especially *Pseudomonas* spp. which have plasmid-encoded enzymes for opening of the rings. However, some of the lignin-degrading fungi can open equivalent aromatic rings during wood decay (see Chapter 10), and an interesting point in both cases is that the organism gains little if any energy benefit from this process. Pesticide breakdown therefore occurs most efficiently in the presence of more readily utilizable nutrients; lignin breakdown similarly requires the presence of readily utilizable substrates. Some common soil fungi such as *Aspergillus*, *Penicillium* and *Trichoderma* spp. can partly metabolize pesticides in the presence of other substrates. Plastics present a different type of problem to pesticides. Their main structural components are long-chain polymers such as polyethylene $[(CH_2-CH_2)_n]$ and polyvinylchloride $[(CH_2-CHCl)_n]$. These components are nonbiodegradable, even though they closely resemble long-chain fatty acids, which fungi can degrade readily by the β-oxidation sequence (Fig. 5.5). The reason is that β-oxidation is an internal cellular process, and plastics are insoluble so they cannot enter the cell. Despite this, several fungi can degrade the other components of commercial plastics, such as plasticizers (glycol esters and phthalates), and in doing so they facilitate the breakdown of plastics by photo-oxidation. If starch-based plastics, already developed experimentally, were to become commonplace then fungi would play a major role in their degradation because many common soil fungi produce amylases.

Cellulose decomposition: a case study of extracellular enzymes

Cellulose is the most abundant natural polymer, representing about 40% of plant wall material, and fungi play the pre-eminent role in its breakdown. We therefore consider this in detail, and use it as a model of the roles of depolymerase enzymes.

Cellulose has a simple chemical structure, consisting of straight chains of about 3000 glucose residues linked by β-1,4-bonds (Fig. 5.6). However, despite this simplicity its breakdown requires the actions of three types of enzyme which together are termed **cellulase** or the cellulase enzyme complex. As shown in Fig. 5.6, they are:

1 an exo-acting enzyme termed **cellobiohydrolase**, which cleaves successive disaccharide units (cellobiose) from the non-reducing ends of the chains;

2 an endo-acting enzyme termed **endoglucanase**, which attacks the centres of the chains at random, breaking the molecule into successively smaller fragments;

3 **β-glucosidase** (or cellobiase) which cleaves cellobiose to glucose.

There can be multiple forms of these enzymes, especially endoglucanase which ranges from about 11 000–65 000 Da, but the exo-enzyme seems

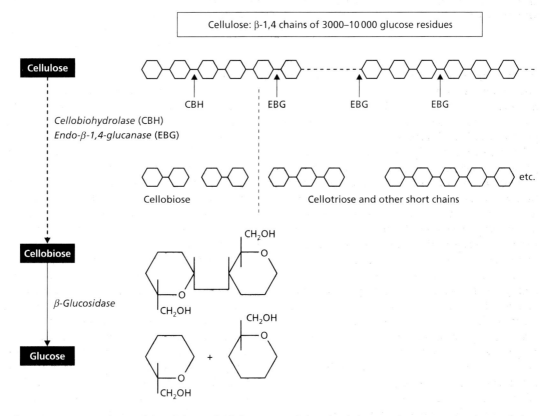

Fig. 5.6 Structure and enzymic breakdown of cellulose. The enzyme cellobiohydrolase (CBH) cleaves disaccharide residues (cellobiose) from the non-reducing end of the cellulose chain. The enzyme endo-β-1,4-glucanase (EBG) cleaves the chain at random to generate shorter chains, providing further ends for the action of CBH. The enzyme β-glucosidase cleaves cellobiose to two glucose residues for uptake by the fungus.

to be more uniform, from 50 000 to 60 000 Da (Mandels, 1981).

The three enzymes act synergistically to degrade natural cellulose or its processed forms such as cotton fabrics or filter paper. Not all aspects of this synergism are fully understood, but at least three factors seem to be involved.

1 The cellulose chains often are aligned to one another and held together by hydrogen bonds and van der Waal's forces to form cellulose microfibrils (so-called crystalline cellulose). This tight packing can preclude the access of enzymes to the centres of the individual cellulose chains, and in this respect the smaller types of endoglucanase may be especially important in the early stages of attack.

2 The endoglucanases progressively generate non-reducing ends on which the exo-acting enzymes can work.

3 The intermediate products of breakdown, especially cellobiose, can bind to the active site of the exo-acting enzyme, causing competitive inhibition of the enzyme action.

So, cellobiose must be cleaved to glucose by β-glucosidase in order for cellulose breakdown to continue. β-Glucosidase is usually wall-bound, like invertase and some other enzymes that act on soluble substrates; there is no advantage in releasing a large enzyme molecule, when the substrate can diffuse faster towards the cell. One of the limitations of commercially available cellulase enzyme mixtures, like those from culture filtrates of *Aspergillus niger* or *Trichoderma reesei*, is that they have a relatively low β-glucosidase activity, so they seldom are as effective as the fungi themselves.

The production of cellulase enzymes is tightly regulated, ensuring that the enzymes are secreted only when needed. Part of this regulation is achieved by the common feedback system termed **catabolite repression**, whereby the genes encoding the enzymes are repressed when readily utilizable substrates like glucose or cellobiose are available. In addition to this, the presence of cellulose will induce the synthesis of enzymes in the absence of catabolite repression. It is difficult to imagine that cellulose itself could be the inducer because it is insoluble. Instead, it is thought that the fungus produces low (constitutive) levels of cellulase enzymes in the absence of glucose, so that the breakdown products such as cellobiose can act

as signals for induction. Cellobiose has been found to act as a weak inducer of cellulases in some fungi, and a derivative of cellobiose, sophorose, is a much stronger inducer for *T. reesei*, the fungus used commonly as a model organism for studies of cellulase action. Thus, by a combination of **gene repression** (by high levels of glucose or cellobiose), **competitive inhibition** of enzyme action (by cellobiose) and **gene induction** (in the absence of glucose but presence of low levels of cellobiose or its derivatives) the rate of cellulose breakdown is matched closely to the rate at which the fungus can use the sugars released from the substrate.

Some anomalies, some developments

The ability to degrade 'crystalline' cellulose (cotton fabrics, etc.) is found in relatively few fungi, but several others can degrade the 'soluble celluloses' such as **carboxymethyl cellulose** which form gels in water and are produced commercially for wallpaper pastes and as thickeners in emulsion paints, processed foods, etc. The fungi that degrade these products secrete an endoglucanase but not cellobiohydrolase; they include the common leaf-surface saprotrophs such as *Cladosporium* spp. and *Aureobasidium pullulans* (see Chapter 1, Fig. 1.9), which also grow on damp kitchen and bathroom walls. This raises the question of why some fungi produce only part of the cellulase enzyme complex if they cannot degrade natural forms of cellulose. A likely explanation was discovered by Taylor and Marsh (1963) who found that several *Pythium* spp., which are not generally considered to be cellulolytic fungi, could degrade cotton fibres taken from unopened cotton bolls but not when the cotton bolls had opened naturally on the plants. When the fibres from unopened bolls were allowed to air-dry they became resistant to breakdown, and simple rewetting did not change this. However, the fibres became degradable again if they were 'swollen' by treatment with potassium hydroxide. The implication is that the cellulose in natural, moist plant cell walls may be susceptible to attack, but it shrinks and the cellulose chains bond together when the tissues have dried. As we shall see in Chapter 10, there are probably two broad groups of cellulose-degrading fungi in nature: (i) those like the leaf-surface residents that can grow on plant tissues before they dry; and (ii) the more 'typical' cellulose-degraders that decompose the fallen, dried plant remains.

Among the typical cellulose-degraders is *T. reesei*, originally isolated from rotting cotton fabric. By routine, repeated selection in laboratory conditions, some strains of this fungus have been obtained that can release as much as 30 g dry weight of cellulase enzymes per litre of culture broth. The gene for a cellobiohydrolase of this fungus has been cloned and engineered into *S. cerevisiae*, under the control of a constitutive promoter. The yeast then secretes large amounts of a functional cellulase, even in the absence of the substrate; it can represent well over 50% of the total protein production by the cultured cells (Beguin, 1990; Penttila *et al.*, 1991). This raises the prospect of producing cellulase enzymes for commercial use from genetically engineered yeast, but we shall see in Chapter 10 that there are further limitations to be overcome because cellulose in crop wastes is often complexed with lignin (so-called **lignocellulose**) which limits the actions of cellulase enzymes.

Nitrogen, phosphorus and iron

Fungi need many mineral nutrients in at least trace amounts (see Table 5.1), but nitrogen, phosphorus and iron merit special mention because of their significance for fungal activities and fungal interactions in nature.

Nitrogen

Of all the mineral nutrients, nitrogen is required in the largest amounts so it can be the limiting factor for fungal growth in natural habitats (see Chapter 10). Fungi do not fix atmospheric nitrogen as some bacteria do, but they can use many other forms of nitrogen and we can understand this most easily by considering the normal pathway for assimilation of nitrogen which is similar for all fungi:

$$NO_3^- \xrightarrow[\text{reductase}]{\text{nitrate}} NO_2^- \xrightarrow[\text{reductase}]{\text{nitrite}} NH_4^+$$

(nitrate)   (nitrite)   (ammonium)

$$\xrightarrow[\text{dehydrogenase}]{\text{glutamate}} \text{glutamate} \xrightarrow[\text{synthetase}]{\text{glutamine}} \text{glutamine}$$

Clearly, if a fungus can use nitrate as a nitrogen source then it can use all the other forms of nitrogen further along this pathway, and so the nitrogen requirements of fungi can be reduced to a few basic rules.

1 All fungi can use amino acids as a nitrogen source. Often they need only one amino acid such as glutamic acid or asparagine, and from this they can produce the other amino acids by **transamination** reactions. For example:

$$HOOCCH_2CH(NH_2)COOH + CH_3COCOOH \leftrightarrow$$
(glutamic acid)   (pyruvic acid)
$$HOOCCH_2COCOOH + CH_3CH(NH_2)COOH$$
(α-ketoglutaric acid)   (alanine)

2 Most fungi can use ammonium as sole nitrogen source. After uptake it is combined with organic acids, usually to produce either glutamic acid (from α-ketoglutaric acid) or aspartic acid; then the other amino acids can be formed by transamination, as above. The fungi that cannot use ammonium and thus depend on organic nitrogen sources include some water moulds (*Saprolegnia* and *Achlya* spp.), several basidiomycota and the mycoparasitic *Pythium* spp. such as *P. oligandrum* (see Chapter 11). However, ammonium is not an ideal nitrogen source, even for many of the fungi that can use it. The reason is that ammonium is taken up in exchange for H+ which can rapidly lower the pH of a culture medium to 4.0 or less, inhibiting the growth of many fungi.

3 Several fungi can use nitrate as sole nitrogen source, converting it to ammonium by the enzymes **nitrate reductase** and **nitrite reductase**. The fungi that cannot use nitrate either lack one of these enzymes or have a mutation in one of them. In fact, nitrate-non-utilizing mutants are commonly used as tools in fungal genetics. They are easily selected by growing fungi on media containing chlorate, because this is toxic to the wild-type strains, either because chlorate is reduced to toxic chlorite by nitrate reductase or because chlorate shuts down the nitrogen metabolism pathway. Thus, only *nit-* strains, lacking nitrate reductase activity or some of the regulatory components, are able to grow.

Despite the apparent simplicity of the nitrogen assimilation pathway, there are complex regulatory controls so that nitrogen sources are not necessarily used in the ways we might expect. For example, if a fungus is supplied with a mixture of nitrogen sources, then ammonium is taken up in preference to either nitrate or amino acids. The reason is that ammonium, or glutamine which is one of the first amino acids formed from it, prevents the synthesis of membrane-uptake proteins for other nitrogen sources, and also prevents the synthesis of enzymes involved in nitrate utilization.

Phosphorus

All organisms need phosphorus in significant amounts for production of sugar phosphates, nucleic acids, ATP and membrane phospholipids, etc. But, phosphorus can be poorly available in natural environments because it occurs as highly insoluble organic phosphates or inorganic (calcium and magnesium) phosphates. Plant roots, in particular, have difficulty in extracting phosphorus from soil, because they deplete the small pool of soluble phosphate in their immediate vicinity and then have to depend on the slow solubilization and diffusion of phosphate from further afield. In contrast, fungi are highly adept at obtaining phosphorus, and they achieve this in several ways (Jennings, 1989). They respond to critically

low levels of available phosphorus by increasing the activity of their phosphorus-uptake systems; they release phosphatase enzymes that can cleave phosphate from organic sources; they solubilize inorganic phosphates by releasing organic acids to lower the external pH; and their hyphae, with a high surface area : volume ratio, extend continuously into fresh zones. In addition to these points, fungi accumulate and store phosphates in excess of their immediate requirements, typically as polyphosphates in the vacuoles. It is therefore not surprising that plants form mycorrhizal associations with fungi. Despite the cost to the plant in maintaining a fungal network, the plant can benefit by obtaining phosphorus from a fungus, and the cost in terms of photosynthate supply to the fungus would be far less than the cost of producing a continuously expanding and relatively inefficient feeder-root system (see Chapter 12).

Iron

Iron is needed in relatively small amounts but is essential as a donor and acceptor of electrons in cellular processes, notably in the cytochrome system for aerobic respiration (see Chapter 6). Iron normally occurs in the ferric (Fe^{3+}) form, insolubil-

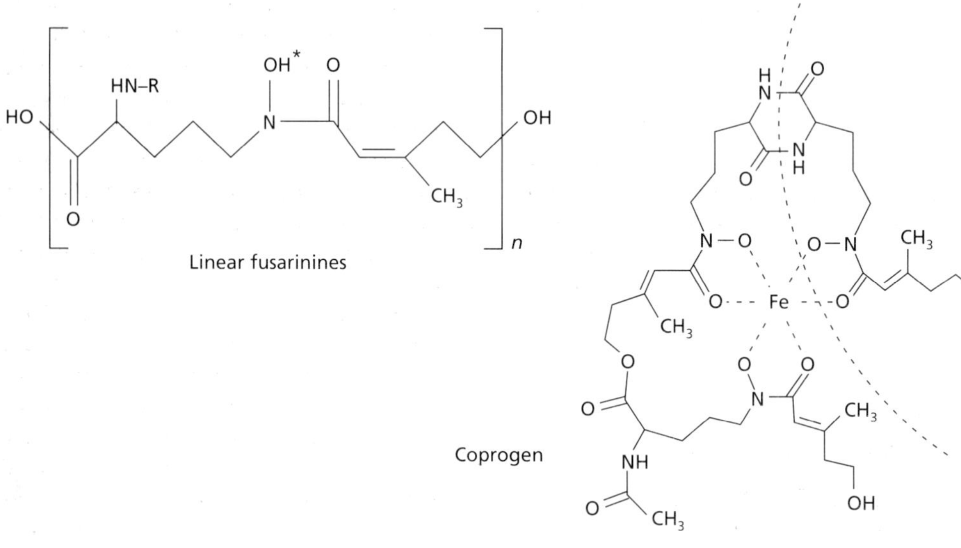

Fig. 5.7 The hydroxamate siderophores of fungi, used for capture of ferric iron (Fe III) from the external environment. The simplest types are linear fusarinines, widely found in *Fusarium*, *Gliocladium* and *Penicillium* species, etc. They sequester iron at the position marked by the asterisk. More complex siderophores such as coprogen have fusarinine units (one shown by broken line) in a ring-like structure.

ized as ferric oxides or hydroxides at pH above 5.5, and it is taken up by a different process compared with other mineral nutrients. Iron must be 'captured' from the environment by the release of iron-chelating compounds termed **siderophores**. These compounds chelate Fe^{3+}, then they are reabsorbed through a specific membrane protein and Fe^{3+} is reduced to Fe^{2+} within the cell, causing its release because the siderophore has a lower affinity for Fe^{2+} than for Fe^{3+}. Finally, the siderophore is exported again to capture a further ion. Siderophores and their specific membrane proteins are produced only in response to iron-limiting conditions.

The siderophores of fungi are of the **hydroxamic acid** type (Fig. 5.7). Despite their high affinity for Fe^{3+}, they have much lower affinity than do the siderophores of fluorescent pseudomonads, such as pseudobactin and pyoverdine, so that fluorescent pseudomonads can be used to control plant-pathogenic fungi in the root zone of crops (Leong, 1986). For example, pseudobactin-producing pseudomonads can suppress germination of the chlamydospores of *Fusarium oxysporum* on low-iron media, whereas mutant pseudomonads, deficient in siderophore production, are ineffective. However, we shall see in Chapter 14 that competition for iron is only one of several ways in which *Pseudomonas* spp. can control plant-pathogenic fungi.

Efficiency of substrate utilization

Industrial microbiologists are specially concerned with the efficiency of substrate conversion into microbial cells or cell products. Microbial ecologists have been less concerned with this, although it has important implications for environmental processes. The two commonly used criteria of efficiency are the **economic coefficient** and the **Rubner coefficient**, obtained as follows:

Economic coefficient (expressed as percentage)

$$= \frac{\text{dry weight of biomass produced}}{\text{dry weight of substrate consumed}}$$

Rubner coefficient (expressed as a proportion)

$$= \frac{\text{heat of combustion of biomass produced}}{\text{heat of combustion of substrate consumed}}$$

Typical values for the economic coefficient range from 20 to 35 if a fungus is grown in a dilute medium, but the values fall markedly if the medium is made progressively richer—fungi seem to be 'wasteful' if supplied with excess substrate. Values for the Rubner coefficient typically are higher than for the economic coefficient; for example *Aspergillus niger* had a Rubner coefficient of 0.55–0.61 (equivalent to 55–61%) over a range of cultural conditions, compared with an economic coefficient of 35–46. The difference is explained mainly by the synthesis of lipid storage reserves during growth, because lipids have higher calorific values than the carbohydrates used as substrates in culture media. However, the important point revealed by all these values is that a considerable proportion of the substrate supplied to a fungus is consumed in energy production rather than being converted into biomass.

Conversion efficiencies are difficult to obtain in natural systems where a mixture of potential substrates is available. However, Adams and Ayers (1985) did this by collecting all the spores produced by the mycoparasite *Sporidesmium sclerotivorum* when grown on sclerotia of its main fungal host, *Sclerotinia minor*, in laboratory conditions. The reported efficiency was exceptionally high: an economic coefficient of 51–60 and a Rubner coefficient of 0.65–0.75, assuming that the whole sclerotial material was 'substrate' in the equations given above. This mycoparasite also has a remarkable way of parasitizing the host sclerotia: it penetrates some of the sclerotial cells initially but then grows predominantly between the sclerotial cells, scavenging small amounts of soluble nutrients that leak from them and creating a nutrient stress. The host cells respond by converting energy storage reserves (principally glycogen) to sugars, which leak from the cells to support further growth of the mycoparasite. Recent studies suggest that this essentially non-invasive mode of parasitism is also employed by **endophytic fungi** in plants (see Chapter 12). They grow slowly and sparsely, between or within the plant cell walls, exploiting nutrients that leak from the host cells. However, there seem to be no reports on the substrate conversion efficiencies of these fungi.

In terms of substrate efficiencies we should note that fungi can grow by **oligotrophy**, using extremely low levels of nutrients (*oligo* = few) on silica gel or glass. They seem to grow by scavenging trace amounts of volatile organic compounds from the atmosphere (Wainwright, 1993).

Fungi that cannot be cultured

To close this chapter we should record that several fungi still cannot be grown in laboratory culture so their nutrient requirements remain largely unknown. They have been termed **obligate parasites** but now more commonly are termed **biotrophic parasites** (see Chapter 12). Many of these fungi are extremely important in environmental and economic terms. They include the ubiquitous arbuscular mycorrhizal fungi (zygomycota), rust fungi (basidiomycota), powdery mildew fungi (ascomycota) and downy mildew fungi (oomycota), all of which form nutrient-absorbing **haustoria** or equivalent structures in the host cells (see Chapter 12). Other obligate parasites include the intracellular parasites of plants (e.g. *Plasmodiophora brassicae* (see Chapter 1, Fig. 1.4) and *Olpidium brassicae*) and the entomophthoran parasites of insects (see Chapter 1, Fig. 1.6).

It remains to be seen if some of these fungi will ever be grown in **axenic** culture (i.e. separate from their hosts); they might have lost the capacity for independent existence during their long evolutionary association with their hosts. However, significant progress has been made in culturing the rust fungi, starting with *Puccinia graminis* (black stem rust of wheat) and now many other rust species (Maclean, 1982). *P. graminis* was found to grow slowly and only after a prolonged lag phase. Its linear extension rate on agar ranged from 30 to 300 μm day^{-1}, compared with rates of 1–50 mm day^{-1} for fungi that are commonly grown in laboratory conditions. *P. graminis* tends to leak vital nutrients, such as cysteine, into the growth medium, so a high spore density is needed in the inoculum to minimize the diffusion of nutrients away from individual sporelings. Also, *P. graminis* produces self-inhibitors which can result in 'staling' of the cultures. This is true of several fungi if they are grown on sugar-rich media, and it causes a progressive reduction and eventual halting of growth. It can be overcome by using relatively weak media. Some other rust fungi need relatively high concentrations of carbon dioxide—

for example, *Melampsora lini* which causes flax rust. If attention is paid to all these points then several rust fungi can be maintained in laboratory culture. However, in this process they change irreversibly to a 'saprotrophic' form that cannot reinfect plants, so the breakthrough in culturing them has not proved as useful as was hoped.

References

Adams, P.B. & Ayers, W.A. (1985) Energy efficiency of the mycoparasite *Sporidesium sclerotivorum in vitro* and in soil. *Soil Biology and Biochemistry*, **17**, 155–8.

Beguin, P. (1990) Molecular biology of cellulose degradation. *Annual Review of Microbiology*, **44**, 219–48.

Booth, C. (1971) *Methods in Microbiology*, Vol. 4. Academic Press, London.

Burton, R.J. & Coley-Smith, J.R. (1993) Production and leakage of antibiotics by *Rhizoctonia cerealis, R. oryzae-sativae* and *R. tuliparum*. *Mycological Research*, **97**, 86–90.

Cho, C., Harold, F.M. & Scheurs, W.J.A. (1991) Electrical and ionic dimensions of apical growth in *Achlya* hyphae. *Experimental Mycology*, **15**, 34–43.

Garraway, M.O. & Evans, R.C. (1984) *Fungal Nutrition and Physiology*. Wiley, New York.

Garrill, A. (1995) Transport. In: *The Growing Fungus* (eds N.A.R. Gow & G.M. Gadd), pp. 163–81. Chapman & Hall, London.

Gow, N.A.R. (1984) Transhyphal electric currents in fungi. *Journal of General Microbiology*, **130**, 3313–18.

Griffin, D. (1994) *Fungal Physiology*, 2nd edn. Wiley-Liss, New York.

Jennings, D.H. (1989) Some perspectives on nitrogen and phosphorus metabolism in fungi. In: *Nitrogen, Phosphorus and Sulphur Utilization by Fungi* (eds L. Boddy, R. Marchant & D.J. Read), pp. 1–31. Cambridge University Press, Cambridge.

Kropf, D.L., Caldwell, J.H., Gow, N.A.R. & Harold, F.M. (1984) Transhyphal ion currents in the water mould *Achlya*. Amino acid proton symport as a mechanism of current entry. *Journal of Cell Biology*, **99**, 486–96.

Leong, J. (1986) Siderophores: their biochemistry and possible role in the biocontrol of plant pathogens. *Annual Review of Phytopathology*, **24**, 187–209.

Maclean, D.J. (1982) Axenic culture and metabolism of rust fungi. In: *The Rust Fungi* (eds K.J. Scott & A.K. Chakravorty). Academic Press, London.

Mandels, M. (1981) Cellulases. *Annual Reports on Fermentation Processes*, **5**, 35–78.

Penttila, M., Teeri, T.T., Nevalainen H. & Knowles, J.K.C. (1991) The molecular biology of *Trichoderma reesei* and its application in biotechnology. In: *Applied Molecular Genetics of Fungi* (eds J.F. Peberdy, C.E. Caten, J.E.

Ogden & J.W. Bennett), pp. 85–102. Cambridge University Press, Cambridge.

Taylor, E.E. & Marsh, P.B. (1963) Cellulose decomposition by *Pythium. Canadian Journal of Microbiology*, **9**, 353–8.

Wainwright, M. (1993) Oligotrophic growth of fungi — stress or natural state? In: *Stress Tolerance of Fungi* (ed. D.H. Jennings), pp. 127–44. Academic Press, London.

Wessels, J.G.H. (1990) Role of cell wall architecture in fungal tip growth. In: *Tip Growth in Plant and Fungal Cells* (ed. I.B. Heath), pp. 1–29. Academic Press, New York.

Chapter 6
Metabolism

This chapter outlines the basic metabolic pathways of fungi, for understanding how fungi grow on different types of substrate and in different environmental conditions. It also covers some distinctive and unusual aspects of fungal metabolism, including the production of secondary metabolites such as antibiotics and mycotoxins.

The structure of the chapter will be based on Fig. 6.1 which gives an overview of the relationships between the main components of fungal metabolism. It is seen that the energy-yielding pathways occupy a central position in metabolism. In addition to their obvious role in generating adenosine triphosphate (ATP), they provide the precursor metabolites for many biosynthetic reactions.

Energy production

The initial breakdown of sugars follows two major routes, the **Embden–Meyerhof** (EM) pathway and the **pentose-phosphate** (PP) **pathway**. As shown in Fig. 6.2, these pathways have the same end-product, glyceraldehyde-3-phosphate, but they are used for different purposes. The EM pathway is the major route for energy production, whereas the PP pathway serves mainly for biosynthesis — it generates intermediates such as ribose-5-phosphate for synthesis of nucleic acids and erythrose-4-phosphate for synthesis of aromatic amino acids.

The EM pathway starts from a six-carbon (6C) sugar such as glucose. This is phosphorylated to fructose-1,6-diphosphate at the expense of two molecules of ATP. Then, fructose-1,6-diphosphate is cleaved to two 3C compounds which, in turn, are converted to two molecules of pyruvic acid. Four molecules of ATP are generated in this process, so the net energy gain is two molecules of ATP from each molecule of glucose metabolized through to pyruvate. All these reactions occur in the cytosol.

In oxidative conditions the two pyruvic acid molecules enter a mitochondrion and are converted to two molecules of acetyl-coenzyme A (CoA) (2C), with the release of two molecules of carbon dioxide. Acetyl-CoA then combines with oxaloacetate (4C) to form citric acid (6C) and this is metabolized sequentially to oxaloacetate by the reactions of the tricarboxylic acid (TCA) cycle. Another two molecules of ATP are synthesized during this process (one in each turn of the cycle).

The reactions of the EM pathway and two turns of the TCA cycle cause the sugar to be oxidized completely, producing six molecules of carbon dioxide. However, there must be a corresponding reduction of other compounds, and this role is served by three nucleotides: nicotinamide adenine dinucleotide (NAD+), NADP+ and flavin adenine dinucleotide (FAD), which are reduced to NADH, NADPH and $FADH_2$, respectively. These nucleotides then need to be reoxidized for the whole process to continue, and this is achieved by passing their electrons along an **electron transport chain**, where oxygen is the **terminal electron acceptor**. As shown in Fig. 6.3, at three steps in this chain from NADH or NADPH the free energy released is sufficient to generate a molecule of ATP (there are two such steps from the reoxidation of $FADH_2$). So, the net result of the whole process can be represented by the following empirical equation, where oxygen (the terminal electron acceptor) has been reduced to water:

$$C_6H_{12}O_6 + 6O_2 \rightarrow 6CO_2 + 6H_2O$$

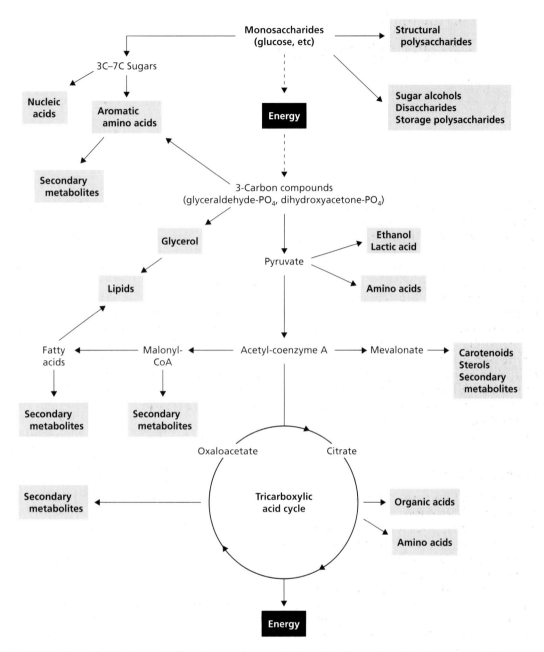

Fig. 6.1 Overview of the basic metabolic pathways of fungi, showing how the main energy-yielding pathways (spine of the diagram) provide the precursors for products used in growth and biosynthesis (boxes).

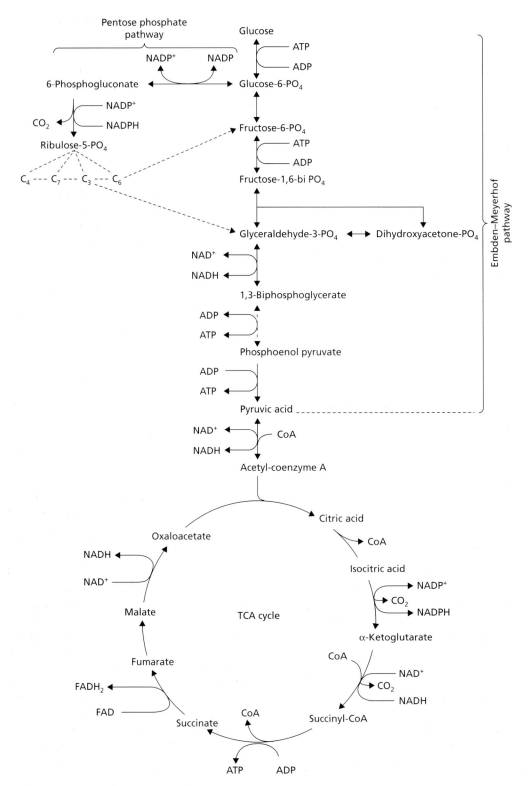

Fig. 6.2 Outline of the Embden–Meyerhof pathway and tricarboxylic acid cycle which provide the major means of generating energy in fungi. Also shown (top left) is an overview of the pentose-phosphate pathway which can provide some energy but is used mainly for biosynthesis.

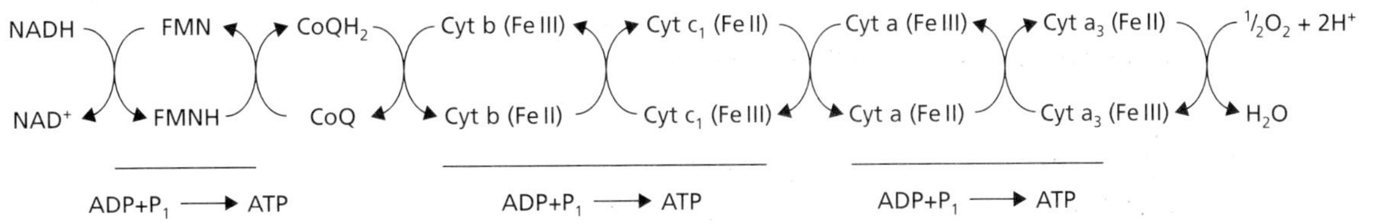

Fig. 6.3 Outline of the respiratory electron-transport chain whereby reduced nicotinamide nucleotides such as NADH (see Fig. 6.2) are reoxidized and their electrons are transferred sequentially to oxygen (the terminal electron acceptor). At three steps along this chain the energy differential is sufficient to produce a molecule of ATP. FMN, flavin mononucleotide; CoQ, coenzyme Q; Cyt, cytochrome.

The oxidation of each molecule of glucose gives a theoretical yield of 38 molecules of ATP, derived as:
two ATPs from the EM pathway;
two ATPs from the TCA cycle;
30 ATPs from the reoxidation of 10 pyridine nucleotides;
four ATPs from the reoxidation of two flavin nucleotides.
However, the yield is usually lower than this because intermediates are drawn continuously from these pathways for biosynthetic reactions, discussed later.

Also, at this stage we should mention the special role of the PP pathway: it can be used as an alternative to the EM pathway for generating energy from sugars (giving one ATP instead of two from the EM route). But, its major role is in biosynthesis, and the NADPH (instead of NADH) produced from it is used normally to provide the reducing power for biosynthetic reactions.

What happens when oxygen is limiting?

Fungi often encounter oxygen-limiting conditions (see Chapter 7), placing severe constraints on the energy-yielding pathways. Without a terminal electron acceptor, the electron transport chain cannot operate, the pool of reduced nucleotides (NADH, etc.) cannot be reoxidized, and the EM pathway and TCA cycle would shut down. However, this problem can be overcome in two ways. Usually, pyruvic acid is converted to either ethanol or lactic acid as shown in the following equations. The end-products of these reactions are more reduced than pyruvate, so the reactions are coupled with the reoxidation of pyridine nucleotides:

(1) $CH_3COCOOH + NADH \xrightarrow{\text{lactic dehydrogenase}}$
   (pyruvic acid) $CH_3CHOHCOOH + NAD^+$
              (lactic acid)

(2) pyruvic acid $\xrightarrow{\text{pyruvic dehydrogenase}}$
                 acetaldehyde $+ CO_2$

then:

$CH_3CHO + NADH \xrightarrow{\text{alcohol dehydrogenase}}$
(acetaldehyde) $CH_3CH_2OH + NAD^+$
                   (ethanol)

Clearly, the TCA cycle cannot operate because pyruvate has been diverted to other products. However, the EM pathway can continue to operate, and ethanol or lactic acid is released into the surrounding medium. Most yeasts and mycelial fungi produce ethanol; indeed, this is the basis of the alcoholic drinks industry (*Saccharomyces cerevisiae*). But, several chytridiomycota produce lactic acid (e.g. *Allomyces*, *Blastocladiella*). Energy-yielding processes of this type, where the terminal electron acceptor during oxidation of a substrate is an organic molecule, are termed **fermentations**. This is the strict biochemical definition of the term fermentation, although industrial microbiologists often use the term more loosely to refer to any production process such as the 'penicillin fermentation'. In terms of energy yield, fermentation is very inefficient because only two ATPs are produced from every molecule of sugar metabolized. A fungus therefore needs an abundant supply of sugars for growth in these conditions. We noted in Chapter 5 that fungi also need other nutrients such as vitamins in anaerobic conditions, because the TCA cycle and several other reactions do not operate, or operate too slowly, to provide the precursors for biosynthesis.

At least some fungi (e.g. *Neurospora crassa*, *Aspergillus nidulans*) have an alternative means of coping with anaerobic conditions: they use nitrate in place of oxygen as the terminal electron acceptor in the electron transport chain, so that the full TCA cycle operates. There is a theoretical yield of 26 ATPs from each molecule of glucose metabolized, using nitrate as terminal electron acceptor. This is lower than with oxygen because the energy differential along the electron transport chain from NADH/NADPH to nitrate is sufficient to generate only two ATPs (and one from $FADH_2$ to nitrate) rather than three ATPs between NADH/NADPH and oxygen. Energy-yielding processes of this type, where substrates are oxidized using an inorganic substance as a terminal electron acceptor are termed **respirations**. This is potentially confusing because the term respiration is inextricably linked with oxygen in everyday use. However, in biochemical terms, organisms can gain energy either by **aerobic respiration** or by **anaerobic respiration**, and both of these processes are distinct from fermentation.

Energy from non-sugar substrates

From Fig. 6.2 it is clear that many types of substrate are potential energy sources, provided that they can be fed into one of the energy-yielding pathways. For example, pentose sugars such as xylose (a major component of hemicelluloses) can be fed into the PP pathway and metabolized to give energy. An amino acid like glutamic acid can be deaminated to an organic acid (α-ketoglutaric acid in this case) which will feed into the TCA cycle and yield a theoretical nine ATPs during its conversion to oxaloacetate. Similarly, if acetate is supplied as a substrate it can be converted to acetyl-CoA and then combined with oxaloacetate (giving citrate) to yield 12 ATPs via the TCA cycle. Fatty acids can be degraded to acetyl-CoA by β-oxidation (see Chapter 5, Fig. 5.5) and then be metabolized in the same way. The only limitation in all cases is that oxygen (or perhaps nitrate) is required for the TCA cycle to operate. The only substrates that can supply energy through *fer-*

mentation are sugars and sugar derivatives, through the EM pathway.

Coordination of metabolism: balancing the pathways

Several intermediates of the basic energy-yielding pathways serve as precursor metabolites for biosynthesis (see Fig. 6.1). For example, dihydroxyacetone phosphate (EM pathway) can be converted to glycerol for lipid synthesis; acetyl-CoA is used for synthesis of fatty acids and sterols; α-ketoglutaric acid is used to produce amino acids of the important glutamate 'family' (glutamate, proline, arginine) and oxaloacetate is used for the equally important aspartate family (aspartate, lysine, methionine, threonine, isoleucine). Therefore, the question arises, how can the pathways for energy production continue when intermediates are removed?

The main problem would arise when intermediates of the TCA cycle are removed, because this would break the cycle. The problem is overcome by special **anaplerotic** reactions (literally 'filling-up' reactions) which replenish the missing intermediates. One such reaction sequence is the glyoxylate cycle, described later in a different context. Another type of anaplerotic reaction involves the addition of carbon dioxide to pyruvic acid, to give oxaloacetate, as follows:

$$\text{pyruvate} + \text{ATP} + \text{HCO}_3^- \xrightarrow{\textit{pyruvate carboxylase}} \text{oxaloacetate} + \text{ADP} + \text{P}_i$$

where P_i is inorganic phosphate. Biotin serves a crucial role as the cofactor of pyruvate carboxylase, acting as a donor or acceptor of carbon dioxide. Biotin is also the cofactor in other carboxylation reactions such as those during the normal synthesis of fatty acids. This explains why biotin must be supplied in a growth medium if a fungus cannot synthesize it (see Chapter 5).

How are sugars generated from non-sugar substrates?

Sugars always are needed for wall synthesis and for the synthesis of nucleic acids, so we must ask how they are produced when a fungus is growing on non-sugar substrates. The process is termed **gluconeogenesis** and is shown in Fig. 6.4. Many of the steps are simply a reversal of the normal EM pathway, but the step between phosphoenolpyruvate (PEP) and pyruvate in the EM pathway is irreversible and so must be bypassed.

Consider the case of a fungus growing on acetate (2C) as the sole carbon source. After uptake, acetate can be converted to acetyl-CoA and used to generate oxaloacetate by the **glyoxylate cycle**—a short-circuited form of the TCA cycle (Fig. 6.4). The first reaction step is the cleavage of isocitrate (a 6C intermediate of the TCA cycle) to yield succinate (4C) and glyoxylate (2C). Then, glyoxylate is condensed with acetyl-CoA to yield malate and thence oxaloacetate. From this point, oxaloacetate is decarboxylated to yield PEP, shown below, and sugars are formed from PEP by reversal of the rest of the EM pathway:

$$\text{Oxaloacetate} + \text{ATP} \xrightarrow{\textit{phosphoenolpyruvate carboxykinase}}$$
$$\text{phosphoenolpyruvate} + CO_2 + \text{ADP} + P_i$$

The central role played by the glyoxylate cycle is evidenced by the fact that its enzymes increase more than 20-fold when acetate is supplied as the carbon source, and mutants that cannot grow on acetate often have a disruption in this pathway. The same processes as described above could be used for generating sugars from other substrates like fatty acids, organic acids, amino acids, ethanol, etc.

Secretion of organic acids

Fungi are important commercial sources of organic acids (see Chapter 1). For example, if *Aspergillus niger* is grown at high glucose levels

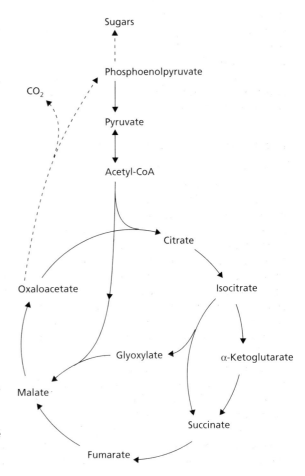

Fig. 6.4 Role of the glyoxylate cycle in generating sugars for biosynthetic processes when fungi are grown on non-sugar substrates such as acetate or organic acids. The conversion of phosphoenolpyruvate (PEP) to pyruvate during energy generation from sugars (see Fig. 6.2) is effectively irreversible, so the glyoxylate cycle generates PEP by an alternative route, then sugars are synthesized from PEP by reversal of most steps in the Embden–Meyerhof pathway (Fig. 6.2).

(15–20%) and low pH (about 2.0) it will convert up to 95% of the sugar supplied to citric acid and release this into the culture medium. Large amounts of oxaloacetate must be generated for this, by the carboxylation of pyruvic acid discussed earlier. Similarly, *Rhizopus nigricans* produces large amounts of fumaric acid, another TCA cycle intermediate. In both cases the growth of the fungus is severely (and purposefully) restricted by low pH or some other factor, and yet the basic metabolic pathways continue to operate. The fungus behaves like a car with its engine ticking over: the fuel (substrate) is not used for growth, so a convenient metabolic intermediate is released as a kind of exhaust product. We shall see later that secondary metabolites such as antibiotics are produced in a similar way. In bacteria these types of process are termed energy spillage or **energy slippage**; the fact that they occur at all indicates that they are necessary, perhaps because microorganisms need to keep their normal metabolic processes operating during periods when growth is temporarily halted.

Translocation and storage compounds

The compounds used for energy storage and translocation in fungi are quite different from those of plants but are strikingly similar to those of insects. The main energy-storage compounds are lipids, **glycogen** (an α-linked polymer of glucose) and **trehalose** (a non-reducing disaccharide; Fig. 6.5). The main translocated carbohydrates are trehalose and straight-chain sugar alcohols (**polyols**) such as **mannitol** (Fig. 6.5) and arabitol, or ribitol in zygomycota. Oomycota have none of these typical 'fungal carbohydrates'. Instead, their storage compounds are lipids and soluble mycolaminarins (β-linked polymers of glucose), and they translocate glucose or similar sugars.

All the fungal carbohydrates can be derived from common sugars taken into the cells; for example, trehalose is composed of two glucose residues, and mannitol is derived from fructose by a polyol dehydrogenase. These compounds are also readily interconverted in hyphae, as Brownlee and Jennings (1981) showed for the dry-rot fungus, *Serpula lacrymans* (Fig. 6.6). Wood blocks previously colonized by *S. lacrymans* were placed on Perspex so that the fungus grew across the

Perspex as a broad mycelial front supplied by mycelial cords behind this (see Chapter 4). When ^{14}C-glucose was placed on the wood blocks it was absorbed by the fungus, and the label was translocated to the colony front. Analysis of the soluble carbohydrates in different zones showed that trehalose predominated in the mycelial cords but a mixture of trehalose and arabitol was present nearer the colony margin. The conversion of trehalose to arabitol would not be difficult, because arabitol is formed from glucose-6-phosphate. The consequence would be an increased solute level near the growing colony margin where the disaccharide is cleaved, helping to draw water and solutes along the mycelial cord.

Sugar alcohols and trehalose have been implicated in host–parasite interactions. The initial studies on this involved ectomycorrhizal fungi of trees, because these fungi produce a sheath of tissue around the root tips, and it can be dissected away for separate chemical analysis (see Chapter 12). When leaves of trees were exposed to $^{14}CO_2$, the label was found mainly as sucrose in the leaf, stem and root tissues, but as mannitol and trehalose in the fungal sheaths. Similar findings have been made for plants infected by rust or powdery mildew fungi; when the leaves were exposed to labelled carbon dioxide they contained the label in the typical plant sugars (sucrose, glucose), but spores collected from the same leaves had the label in fungal carbohydrates. Most plants do not metabolize these compounds (except in seeds and fruits), so fungal parasites that grow in intimate association with plant cells are suggested to

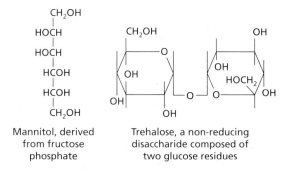

Mannitol, derived from fructose phosphate

Trehalose, a non-reducing disaccharide composed of two glucose residues

Fig. 6.5 Mannitol (a polyol) and the disaccharide trehalose, two common forms in which carbohydrates are translocated in fungal hyphae.

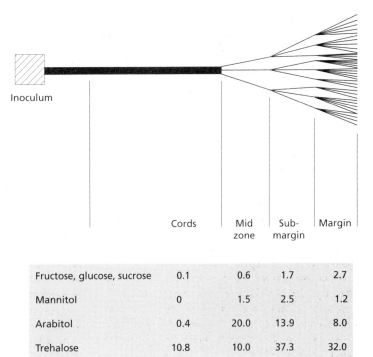

	Cords	Mid zone	Sub-margin	Margin
Fructose, glucose, sucrose	0.1	0.6	1.7	2.7
Mannitol	0	1.5	2.5	1.2
Arabitol	0.4	20.0	13.9	8.0
Trehalose	10.8	10.0	37.3	32.0

Fig. 6.6 Soluble carbohydrate contents of different regions of the mycelial system of *Serpula lacrymans* growing from a wood block across a sheet of Perspex (data are per cent of dry weight). (From Brownlee & Jennings, 1981.)

benefit from this 'metabolic valve' — a one-way flow of nutrients from the host. However, the value of this has been questioned because the downy mildew pathogens (oomycota) grow in intimate association with plant cells and yet do not have the typical fungal carbohydrates. Farrar and Lewis (1987) reviewed the mechanisms of nutrient transfer in all these biotrophic interactions (see Chapter 12) and suggested that the important feature is the ability of a fungus to maintain a concentration gradient of sugars towards itself by rapidly translocating sugars away from the infection site and using them for sporulation or hyphal growth. In a related context, the green algae in lichen associations synthesize mannitol, which the fungi can use, perhaps helping in the stability of these relationships. Also, the orchids can utilize trehalose as a carbohydrate source, and this helps them to obtain sugars from their mycorrhizal fungi (see Chapter 12).

The actual pathways of nutrient translocation in fungi are largely unknown but nutrients can move both forwards and backwards in hyphae, from regions of relative abundance to relative shortage. In this respect the tubular vacuolar system

described in Chapter 2 may be significant because it can transport fluorescent dyes against the general flow of cytoplasm. It may also be significant that mannitol is a common constituent of fungal vacuoles, where it is thought to have a role in regulating cellular pH. So, mannitol might have a multiplicity of functions — as a storage compound, a metabolic regulator and a translocated nutrient within a tubular vacuolar system.

Chitin synthesis

Polysaccharide synthesis in fungi is similar to that in other organisms, and can be represented by the following general equation, where the energy required for synthesis is contained in the donor–sugar bond:

[donor + sugar unit] + acceptor
→ donor + [acceptor + sugar unit]

For example, in the synthesis of chitin (Fig. 6.7), fructose-6-phosphate (from the EM pathway) is initially converted to *N*-acetylglucosamine (GlcNAc) by successive additions of an amino group (from the amino acid glutamine) and an

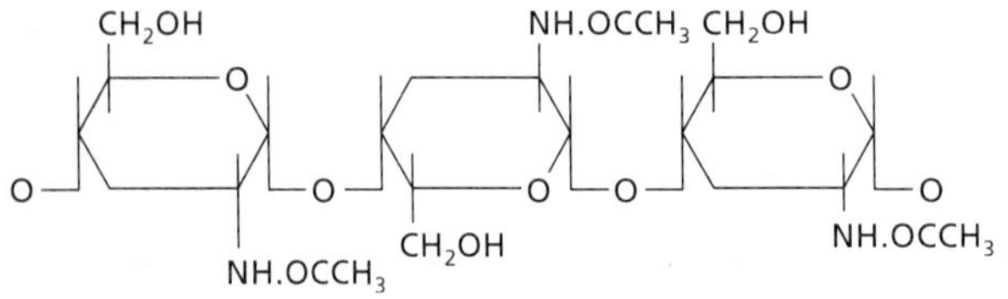

Chitin, a β-1,4 linked polymer of N-acetylglucosamine

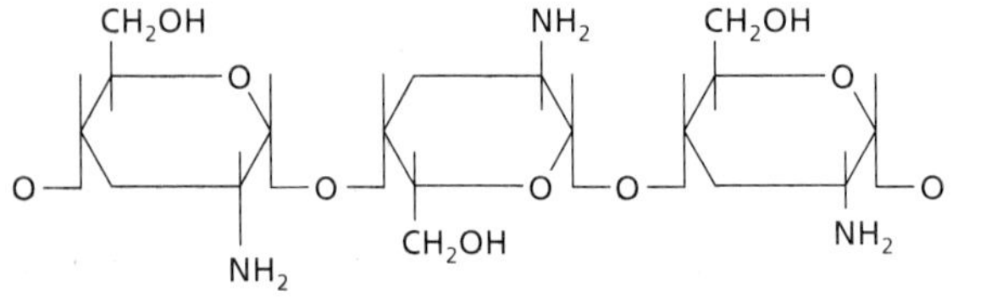

Chitosan, a poorly or non-acetylated derivative of chitin

Fig. 6.7 Structure of chitin and its derivative, chitosan.

acetyl group from acetyl-CoA. Then, GlcNAc reacts with uridine triphosphate (UTP) to form uridine diphosphate (UDP)–GlcNAc (donor–sugar unit) plus inorganic phosphate, and this molecule is added stepwise to the elongating chitin chain:

$$UTP + GlcNAc \rightarrow UDP\text{–}GlcNAc + P_i$$

$$UDP\text{–}GlcNAc + n[GlcNAc] \rightarrow$$
$$UDP + n + 1[GlcNAc]$$

Chitin is such a characteristic component of fungal walls that it has been used for assessing the fungal content of natural materials such as decomposing organic matter, infected plants or even the infected tissues of animals (*post mortem*!). For this, the material is ground thoroughly then treated with strong alkali to remove the acetyl groups from chitin, leaving chitosan (Fig. 6.7). Then, the chitosan is treated with nitrous acid to remove the amine groups and produce an aldehyde, 2,5-anhydromannose, which can be complexed with a dye and assessed colorimetrically. Standards containing known weights of fungal mycelium enable the colorimetric readings to be converted to fungal content. However, there are problems of interpretation, not least the inability of the method to distinguish between fungal and insect chitin, or living and dead hyphae. Alternative methods include the analysis of ergosterol content of materials, because this is an exclusively fungal sterol (see Chapter 14). This topic was reviewed by Newell (1992).

Lysine biosynthesis

Unlike other amino acids (see Chapter 5), lysine is synthesized by two specific pathways that are distinct from one another, with no enzymes in common. They are termed the **DAP** and **AAA** pathways after their characteristic intermediates, α-diaminopimelic acid and α-aminoadipic acid (Fig. 6.8). The AAA pathway is found only in the chitin-containing fungi and some euglenids. All other organisms that synthesize lysine—the plants, bacteria and oomycota—use the DAP pathway (animals do not synthesize lysine and need to be supplied with it). Such a major difference between fungi and other organisms indicates an ancient evolutionary divergence.

Secondary metabolism

The term secondary metabolism was coined to describe a wide range of reactions whose products are not directly involved in normal growth. In this respect, secondary metabolism differs from intermediary metabolism (the normal cellular metabolic pathways). Some thousands of secondary metabolites have been described from fungi (Turner, 1971; Turner & Aldridge, 1983), including antibiotics, plant growth hormones, mycotoxins, etc. The only features that they have in common are:

1 they tend to be produced at the end of the growth phase in batch culture or when growth is substrate-limited in continuous culture (see Chapter 3);

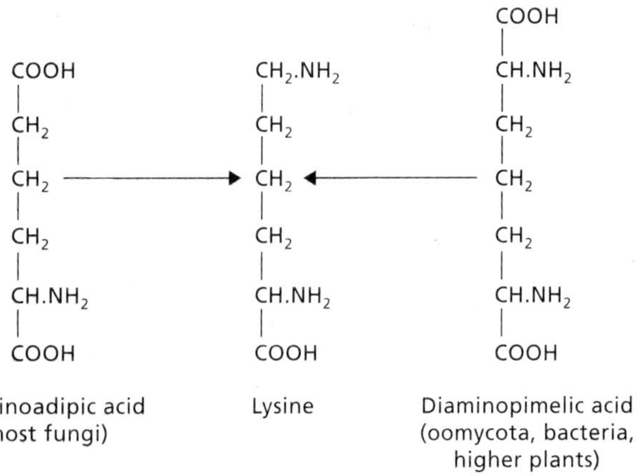

Fig. 6.8 The amino acid lysine and the characterisitic intermediates of the two distinct lysine-biosynthetic pathways.

α-Aminoadipic acid (most fungi)

Lysine

Diaminopimelic acid (oomycota, bacteria, higher plants)

2 they are produced from common metabolic intermediates but by special enzymic pathways encoded by specific genes;

3 they are not essential for growth or normal metabolism;

4 their production tends to be genus-, species- or even strain-specific.

Interest in these compounds stems mainly from their commercial or environmental significance. All the antibiotics used in medicine are secondary metabolites, including penicillin, cephalosporin and griseofulvin which are produced by fungi (see Chapter 1, Table 1.4). Fungi also produce many toxic secondary metabolites, such as aflatoxins, ochratoxins and ergot alkaloids in food products. Some other secondary metabolites are involved in differentiation, examples being the melanins in spore and hyphal walls and the sex hormones discussed in Chapter 4. Nevertheless, the vast majority of secondary metabolites of fungi have no obvious role, and mutants that do not produce them grow as well as the wild-type strains in culture. This raises the question of why they are produced.

One of the most plausible suggestions is that the process of secondary metabolism is necessary, regardless of the end-products. According to this view, secondary metabolism acts as an overspill or escape valve, to remove intermediates from the basic metabolic pathways when growth is temporarily restricted. This is similar to the explanation we noted earlier for the overproduction of organic acids: an organism might need to maintain the basic metabolic pathways during periods

when growth is restricted, but the common metabolic intermediates cannot be allowed to accumulate because they would disrupt metabolic control, so these intermediates are shunted into secondary metabolites which are either exported from the cell or accumulate as inactive compounds. It could then be argued that the genes encoding the secondary metabolic pathways are free to mutate (certainly more so than those encoding basic metabolism) and selection pressure would favour mutations that lead to useful products. Thus, antibiotics could have been useful in defence of territory, mycotoxins as insect antifeedants (see later), melanin for protection against ultraviolet (UV) damage (see Chapter 7), sex hormones for attracting partners (see Chapter 4) and flavour or odour components of toadstools for attracting insects for spore dispersal. If this view of secondary metabolism is correct, then it seems that many fungi have still to find useful roles for their secondary metabolites, or we have yet to find them!

In any case, secondary metabolism is under tight regulatory control. As noted above, secondary metabolites typically are produced towards the end of the exponential growth phase in batch cultures, when growth is limited by a critical nutrient but when other nutrients are still available. In continuous culture systems, however, secondary metabolites can be produced throughout the exponential growth phase; the critical factor is that the genes encoding secondary metabolism are repressed by high levels of particular nutrients, and in chemostats the culture medium can be designed so that these repressor substrates are the growth-limiting nutrients, always present at low concentration because they are utilized as soon as they enter the chemostat (see Chapter 3).

The pathways and precursors of secondary metabolism

The principal pathways and precursors of secondary metabolism in fungi are listed in Table 6.1. The single most important pathway is the **polyketide pathway**, which seems to have no other role except in secondary metabolism. It is shown in simplified form in Fig. 6.9. The precursor is acetyl-CoA, which is carboxylated to form malonyl-CoA (a normal event in the synthesis of fatty acids), then three or more molecules of malonyl-CoA condense with acetyl-CoA to form a chain. This chain undergoes cyclization, then the ring systems are modified in different ways to give a wide range of products. These include the antibiotic griseofulvin

(from *Penicillium griseofulvum*) which is used for treatment of dermatophyte infections of humans (see Chapter 14), the potent aflatoxins (from *Aspergillus flavus* and *A. parasiticus*) discussed later, and the ochratoxins from various *Penicillium* and *Aspergillus* spp.

Another important pathway in fungi is the **isoprenoid pathway** (Fig. 6.10), used normally for the synthesis of sterols. Again, acetyl-CoA is the precursor, but three molecules of this condense to form mevalonic acid (a 6C compound) which is then converted to a 5C **isoprene** unit. The isoprene units condense head-to-tail to form chains of various lengths, then the chains undergo cyclization and further modifications. The products of this pathway include the mycotoxins of *Fusarium* spp. that grow on moist grain, such as T-2 toxin and the trichothecenes.

The **shikimic acid pathway**, used normally for the production of aromatic amino acids, provides the precursors for the hallucinogenic secondary metabolites, lysergic acid and psilocybin in toad-

Table 6.1 Examples of secondary metabolites derived from different pathways and precursors.

Precursor	Pathway	Metabolites, representative organisms (chapter reference)
Sugars	—	Few, e.g. muscarine (*Amanita muscaria*), kojic acid (*Aspergillus* spp.) (1)
Aromatic amino acids	Shikimic acid	Some lichen acids
Aliphatic amino acids	Various, including peptide synthesis	Penicillin (*Penicillium notatum, P. chrysogenum*) (6) Fusaric acid (*Fusarium* spp.) (12) Ergot alkaloids (*Claviceps, Acremonium* spp.) (6) Lysergic acid (*Claviceps purpurea*) Sporidesmin (*Pithomyces chartarum*) (10) Beauvericin (*Beauveria bassiana*) (13) Destruxins (*Metarhizium anisopliae*) (13)
Organic acids	TCA cycle	Rubratoxin (*Penicillium rubrum*) (6) Itaconic acid (*Aspergillus* spp.) (1)
Fatty acids	Fatty acid	Polyacetylenes (basidiomycota mycelia and fruitbodies)
Acetyl-CoA	Polyketide	Patulin (*Penicillium patulum*) (6) Usnic acid (many lichens) Ochratoxins (*Aspergillus ochraceus*) (6) Griseofulvin (*Penicillium griseofulvum*) (14) Aflatoxins (*Aspergullus flavus, A. parasiticus*) (6)
Acetyl-CoA	Isoprenoid	Trichothecenes (*Fusarium* spp.) (6) Fusicoccin (*Fusicoccum amygdali*) (7) Sirenin, trisporic acids, oogoniol, antheridiol (4) Cephalosporins (*Cephalosporium* spp.) Viridin (*Trichoderma virens*) (11)

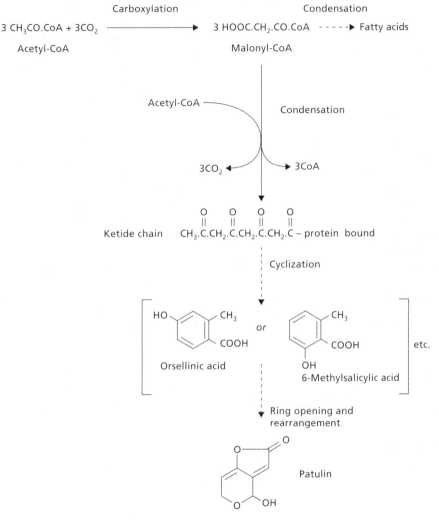

Fig. 6.9 Outline of the polyketide pathway, one of the main pathways for production of fungal secondary metabolites, such as the mycotoxin (and antibiotic) patulin. Various intermediates, shown in brackets, can be formed during cyclization, leading to different final products. Patulin is derived from 6-methylsalicylic acid. Longer ketide chains than the one shown here can give multiple ring systems, leading to products such as the aflatoxins (see Fig. 6.12).

stools of *Psilocybe*, and the toxin muscarine in toadstools of *Amanita muscaria*. Yet other pathways lead from aliphatic amino acids to the penicillins (see later) and the amatoxins and phallotoxins (in fruitbodies of *Amanita phalloides*), from fatty acids to the volatile polyacetylenes of mycelia and fruitbodies of several basidiomycota, and from intermediates of the TCA cycle to the rubratoxins of *Penicillium rubrum*. The important general point is

that all the precursors of these pathways are key metabolic intermediates (acetyl-CoA, etc.), so if secondary metabolism is an overspill device then the escape valves are located at crucial points in the normal metabolic pathways.

Against this background, we now consider three examples of secondary metabolites of special interest — the penicillins, aflatoxins and ergot-related alkaloids.

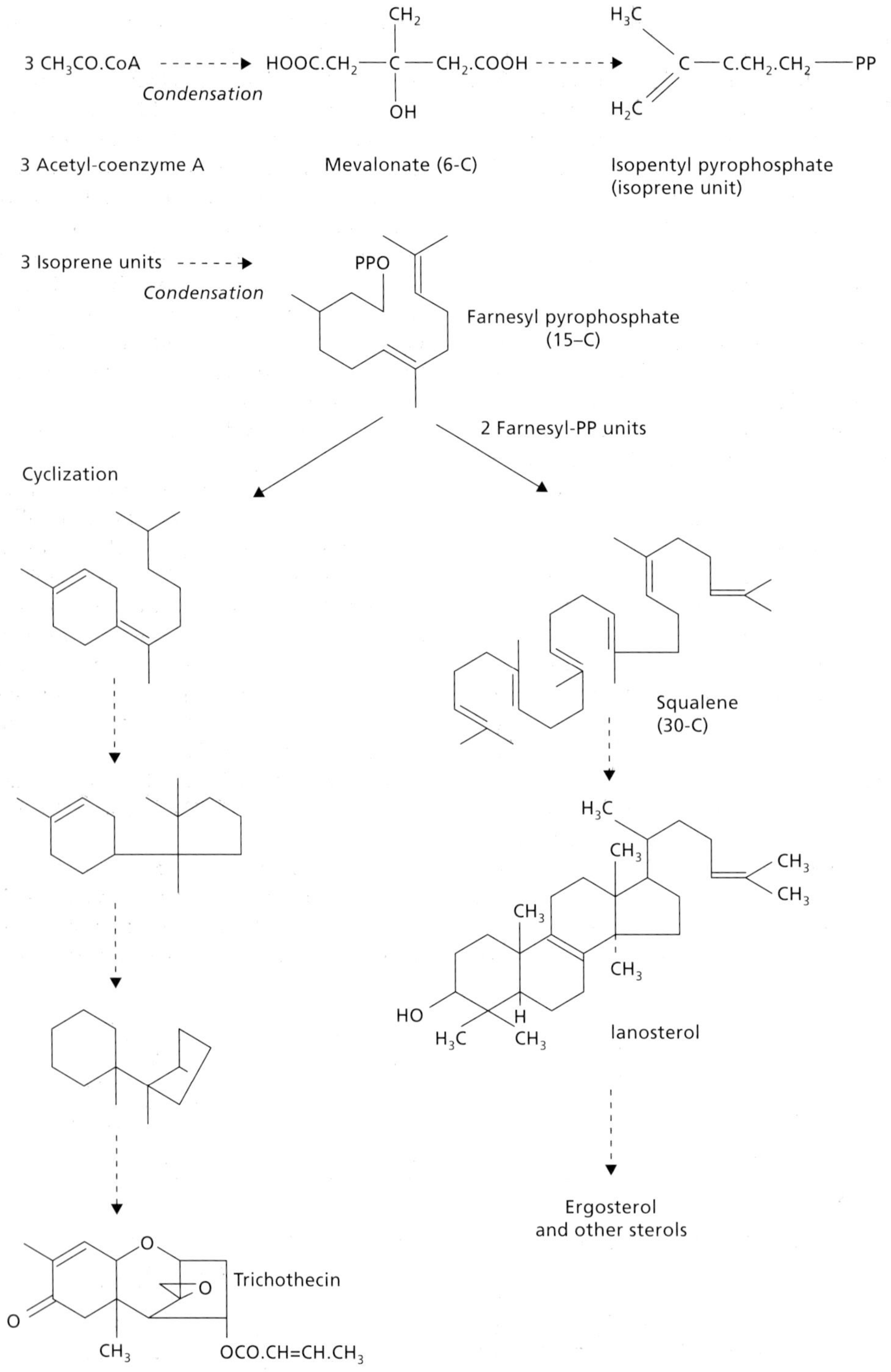

Fig. 6.10 Outline of the isoprenoid pathway for synthesis of sterols, several fungal pheromones and several mycotoxins such as trichothecin, one of the trichothecene group of toxins.

Penicillins

Penicillin was discovered by Alexander Fleming in 1929 as a metabolite of *Penicillium notatum*. However, Fleming could not purify the compound in stable form, and this development together with the design of large-scale fermentation systems did not occur until the early 1940s. Then, penicillin rapidly became the 'wonder drug', a reputation that it fully deserves because it has saved literally millions of lives. It is most active against Gram-positive bacteria, preventing the cross-linking of peptides during the synthesis of peptidoglycan in bacterial cell walls, so that the cells do not grow properly and are susceptible to osmotic lysis. In recent years the role of penicillin in chemotherapy has declined because of the widespread development of penicillin resistance among the target bacteria, but it is still used widely, especially for infections by *Streptococcus* spp. Penicillin is now produced commercially from high-yielding strains of *P. chrysogenum*. These strains have been mutated and selected by traditional methods to produce over 30 times as much antibiotic as wild-type strains; the basis of this increased production is now known to be due to multiple copies of the biosynthetic genes.

Figure 6.11 shows the basic structure of penicillin. It is a ring system derived from two amino acids, L-cysteine and D-valine, but is synthesized from a tripeptide precursor (α-aminoadipic acid–cysteine–valine) by the replacement of α-aminoadipic acid with an acyl group (R in Fig. 6.11). This step is catalysed by the enzyme acyl transferase. In the early years of commercial production, it was discovered that modification of the culture medium would produce penicillins with different acyl groups, conferring different properties on the molecule. So, a range of penicillins were produced commercially by carefully controlling the supply of acyl precursors in the culture vessels. However, it was then found that several bacteria produce the enzyme **penicillin acylase**, which removes the acyl side chain and leaves the basic molecule, **6-aminopenicillanic acid** (6-APA). Chemists could then attach any desired side chain to this molecule, with a high degree of precision. So, the modern production method for penicillins involves three stages.

1 Culture of the fungus to produce maximum amounts of any type of penicillin, usually penicillin G which was the type first discovered (Fig. 6.11). This is done by **fed-batch** culture (see Chapter 3) in which glucose is added in stages to prevent suppression of the secondary metabolic genes, while the precursor amino acids are supplied in excess. Also, the pH and aeration are carefully controlled because the penicillin molecule dissociates above pH 7.5.

2 Recovery of penicillin for the culture filtrate and treatment with penicillin acylase to produce 6-APA.

3 Chemical addition of specific acyl groups to produce a range of 'semi-synthetic' penicillins which have different specific properties. For example, **oxacillin** and closely related compounds

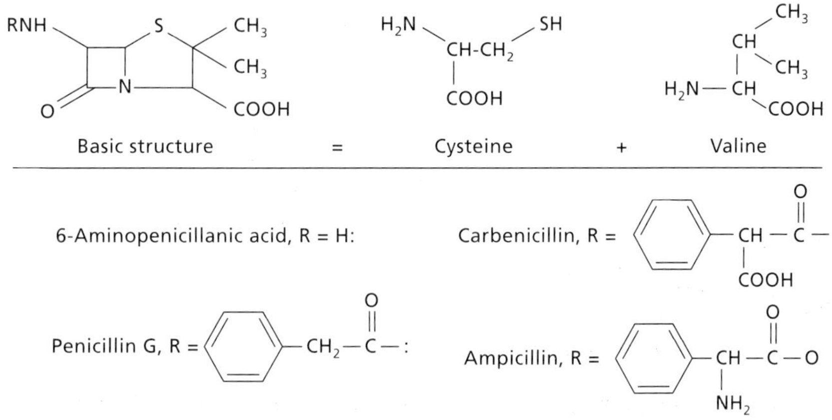

Fig. 6.11 Structure of penicillins.

are resistant to some bacterial β-lactamases — the enzymes that cleave the ring of penicillin G and thus inactivate the antibiotic. **Ampicillin** has significant activity against some Gram-negative bacteria, including enteric bacteria, whereas the natural penicillins act mainly against Gram-positive cocci like *Streptococcus*. **Penicillin V** has enhanced resistance to degradation by stomach acid, so it is one of the best for oral administration, whereas penicillin G is largely destroyed by stomach acid.

Despite these advances, all the penicillins are susceptible to a plasmid-encoded β-lactamase of some enteric bacteria, and they can cause allergic reactions in some patients. These problems are being approached by development of a closely related group of antibiotics, the **cephalosporins**, which were originally discovered as products of the fungus *Cephalosporium acremonium*, but now are produced commercially from strains of actinomycetes, *Streptomyces* spp.

Mycotoxins

Mycotoxins are a diverse range of compounds from different precursors and pathways, but grouped together on the basis of their toxicity to humans or higher animals. Some examples are shown in Table 6.2. A few of these toxins have been known for a long time, especially the toxins in toadstools (e.g. phallotoxins, amatoxins) and the alkaloids in sclerotia (ergots) of *Claviceps purpurea* which develop in place of the grain in infected cereals. But, the wider significance of mycotoxins was not appreciated until 1960 when 100 000

young turkeys died in Britain after being fed on fungus-contaminated cottonseed meal. The cause was traced to a mycotoxin, termed **aflatoxin**, and it raised awareness that even microscopic fungi can produce toxins in stored food or feed products. In fact, the aflatoxins are among the most potent known carcinogens; they have been shown to cause hepatomas (liver cancer) in experimental animals and are implicated strongly in some cases of liver cancer in humans. Table 6.3 illustrates the scale of the potential mycotoxin problem in food and feed products in general. For some of the most toxin-prone foodstuffs such as peanuts (aflatoxin) and recently in apple juice (**patulin** produced by the common apple-rot fungus *Penicillium expansum*) there are strict regulations or guidelines on the maximum permitted levels of contamination.

Aflatoxins are produced mainly by *Aspergillus flavus* (hence the name aflatoxin) and *A. parasiticus*, which are common in tropical and subtropical conditions. For example, they grow on the root systems and crop debris of groundnuts (peanuts), providing inoculum for colonization of the subterranean groundnut fruits. *A. flavus* can be isolated from the shells of almost any peanuts bought from a grocery store—it is seen as grey or black patches inside the shells. However, most strains do not produce aflatoxin, and even the strains that do so require specific conditions (see Chapter 7). Oil-rich materials such as peanut fruits and cottonseed are especially favourable for toxin production, consistent with the finding that aflatoxin production in laboratory culture is stimulated by lipids. After breakdown of the lipids by lipases, the fatty acids

Table 6.2 Some environmentally important mycotoxins.

Toxin	Fungi characteristically involved	Principal food/feed	Effects
Aflatoxins	*Aspergillus flavus* *A. parasiticus*	Peanuts, oilseeds	Liver damage
Sterigmatocystin	*A. versicolor*	Grain, oilseeds	Liver damage
Ochratoxins	*A. ochraceus*	Grain, oilseeds	Liver damage
Citrinin	*Penicillium citrinum*	Peanut, cereals	Kidney damage
Penicillic acid	*P. cyclopium*	Cereals	Cardiac toxin
Rubratoxins	*P. rubrum*	Seeds	Haemorrhage
Trichothecenes	All *Fusarium* spp.	Cereals	Various (oestrogenic, abortive)
Patulin	*A. clavatus*	Seeds	Neurotoxin
	P. expansum	Apples	

Table 6.3 Occurrence of ochratoxin A in tested samples of potential food and feed products. (From WHO, 1979.)

Product	Country	Number of samples tested	Per cent contaminated	Contamination level ($\mu g\,kg^{-1}$)
Wheat, oats, barley, rye for feed	Canada	32	56	30–27 000
Barley, oats	Denmark	33	58	27–27 500
Maize	France	463	3	15–200
Maize	USA	293	1	83–166
Wheat	USA	577	2	5–115
Barley, oats for feed	Sweden	84	9	16–409
Coffee beans	USA	267	7	20–360

are metabolized to acetyl-CoA by β-oxidation (see Chapter 5), then aflatoxins can be produced from acetyl-CoA by the polyketide pathway (see Fig. 6.9).

Aflatoxins are detected in extracts of contaminated foods by thin-layer chromatography, because they show natural green or blue fluorescence under UV illumination. This colour difference is related to the ring structure (Fig. 6.12) and has led to the distinction between aflatoxin G (green-fluorescing) and aflatoxin B (blue). It is also linked to toxicity, because aflatoxin B_1 is much

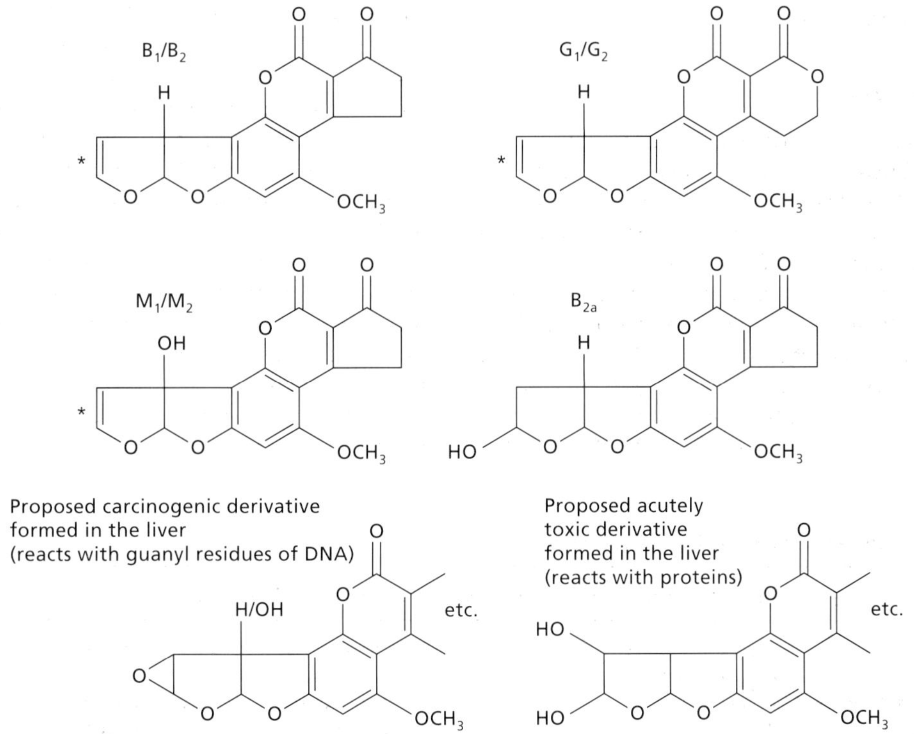

Fig. 6.12 Structure of some common aflatoxins. The most toxic compounds (aflatoxins B_1, G_1 and M_1) have a double bond at the position marked *. The less toxic compounds (B_2, G_2, M_2) lack the double bond at this position. Aflatoxins M_1 and M_2 are found in the milk of cows fed with B_1 and B_2. The non-toxic compound B_{2a} is produced by acid treatment of B_1. Aflatoxins B_1, G_1 and M_1 are metabolized in the liver to produce a highly carcinogenic derivative with an epoxide bridge at * and an acutely toxic compound with two hydroxy substituents.

more toxic than aflatoxin G_1. However, the toxicity of all these compounds depends on the presence of a double bond on the end furan ring (marked * in Fig. 6.12). Aflatoxins B_1 and G_1 have this double bond and they are carcinogens, whereas aflatoxins B_2 and G_2 lack the bond and are only weakly toxic. A similar difference is found between aflatoxin M_1 and M_2 which are found in the milk of cows fed with aflatoxin B_1 or B_2, respectively; roughly 5% of the aflatoxin intake by a cow can be found as aflatoxin M in the milk. The aflatoxins are absorbed from the gut and then pass to the liver where they are metabolized to a highly reactive, but unstable molecule with an epoxide bridge at the position of the double bond on the end furan ring (Fig. 6.12). This epoxide group enables the toxin to bind to deoxyribonucleic acid (DNA) and depurinate it, causing genomic lesions that can lead to hepatomas. In addition, the epoxide can give rise to dihydroxyaflatoxin, which is acutely toxic to some animals.

The toxins of endophytic fungi

Endophytes are fungi that grow inconspicuously within the tissues of functioning plants and usually cause no damage (see Chapter 12). Indeed, they have tended to go unnoticed because in many cases they can be detected only by staining and close microscopical inspection of plant tissues to observe their sparse hyphae in or between the plant cell walls. However, recent studies show that they can be significant in many ways. The endophytes of some pasture grasses, in particular, can increase the productivity of a grassland, increase the stress-tolerance of the plants, and significantly reduce insect damage (Clay, 1989). Their role as insect antifeedants is linked to the production of mycotoxins, some of which also cause problems in grazing animals, such as 'ryegrass staggers' of ruminants. The grass-associated endophytic fungi are a taxonomically difficult

group. Many of them grow extremely slowly in laboratory culture and do not produce sporing stages; in fact, some of them might have lost this ability altogether because they always grow within a plant and can be transferred from generation to generation by growing into the seed coat. The few isolates that have been induced to sporulate in culture produce a simple conidial stage classified as *Acremonium* and resembling the conidial stage of the ergot fungus *Claviceps purpurea*. Thus, *A. coenophialum* is commonly found in the fescue grasses, and *A. lolii* in ryegrasses. Like *Claviceps*, these clavicipitaceous endophytes produce alkaloids and related compounds in culture — the ergopeptine alkaloids (*A. coenophialum*) and the lolitrem alkaloids (*A. lolii*). Presumably they also produce these compounds in the plant tissues, where the slow, substrate-limited growth would be consistent with the production of secondary metabolites.

References

Brownlee, C. & Jennings, D.H. (1981) The content of carbohydrates and their translocation in mycelium of *Serpula lacrymans*. *Transactions of the British Mycological Society*, **77**, 615–19.

Clay, K. (1989) Clavicipitaceous endophytes of grasses: their potential as biocontrol agents. *Mycological Research*, **92**, 1–12.

Farrar, J.F. & Lewis, D.H. (1987) Nutrient relations in biotrophic infections. In: *Fungal Infection of Plants* (eds G.F. Pegg & P.G. Ayres), pp. 92–132. Cambridge University Press, Cambridge.

Newell, S.Y. (1992) Estimating fungal biomass and productivity in decomposing litter. In: *The Fungal Community: its Organization and Role in the Ecosystem* (eds G.C. Carroll & D.T. Wicklow), pp. 521–61. Marcel Dekker, New York.

Turner, W.B. (1971) *Fungal Metabolites*. Academic Press, London.

Turner, W.B. & Aldridge D.C. (1983) *Fungal Metabolites. II.* Academic Press, London.

WHO (1979) Environmental Health Criteria 11. *Mycotoxins*. World Health Organization, Geneva.

Chapter 7

Environmental conditions for growth, and tolerance of extremes

In this chapter we consider the effects of the major environmental factors on fungal growth, especially temperature, pH, aeration, water and light. We also consider environmental extremes, because some of the most interesting activities of fungi are seen in these conditions. However, a few introductory points must be made to put the content of this chapter into perspective.

First, we will be concerned with growth rather than survival. All organisms survive to at least some degree in conditions that do not support their growth, but the adaptations for this have been covered already (see Chapter 4) or will be discussed later (e.g. see Chapter 9). Second, most fungi can grow over a wider range of conditions than will support their full life cycle. This should be borne in mind when extrapolating from the laboratory to natural environments. Third, fungi often will tolerate one suboptimal factor if all others are near optimum, but a *combination* of suboptimal factors can prevent fungal growth. For example, several fungi can grow at low pH (less than 4.0), and several can grow in anaerobic conditions, but few, if any, fungi can grow when low pH is combined with anaerobiosis. This is why peat deposits have accumulated over thousands of years in boggy areas, and why these peat deposits are now disappearing at an alarming rate when the land is drained for agriculture or forestry, allowing fungi to degrade the acidic organic matter in aerobic conditions, whereas previously they were prevented from doing so by anaerobiosis. Humans are the culprits, but fungi are the agents.

Finally, the effect of any factor on a fungus in culture can be a poor predictor of the effect in nature, where competitive interactions come into play. A classic example of this is shown in Fig. 7.1.

Wheat plants were grown in sterilized soil and inoculated with the take-all fungus (*Gaeumannomyces graminis*), an aggressive root pathogen (see Chapter 12). The fungus caused progressively more disease as the soil temperature was raised from 13 to 23 or 27°C (its optimum for growth on agar). But, in natural, unsterile soil the amount of disease declined as the temperature was raised above 18°C. The main reason for this is that higher temperatures favour other micro-organisms even more than they favour the take-all fungus; these other microbes include fluorescent pseudomonads that inhibit the take-all fungus by antibiosis (see Chapter 14). Thus, we see that many interacting factors must be taken into account in natural conditions—the subject of Chapters 10 and 11.

Temperature

Fungi can be grouped into three broad categories in terms of their temperature requirements for growth: **psychrophiles** (cold-loving), **mesophiles** (growing at moderate temperatures) and **thermophiles** (heat-loving). However, the temperature ranges of these groups differ from those of bacteria, because relatively few fungi grow at 37°C (human body temperature) and the upper limit for growth of any fungus is 62–65°C. In contrast, some bacteria thrive at 70–80°C, and some archaea are hyperthermophiles, growing at over 100°C. The spectrum of the temperature ranges of some representative fungi is shown in Fig. 7.2.

Most fungi are mesophilic, growing within the range 10–35°C, although with different tolerances within this range, and with optima between 20 and 30°C. For routine purposes these fungi can be grown at room temperature (22–25°C). It is worth noting that many bacteria of natural environments

121

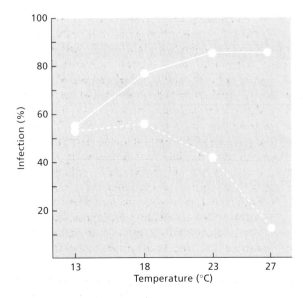

Fig. 7.1 Effect of temperature on infection of wheat by the take-all fungus *Gaeumannomyces graminis* in sterilized soil (solid line) and unsterilized soil (broken line). (From Henry, 1932.)

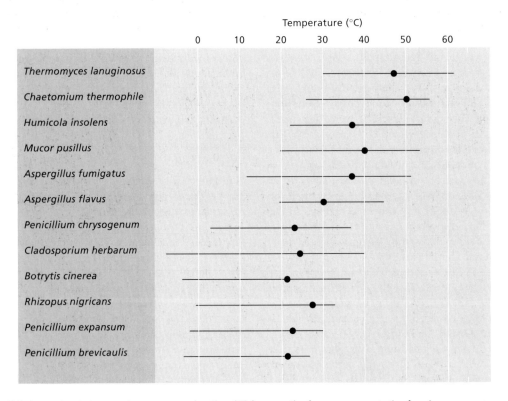

Fig. 7.2 Approximate temperature ranges and optima (●) for growth of some representative fungi.

have similar ranges to this; for example, both *Rhizobium* and *Agrobacterium* which are common root-associated bacteria have upper limits of about 30°C.

Only about 100 fungi can be considered to be thermophilic, with a minimum temperature of about 20°C, an optimum near 40°C and a maximum extending to 50–60°C. Again, there are variations within this range (Fig. 7.2), but these fungi are common in composts (see Chapter 10), in birds' nests and in sun-heated soils. An interesting and important example is *Aspergillus fumigatus* (Fig. 7.3) which can grow from 12 to 55°C, spanning most of the mesophile–thermophile range. It is found commonly in composts and on moulding grain, but it can also grow on hydrocarbons in aviation kerosene, and it can grow in the lungs after being inhaled as air-borne conidia, or it can colonize surgical wounds and grow within the tissues of transplant patients (see Chapter 13). It is thus a common and potentially dangerous opportunist.

A few fungi are psychrophilic (or psychrotolerant), with the ability to grow at or below 0°C and with upper limits of about 20°C, but higher for the psychrotolerant species. These fungi include several yeasts and mycelial species of permanently cold regions of the world, and a few species that cause spoilage of meat in cold storage. *Cladosporium herbarum* (Fig. 7.3) and *Thamnidium elegans* (Fig. 7.3) are found on cold-stored meat, but their normal habitats are leaf surfaces (*Cladosporium*) or soil and animal dung (*Thamnidium*). Owing to its strong proteolytic activity, *Thamnidium* has even been patented for use in tenderizing steak, but it is hard to find anyone who has used it! In a different context, the 'snow moulds' cause serious damage to cereal crops or grass turf if there is prolonged snow cover. In northern parts of the USA, the sclerotium-producing *Typhula* spp. (ascomycota) can kill up to 50% of the winter-sown cereal crop each year. Psychrophilic *Pythium* spp. cause similar problems in Japan. In Britain, however, it is more common to see cereals or

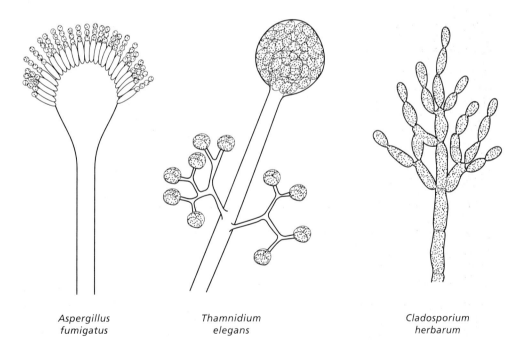

Aspergillus
fumigatus

Thamnidium
elegans

Cladosporium
herbarum

Fig. 7.3 *Aspergillus fumigatus* (deuteromycota) conidiophore with characteristic club-shaped terminal vesicle bearing phialides only on the upper half. *Thamnidium elegans* (zygomycota) sporangiophore which characteristically has a large terminal black sporangium and smaller sporangioles borne on side branches. *Cladosporium herbarum* (deuteromycota) with characteristic branching pattern of the conidiophore and conidial chains.

turf damaged by *Monographella nivalis* (formerly termed *Fusarium nivale*). This fungus is only weakly parasitic, but it invades and rots the plant tissues when their resistance is lowered by prolonged low temperature and low light. Late-season applications of nitrogenous fertilizer can predispose turf to attack because nitrogen promotes lush growth, rendering the plants susceptible to winter damage.

Physiology of temperature tolerance

The ability to grow at extreme temperatures involves total adaptation of an organism, not just the possession of a specific attribute. Probably for this reason, the cellular complexity of all eukaryotes limits their upper temperature to about 60–65°C. Also, for this reason, physiologists have tended to focus on the extreme temperature-tolerance of prokaryotes, so there is little direct information concerning the physiology of temperature tolerance of fungi. Nevertheless, the mechanisms are likely to be similar in different types of organism.

In order for growth to occur, the fluidity of the cell membrane must be maintained within certain limits, and all micro-organisms seem to achieve this by varying the composition of the membrane lipids. Saturated fatty acids (as in butter) are less fluid than unsaturated fatty acids (as in margarine) at any given temperature, and a comparison of nine thermophilic fungi with nine mesophiles of the same genera showed that the thermophiles consistently had a higher ratio of saturated to unsaturated lipids in their membranes. When the thermophilic fungi were grown near their lower temperatures they changed their membrane lipid composition so that the proportion of unsaturated lipids was higher. However, this is not a special feature of thermophiles because even mesophiles will show the same response when grown at different temperatures within their range.

A characteristic feature of thermophiles is that their enzymes and ribosomal components are more heat stable than those of mesophiles when extracted and tested in cell-free systems. This has been shown for thermophilic yeasts as well as for bacteria. The heat stability of enzymes is due to increased bonding between the amino acids near the enzyme active site, including bonds other than hydrogen bonds which are heat labile. Heat-stabilizing factors in the cytosol can also contribute to the thermostability of enzymes. For example, the glutamate synthetase of *Bacillus stearothermophilus* is not inherently heat stable when extracted and tested *in vitro*, but can be stabilized by adding a combination of ammonium (NH_4^+) and glutamate, or adenosine triphosphate (ATP) and glutamate which is probably the case *in vivo*.

The psychrophilic fungi almost certainly have a high proportion of unsaturated fatty acids in their cell membrane. Their ribosomes also differ from those of mesophiles, because ribosomes extracted from the psychrophilic yeast, *Candida gelida*, were found to be less stable than those from the mesophile *C. utilis* as the temperature was raised progressively in cell-free systems. The loss of stability was caused by dissociation of the protein from the ribonucleic acid (RNA) components of the ribosomes. However, the significance of this study has been questioned because none of the ribosomes dissociated at temperatures below 50°C. More relevant was the finding that ribosomes extracted from *C. gelida* lost 70% of their ability to synthesize a polypeptide chain after exposure to 30°C for 5 min, and 100% loss occurred after 5 min at 40°C, whereas the ribosomes of *C. utilis* were wholly unaffected by these treatments. In further comparisons of these two species, the aminoacyl–transfer RNA synthetases (which deliver amino acids to the ribosomes) and pyruvate decarboxylase (which converts pyruvic acid to acetaldehyde during the ethanol fermentation) of *C. gelida* were very heat sensitive. So, this obligately psychrophilic yeast has several temperature-sensitive components which probably set its upper limit for growth.

Recent studies on psychrophily have tended to address the topic from a different angle by asking why the majority of fungi cannot grow at low temperatures. They have sufficient cell solutes to prevent the cytosol from freezing at 0°C, and in any case fungi can increase their solute levels when necessary (see later). Kinetic studies show that chemical reactions still occur at 0°C so that even mesophiles should be able to grow at this temperature. However, it has been found that psychrophiles and mesophiles differ in their abilities

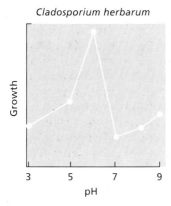

Cladosporium herbarum

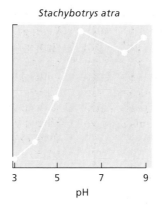

Stachybotrys atra

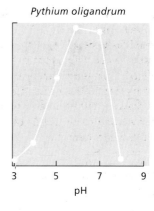

Pythium oligandrum

Fig. 7.4 pH growth response curves of three representative fungi in pure culture.

to assemble polyribosomes. When mesophiles are shifted down below their temperature limit, they continue to synthesize protein for a while but the ribosomes do not reattach to messenger RNA when the current round of protein synthesis is completed. At equivalent temperatures the psychrophiles continue to synthesize proteins. These and other aspects of temperature tolerance of fungi were reviewed by Griffin (1994).

Hydrogen ion (H+) concentration

In buffered culture media, many fungi will grow over the pH range 4.0–8.5, or sometimes 3.0–9.0, and they show relatively broad pH optima of about 5.0–7.0. However, there are variations within this 'normal' range, as shown by the three examples in Fig. 7.4. Several fungi are **acid-tolerant**, including some yeasts which grow in the stomachs of animals and some mycelial fungi (*Aspergillus*, *Penicillium* and *Fusarium* spp.) which will grow at pH 2.0. But, their pH optimum in culture is usually 5.5–6.0. Truly **acidophilic** fungi seem to be rare, and the most-cited example is *Acontium velatum* which will grow in 1.25 M sulphuric acid. It can initiate growth at pH 7.0 but it rapidly lowers the pH of culture media to about 3.0 which is probably close to its optimum. There are many naturally acidic environments where the acid-tolerant or acidophilic fungi can grow. In contrast, there are few strongly alkaline environments, although several fungi will grow up to pH 10–11 in culture (e.g. *Fusarium oxysporum*, *Penicillium variabile*).

In all cases that have been investigated, the fungi that grow at extremes of pH are found to have a cytosolic pH of about 7. The intracellular pH can be assessed crudely in extracts of disrupted cells or by observing the colour of pH-indicator dyes such as neutral red that are absorbed by living hyphae. However, the most accurate methods involve the insertion of pH-sensitive electrodes into hyphae, or loading hyphae with pH-sensitive fluorescent dyes that are permeabilized through the plasma membrane. These dyes show peaks of fluorescence at two wavelengths, and the relative size of the two peaks changes with pH, enabling changes of less than 0.1 pH unit to be measured accurately. The findings suggest that the fungal cytosol has strong buffering capacity; even when the external pH is changed by several units, the cytosolic pH changes by, at most, 0.2–0.3 units. This can be achieved in several ways: (i) by the selective uptake or release of ions; (ii) by exchange of materials between the vacuoles (which normally are acidic) and the cytosol; and (iii) by the interconversion of sugars and polyols such as mannitol (see Chapter 6) which involves the sequestering or release of H+.

Because the cytosolic pH is so tightly regulated, any perturbation of this can act as an intracellular signal leading to differentiation or change of growth polarity, etc. There are several examples of this in plant and animal cells, and a recent study suggests that it is also true for fungi. The cleavage of zoospores in the sporangia of *Phytophthora* and *Pythium* spp. (see Chapter 4) can be induced experimentally by a cold shock. By using a fluores-

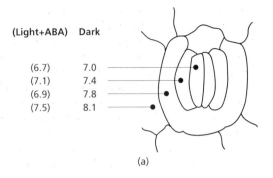
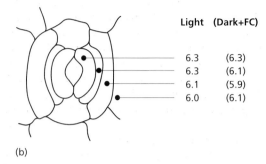
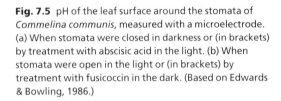

(a) (b)

Fig. 7.5 pH of the leaf surface around the stomata of *Commelina communis*, measured with a microelectrode. (a) When stomata were closed in darkness or (in brackets) by treatment with abscisic acid in the light. (b) When stomata were open in the light or (in brackets) by treatment with fusicoccin in the dark. (Based on Edwards & Bowling, 1986.)

cent pH-indicator dye, Suzaki *et al.* (1996) found that the cytosolic pH was raised transiently from 6.84 to 7.04 by this treatment, and no zoospore cleavage occurred if the sporangia had been microinjected with a buffer of pH 7.0, to prevent any change in cytosolic pH.

Ecological implications of pH

The effects of pH are much easier to investigate in laboratory conditions than in nature, because pH is not a unitary factor — as the pH changes so do many other things, leading to problems of interpretation. For example, pH affects the net charge on membrane proteins, with consequences for nutrient uptake. It also affects the degree of dissociation of mineral salts, and the balance between dissolved carbon dioxide (CO_2) and bicarbonate ions. Soils of low pH can have potentially toxic levels of available trace elements such as aluminium (Al^{3+}), manganese (Mn^{2+}), copper (Cu^{2+}) or molybdenum (Mo^{3+}) ions. Conversely, soils of high pH can have poorly available levels of essential nutrients such as iron (Fe^{3+}), calcium (Ca^{2+}) and magnesium (Mg^{2+}). Some of the anomalous fluctuations in the pH-growth curves in Fig. 7.4 were probably caused by such effects. Nevertheless, in general the pH-response curves in laboratory culture seem to be relevant to natural situations. For example, *Pythium* spp. are gener-

ally intolerant of very low pH but occur in soils above pH 4–5, *Stachybotrys chartarum* is found predominantly in near-neutral and basic soils (see Fig. 7.4), and *Trichoderma* spp. are characteristic of acidic soils.

Fungi can alter the pH around them and thus to some degree create their own environment. The most general method of doing this is by selective uptake and exchange of ions. For example, NH_4^+ is taken up in exchange for H^+, so the external pH can be lowered to 4 or less, leading to growth inhibition of the more acid-sensitive fungi such as *Pythium* spp. Conversely, the uptake of nitrate (NO_3^-) can cause the external pH to rise by 1 unit. Fungi also can release organic acids (see Chapter 6) which lower the external pH. Some of the aggressive tissue-rotting pathogens of plants release large amounts of oxalic acid in culture and seem to exploit this in their pathogenicity. For example, both *Sclerotium rolfsii* and *Sclerotinia sclerotiorum* secrete oxalic acid in plant tissues, lowering the pH to about 4.0. They also secrete pectic enzymes with acidic pH optima, and one of the roles of oxalic acid may be to form complexes with Ca^{2+}, removing this from the pectic compounds in plant cell walls so that the walls are more easily attacked by the pectic enzymes (see Chapter 12).

Environmental pH can help to orientate fungal growth, as Edwards and Bowling (1986) found by mapping the pH of leaf surfaces with microelectrodes. A pH gradient of more than 1 unit was found to occur locally around closed stomata, but little or no gradient was detected around open stomata (Fig. 7.5). This was true when the opening of stomata was controlled naturally by light or darkness and also when it was controlled experimentally by chemicals: the plant hormone abscisic acid causes stomata to close in the light, whereas the fungal metabolite fusicoccin (from the plant pathogen *Fusicoccum amygdali*) causes stomata to open in darkness. As we saw in Chapter 4, several plant pathogens infect through stomata and they can be guided by topographical signals. But, Edwards and Bowling found that pH gradients might also be involved, because germ-tubes of the rust fungus, *Uromyces viciae-fabae*, frequently terminated over open stomata but not over closed stomata. To test this, they made nail-varnish replicas of leaf surfaces with open stomata and placed the leaf replicas (surface pH 6.5) on agar of pH 6.0 or 7.0 so that this was the pH of the stomatal pore. When rust spores germinated on the leaf replicas a significantly higher proportion of the germ-tubes were found to locate the 'stomatal pores' of pH 6 than of pH 7, suggesting that they grew down a pH gradient.

Aeration

Most fungi are strict aerobes, in the sense that they require oxygen in at least some stage of their life cycle. Even *Saccharomyces cerevisiae*, which can grow continuously by fermenting sugars in anaerobic conditions (see Chapter 6), needs oxygen for sexual reproduction. So long as this is recognized we can restrict the discussion to the effects of oxygen on somatic growth. Then, fungi can be grouped in four categories of behaviour (Emerson & Natvig, 1981).

1 Some fungi are **obligate aerobes**. Their growth is reduced markedly if the partial pressure of oxygen (Po_2) is lowered much below that of air (0.21). For example, growth of the take-all pathogen of cereals is reduced even at Po_2 0.18. The thickness of a water film around the hyphae is crucial in these cases because of the slow diffusion of oxygen through water, as we saw for rhizomorphs of *Armillaria mellea* in Chapter 4.

2 Many yeasts and several mycelial fungi (e.g. *Fusarium oxysporum*, *Mucor hiemalis*, *Aspergillus fumigatus*) are **facultative aerobes**; they grow in aerobic conditions but also can grow in the absence of oxygen by fermenting sugars. The energy yield is much lower from fermentation and the biomass production is often less than 10% of that in aerobic culture, as Louis Pasteur first recognized for *S. cerevisiae*. However, the yield of some mycelial fungi can be at least 50% of normal if NO_3^- is available for anaerobic respiration.

3 A few aquatic fungi are **obligately fermentative**, because they lack mitochondria or cytochromes (e.g. *Aqualinderella fermentans*, oomycota) or they have rudimentary mitochondria and low cytochrome content (e.g. *Blastocladiella ramosa*, chytridiomycota). They grow in the presence or absence of oxygen, but always by fermentation. In this respect they resemble the lactic acid bacteria which always ferment. Fungi of this type are found in nutrient-enriched waters, where fermentable substrates are abundant.

4 The few chytridiomycota that grow in the rumen of cows and sheep are **obligate anaerobes** — their somatic cells are killed by exposure to oxygen. *Neocallimastix* spp. have received most study in this respect (Orpin, 1993; Trinci *et al.*, 1994). Their obligately anaerobic zoospores show chemotaxis to plant sugars in culture, and rapidly accumulate on chewed herbage in the rumen, where they encyst preferentially on the exposed ends of xylem vessels. Then, the rhizoids penetrate the plant tissue and release cellulase and other polymer-degrading enzymes, providing the fungus with a source of nutrients. These rumen chytrids are unusual among fungi because they have a **mixed acid fermentation**, the main products of which are formic acid (HCOOH), acetic acid, lactic acid, ethanol, CO_2 and hydrogen (H_2). The proportions of the end-products of these fermentations are variable, because many of the intermediates are interconvertible (Fig. 7.6). It is interesting that these organisms contain **hydrogenosomes**, which are functionally equivalent to the mitochondria of aerobic organisms, and are involved in energy generation by electron transfer, as shown in Fig. 7.6 (Marvin-Sikkema *et al.*, 1994). The end-products of fermentation can be used by other rumen organisms such as the methane-producing bacteria. In laboratory culture the degradation of cellulose by rumen fungi is often enhanced by the presence of these bacteria, so these organisms form a **consortium**, benefiting from one another's activities.

Physiology of oxygen tolerance

The fact that oxygen can be toxic to some organisms may be surprising, but the reason is that

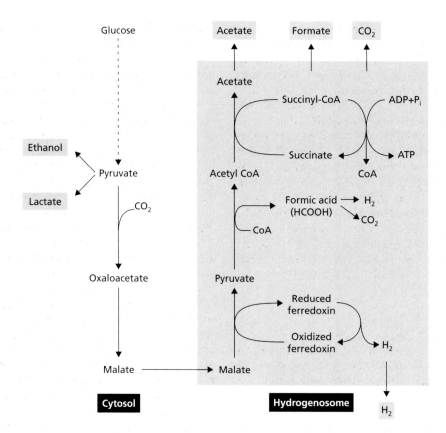

Fig. 7.6 Diagrammatic representation of the mixed-acid fermentation of the rumen chytrid *Neocallimastix*. End products of the fermentation are shown in small boxes. Part of the fermentation occurs in the cytosol, part in the hydrogenosome. Some of the details are known (Orpin, 1993; Marvin-Sikkema *et al.*, 1994); others are assumed here, based on the mixed-acid fermentation of some enteric bacteria.

oxygen can react with cellular components such as flavoproteins to generate hydrogen peroxide (H_2O_2) (commonly used as a disinfectant) and the highly reactive superoxide radical (O_2^-) which is an oxygen atom with an extra, unpaired electron. Superoxide readily donates the extra electron to any compound in its vicinity, causing cellular damage. Thus, all aerobic organisms need mechanisms for coping with these potentially hazardous effects, and they do so in two ways. First, they have an enzyme, **superoxide dismutase**, which converts superoxide to hydrogen peroxide as follows:

$$2O_2^- + 2H^+ \rightarrow H_2O_2 + O_2$$

Then the enzyme **catalase** converts H_2O_2 to water:

$$2H_2O_2 \rightarrow 2H_2O + O_2$$

Obligate anaerobes lack one or both of these enzymes. For example, *Neocallimastix* has superoxide dismutase but not catalase, so its inability to deal with peroxides probably accounts for its failure to tolerate oxygen.

Carbon dioxide

All fungi need CO_2 in at least small amounts for carboxylation reactions that generate fatty acids, oxaloacetate, etc. (see Chapter 6). Fungi that grow in anaerobic conditions often have a high CO_2 requirement, whereas several aerobic fungi can be inhibited by high CO_2. However, the significance of this in natural environments is difficult to judge. CO_2 dissolves in water to form carbonic acid (H_2CO_3), which dissociates to bicarbonate ions in a pH-dependent manner. At pH 8, the equilibrium is approximately 3% CO_2 (equivalent to H_2CO_3) with 97% HCO_3^-, but at pH 5.5 it is approximately 90% CO_2 with 10% HCO_3^-. Studies in laboratory culture at different pH suggest that fungi are more sensitive to the bicarbonate ion than to CO_2 as such. Even so, it can be questioned whether CO_2 (or bicarbonate) is a major growth inhibitor in nature. CO_2 is much more soluble than oxygen in water, and when their different diffusion coefficients are taken into account (the coefficient for CO_2 is actually lower) it can be calculated that CO_2 diffuses about 23 times more rapidly than oxygen in water films. Thus, when a fungus is respiring aerobically, generating 1 mole of CO_2 for every mole of oxygen consumed, the oxygen will become depleted in a water film before the CO_2 level has reached even 1%.

Water availability

All fungi need the physical presence of water for diffusion of nutrients into the cells and for the release of extracellular enzymes. Fungi also need to take up water to maintain their cytoplasm. However, water can be present in the environment and still be unavailable because it is bound by external forces. These forces include the osmotic potential (solute binding forces, φ_π), matric potential (physical binding forces, φ_m), turgor (φ_p) and gravimetric potential (φ_g). Their effects are additive so they are encompassed by the general term **water potential**, denoted φ and defined in terms of energy. Thus, the water potential of an environment is represented by:

$$\varphi = \varphi_\pi + \varphi_m + \varphi_p + \varphi_g$$

In order for a fungus to retain its existing water, it must generate a potential equal to the external φ, and in order to gain water from the environment it must generate a potential greater than φ.

Most fungi are highly adept at gaining water, even when the external forces are high, and so they grow by remaining turgid. However, the water moulds (*Saprolegnia*, *Achlya* spp.) are major exceptions to this; they seem to have little or no ability to maintain their turgor against external forces, probably because they grow only in extremely dilute freshwater environments. Of interest, these fungi grow normally even when they have lost turgor, probably by cytoskeletal extension of the apex (see Chapter 3). But, they are unable to penetrate solid surfaces in these conditions so they are, in effect, crippled (Harold *et al.*, 1996). All other fungi will grow and remain turgid over a range of external potentials. In fact, the fungi as a whole excel at this, and it is one of their most remarkable features.

Before we consider these points, we must introduce some further terms. In the older literature the availability of water was defined by the equilibrium relative humidity (RH). On this basis, an RH of 70% is near the lowest limit for fungal growth, although a few yeasts and mycelial fungi (e.g. *Xeromyces bisporus*, ascomycota) can obtain water at 61–62% RH, which is unrivalled by any other organism. In the food industry the term water activity (a_w) has commonly been used; it is equivalent to RH but expressed in decimal terms, 1.0 for pure water (no forces preventing its availability) down to zero. For most environmental work the term water potential is preferred (see Papendick & Mulla, 1986) and it is measured in megapascals (1 MPa is equivalent to 9.87 atm, or 10 bar). As familiar reference points, normal sea water has a potential about –2.5 MPa, and most plants reach 'permanent wilting point' in soils of about –1.5 MPa. The units are negative because these environments exert a pull on water.

Almost all fungi of soil and other terrestrial habitats can grow readily in media of –2 MPa. If the water-stress is increased beyond this then the aseptate fungi (zygomycota, oomycota) are the first to stop growing (their lowest limit is about –4 MPa). But, many septate fungi will grow at –4 MPa, and are not considered to be particularly stress-tolerant. Some fungi grow down to –5 or –10 MPa, and the most stress-tolerant fungi will grow at near-maximum rates at –20 MPa and make at least some growth at –50 MPa. These highly stress-tolerant fungi include the yeast *Zygosaccharomyces rouxii* and some *Aspergillus* spp., which initiate the spoilage of stored food products (see later). Having said this, we must make one important qualification: the response of a fungus depends on how the external stress is generated. Most fungi tolerate sugar-imposed osmotic potentials better than salt-imposed potentials—they are inhibited by salt toxicity long before they are inhibited by osmotic potential as such. Also, many fungi tolerate sugar-generated osmotic potential more than matric potential generated by adding high-molecular-weight polyethylene glycol to a growth medium.

Physiological mechanisms

Typically, fungi respond to low (negative) external water potentials by generating a lower internal osmotic potential. In some cases this is achieved by uptake and accumulation of ions, such as the uptake of potassium (K+) by the marine fungus *Dendryphiella salina*. However, this is not very satisfactory because high ionic levels are potentially damaging to cells, and even the marine fungi seem to take up K+ primarily as a mechanism for preventing sodium (Na+) toxicity. The more general mechanism is to accumulate sugars or sugar derivatives that do not interfere with the regulation of normal metabolic pathways. For this reason, these osmotically active compounds are termed **compatible solutes**. The most common compatible solute in fungi is glycerol, which is characteristic of the most stress-tolerant fungi (Hocking, 1993). Mannitol, trehalose and arabitol can also con-

tribute to the osmotic potential (see Chapter 6). The oomycota do not synthesize these typical 'fungal carbohydrates' and so (except for water moulds) they seem to accumulate the amino acid proline, as some bacteria also do in response to water-stress.

The compatible solutes of fungi can be generated from storage reserves (e.g. glycogen) or from nutrients taken into the cells. Fungi have at least some ability to change their compatible solutes, depending on the factors that cause the water-stress. Hallsworth and Magan (1994) demonstrated this for insect-pathogenic fungi, grown on media adjusted to different levels of osmotic stress with glycerol, trehalose or other compounds. Analysis of the solutes in the spores produced on these media (Fig. 7.7) showed that glycerol often accumulated when glycerol was used as the external osmoticum, whereas mannitol and other sugar alcohols accumulated in response to glu-

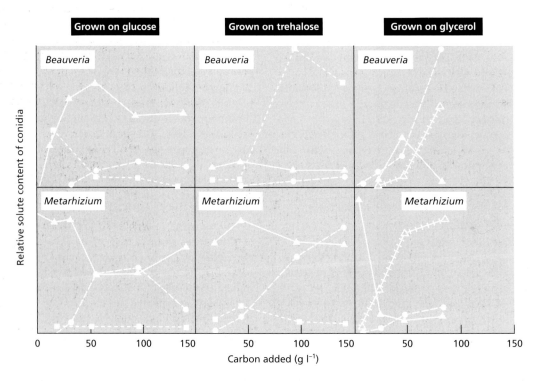

Fig. 7.7 Content of different solutes in the conidia of two insect-pathogenic fungi, *Beauveria bassiana* and *Metarhizium anisopliae*, when conidia were produced on colonies grown with increasing water stress generated by glucose, trehalose or glycerol. The solutes shown are mannitol (▲); erythritol or arabitol (●); trehalose (■); and glycerol (△). (Adapted from Hallsworth & Magan, 1994.)

cose-imposed stress, and trehalose could accumulate when trehalose was applied externally. These authors also have found that spores produced in media of high osmotic potential, and therefore containing high levels of solutes, are better able to germinate and infect insects in relatively dry conditions. This could be significant for use in insect biocontrol because, as we shall see in Chapter 13, the humidity requirement for germination of insect-pathogenic fungi is a major barrier to their exploitation in practice.

Comparisons between stress-tolerant fungi and stress-intolerant fungi have shown that both types produce compatible solutes in response to water-stress but they differ in their ability to retain the solutes in their cells. For example, glycerol is the compatible solute of both *Saccharomyces cerevisiae* (non-tolerant) and *Zygosaccharomyces rouxii* (stress-tolerant), and both fungi produce it to the same degree when subjected to water-stress, but glycerol then leaks from *S. cerevisiae* into the culture medium whereas *Z. rouxii* retains it. This also was true in a comparison of the stress-tolerant species *Penicillium janczewskii* and the non-tolerant species *P. digitatum*. Membrane fluidity thus seems to be implicated, and there is evidence of a higher content of saturated lipids in the membranes of stress-tolerant yeasts.

Ecological and practical aspects

Water-stress-tolerant fungi are economically important in the spoilage of cereal grains and other stored food products. None of these fungi can grow if the grain is dried to 14% moisture level, but this is not always practicable and if the level is even slightly higher (15–16%) then the stress-tolerant *Aspergillus* spp. (sexual stage, *Eurotium*) start to grow. *A. amstelodami* will initiate spoilage at –30 MPa in any slightly moist pocket of a grain store. It generates 'metabolic water' by degrading starch to glucose and respiring this to CO_2 and water. It also generates metabolic heat, causing the water to evaporate and condense elsewhere in the grain mass, so that the moulding spreads progressively and eventually paves the way for the growth of less stress-tolerant fungi. Figure 7.8 illustrates how this succession can occur. The *Aspergillus* and *Penicillium* spp. typically cause post-harvest spoilage, but the most stress-tolerant species (*A. amstelodami* and *A. restrictus*) are the initiators, followed by *A. fumigatus* and *Penicillium* spp. In contrast, *Fusarium* spp. commonly are regarded as 'field fungi'—they initiate spoilage in field conditions if there is a wet harvest season but they are intolerant of severe water-stress. Figure 7.8 also shows how tempera-

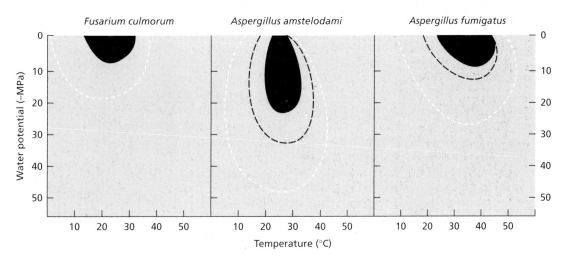

Fig. 7.8 Growth rate isopleths (the combination of temperature and water potential at which different growth rates are seen) for three fungi that cause spoilage of cereal grains. Growth rate of 0.1 mm/day (broken white line), 2.0 mm/day (broken black line) and ≥ 4.0 mm/day (black area). (Data from Ayerst, 1969 for the *Aspergillus* spp., and from Magan & Lacey, 1984 for *Fusarium culmorum*.)

131

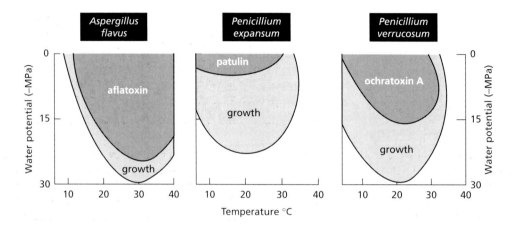

Fig. 7.9 Combination of temperature and water potential over which three spoilage fungi grow (light shading) and produce their mycotoxins (darker shading). (Reproduced from Northolt & Bullerman, 1982.)

ture interacts with water availability to affect the growth of spoilage fungi. The data are based on pure culture studies, but they have predictive value and can be used to design the most cost-effective but safe storage conditions.

Similar laboratory studies on the interaction of water-stress and temperature can be used to predict the conditions for mycotoxin production (Fig. 7.9). For example, *A. flavus* can produce afla-toxin A over most of the environmental range that supports its growth, whereas *Penicillium verruco-sum* produces ochratoxin A over a narrower part of its growth range. The production of patulin (which causes haemorrhaging of the lungs and brains of experimental animals) occurs over an even nar-rower range of conditions than those that support the growth of the apple-rot fungus *P. expansum*.

The fungi that grow commonly as saprotrophs on the surfaces of living leaves (the **phyllosphere**) show a different type of adaptation to water-stress. These fungi, such as *Cladosporium*, *Alternaria* and *Aureobasidium*, do not grow at low water potentials but have a remarkable ability to withstand peri-odic wetting and drying. Park (1982) demon-strated this in a simple and elegant way. He grew these fungi on pieces of transparent cellulose film ('Cellophane') on top of malt extract agar ('low water-stress) then removed the film with the attached fungal colonies and suspended it over saturated solutions of sodium nitrite or potassium

nitrate, giving atmospheres of 66% or 45% RH, respectively. After different times in these condi-tions, the cellulose film was replaced on the malt extract agar and examined for signs of regrowth of the fungi. Even after 2 or 3 weeks of drought, the leaf-surface fungi initiated regrowth within 1 h of being placed on the agar, and this regrowth occurred from the original hyphal tips. In contrast, a range of common soil fungi (e.g. *Fusarium*, *Trichoderma*, *Gliocladium*) or typical food-spoilage fungi (*Penicillium* spp.) never regrew from their original hyphal tips, although many of these fungi could regrow after 24 h from spores or surviving hyphal compartments behind the tips. The phyl-losphere fungi often have melanized hyphae, but their growing tips are hyaline. The severe drought-tolerance of their tips is still unexplained but perhaps involves hydrophobins, discussed in Chapter 4. In any case, these fungi clearly are adapted to the fluctuating moisture conditions of the leaf surface, and the same adaptations prob-ably enable these 'sooty moulds' to grow on kit-chen and bathroom walls in our homes.

Light

Visible light (wavelengths about 380–720 nm) has relatively little effect on somatic growth, although it can cause a zonation of some fungi on agar, so that circular zones of dense, highly branched growth alternate with zones of more normal spreading growth (Fig. 7.10). Lysek (1978) sug-gested that this occurs because light inhibits the extension of tips of the surface-growing hyphae, leading to dense branching, while hyphae within

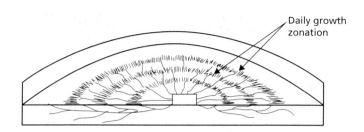

Daily growth zonation

Fig. 7.10 Diurnal growth zones of a fungal colony exposed to light. (Based on Lysek, 1978.)

the agar continue to grow and 'overtake' the surface hyphae, then reach the surface where their own growth is inhibited. A similar zonation can be caused by temperature fluctuations and it even occurs constitutively in some mutants of *Neurospora crassa* and *Ascobolus immersus*, where it is known to be governed by a single gene.

In contrast to somatic growth, light can profoundly affect reproduction or other differentiation events. Several fungi produce circular zones of asexual sporulation on agar plates, corresponding to daily alternations of light and darkness. Both *N. crassa* and *Trichoderma* spp. often show these circular zones of conidiation, while *Podospora anserina* and some other ascomycota show similar zones of perithecium (sexual) development. Sometimes, these zones correspond to earlier zones of somatic growth, described above, but with a delay while the differentiated structures develop, suggesting that these factors are linked. Often, these responses are elicited by near-ultraviolet (NUV, 330–380 nm) or blue light (about 450 nm), implicating a flavin-containing receptor. But, there is considerable variation, perhaps linked to the natural habitat, because *Alternaria* spp. are induced to sporulate by UV irradiation (280–290 nm), and in *Botrytis cinerea* the triggering is by NUV but reversed by blue light (see Chapter 4).

The fruitbodies of many basidiomycota are formed in response to light, but with an additional requirement for a low level of CO_2 in several cases. The process has been studied in *Coprinus congregatus* which produces its 'ink-cap' toadstools quite readily in laboratory culture (Ross, 1985). As shown in Fig. 7.11, agar colonies of this fungus must reach a critical age (about 3 days) before they will respond to a light trigger. Then, even a short exposure to white light causes a temporary halt in growth at the colony margin. After this the fungus regrows, but the site of the 'halt' becomes the

site where fruitbody primordia will develop subsequently if the colony receives a second light trigger (or continuous light after the original stimulus) at least 3-h later. This time interval is thought to be needed for synthesis of a new gene product that forms part of a light–receptor complex. The initial light stimulus is known to cause a physiological change in the hyphae because, after some hours and even in the absence of a second trigger, the hyphae at the 'halt' site become melanized (the colony has grown on in the meantime). The melanization is caused by an enzyme, phenol oxidase, which was originally membrane-bound in the hyphae but is released from the membrane by the light trigger and is then detected in hyphal extracts. Whether it has any role in the subsequent stages of development is unknown, but this membrane-bound enzyme is always located at the hyphal tips, ready for release and localized melanization when the tips are exposed to light. So, the fungus has an elegant method of sensing when it has reached the surface of soil or substrate, for initiating the fruitbodies. However, only the sites at which primordia will develop are determined by the light trigger, and the development of the fruitbody primordia themselves occurs when the conditions restrict further somatic growth—for example, when one critical nutrient becomes growth-limiting or when the colony reaches the edge of an agar plate. Then, the mycelial reserves are diverted to fuel the development of the fruitbodies, as discussed in Chapter 4.

Light has other effects on fungal reproductive structures, notably in eliciting phototropism of the sporangiophores of some zygomycota and of the ascal tips of some ascomycota (see Chapter 9). We end the topic by noting that the light responses of fungi often are clearly related to ecology; the light-responsive fungi almost invariably are those that produce air-borne spores.

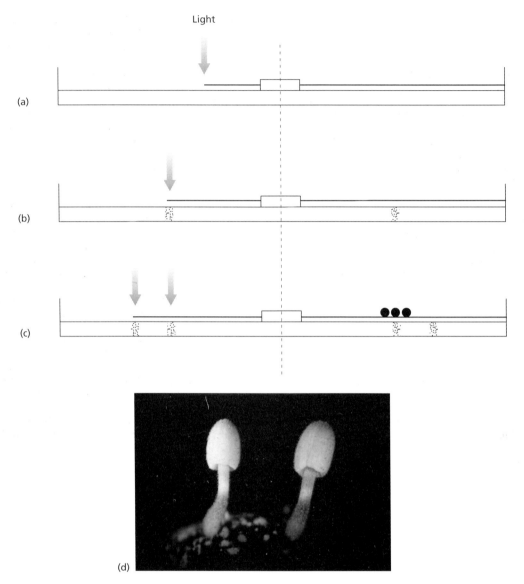

(a)

(b)

(c)

(d)

Fig. 7.11 Diagrammatic representation of effects of light on differentiation of fruitbodies of *Coprinus congregatus*. (Based on Ross, 1985.) The left side of each diagram (a–c) shows the treatment applied at one stage of colony growth; the right side shows the result of that treatment when the colonies were incubated until they reached the edge of the agar plate. (a) Colony exposed to light for less than 3 h when the colony was less than 3 days old: **no effect on growth**. When the colony reached the edge of the agar plate there was **no differentiation.** (b) Colony exposed to light for less than 3 h when the colony was more than 3 days old: **temporary halt in colony growth and melanization of colony margin** (stippled). When the colony reached the edge of the agar plate there was **no differentiation.** (c) Colony given two light exposures at 3 h intervals when more than 3 days old: **temporary halt in colony growth and melanization of colony margin**. When the colony reached the edge of the agar plate it **developed fruitbody primordia where the colony was first exposed to light**. (d) Mature fruitbodies of a related *Coprinus* sp. growing on dung.

References

Ayerst, G. (1969) The effects of moisture and temperature on growth and spore germination in some fungi. *Journal of Stored Products Research*, **5**, 127–41.

Edwards, M.C. & Bowling, D.J.F. (1986) The growth of rust germ tubes towards stomata in relation to pH gradients. *Physiological and Molecular Plant Pathology*, **29**, 185–96.

Emerson, R. & Natvig, D.O. (1981) Adaptation of fungi to stagnant waters. In: *The Fungal Community* (eds D.T. Wicklow & G.C. Carroll), pp. 109–28. Marcel Dekker, New York.

Griffin, D. (1994) *Fungal Physiology*, 2nd edn. Wiley-Liss, New York.

Hallsworth, J.E. & Magan, N. (1994) Effect of carbohydrate type and concentration on polyhydric alcohol and trehalose content of conidia of three entomopathogenic fungi. *Microbiology*, **140**, 2705–13.

Harold, R.L., Money, N.P. & Harold, F.M. (1996) Growth and morphogenesis in *Saprolegnia ferax*: is turgor required? *Protoplasma*, **191**, 105–14.

Henry, A.W. (1932) Influence of soil temperature and soil sterilization on the reaction of wheat seedlings to *Ophiobolus graminis* Sacc. *Canadian Journal of Research*, **7**, 198–203.

Hocking, A.D. (1993) Responses of xerophilic fungi to changes in water activity. In: *Stress Tolerance of Fungi* (ed. D.H. Jennings), pp. 233–56. Academic Press, London.

Lysek, G. (1978) Circadian rhythms. In: *The Filamentous Fungi*, Vol. 3 (eds J.E. Smith & D.R. Berry), pp. 376–88. Arnold, London.

Magan, N. & Lacey, J. (1984) Effect of temperature and pH on water relations of field and storage fungi. *Transactions of the British Mycological Society*, **82**, 71–81.

Marvin-Sikkema, F.D., Driessen, A.J.M., Gottschal, J.C. & Prins, R.A. (1994) Metabolic energy generation in hydrogenosomes of the anaerobic fungus *Neocallimastix*: evidence for a functional relationship with mitochondria. *Mycological Research*, **98**, 205–12.

Northolt, M.D. & Bullerman, L.B. (1982) Prevention of mould growth and toxin production through control of environmental conditions. *Journal of Food Production*, **45**, 519–26.

Orpin, C.G. (1993) Anaerobic fungi. In: *Stress Tolerance of Fungi* (ed. D.H. Jennings), pp. 257–73. Academic Press, London.

Papendick, R.I. & Mulla, D.J. (1986) Basic principles of cell and tissue water relations. In: *Water, Fungi and Plants* (eds P.G. Ayres & L. Boddy), pp. 1–25. Cambridge University Press, Cambridge.

Park, D. (1982) Phylloplane fungi: tolerance of hyphal tips to drying. *Transactions of the British Mycological Society*, **79**, 174–8.

Ross, I.K. (1985) Determination of the initial steps in differentiation in *Coprinus congregatus*. In: *Developmental Biology of Higher Fungi* (eds D. Moore, L.A. Cassleton, D.A. Wood & J.C. Frankland), pp. 353–73. Cambridge University Press, Cambridge.

Suzaki, E., Suzaki, T., Jackson, S.L. & Hardham, A.R. (1996) Changes in intracellular pH during zoosporogenesis in *Phytophthora cinnamomi*. *Protoplasma*, **191**, 79–83.

Trinci, A.P.J., Davies, D.R., Gull, K. *et al.* (1994) Anaerobic fungi in herbivorous animals. *Mycological Research*, **98**, 129–52.

Chapter 8
Genetics

In this chapter we cover the basic and applied genetics of fungi, including the features that continue to make fungi important model organisms for genetical research. The chapter includes recent molecular approaches in a range of fields such as the analysis of pathogenicity determinants of fungi and the development of fungi as 'factories' for foreign gene products. It also covers the most intriguing recent developments, including the roles of extrachromosomal genes in ageing and the effects of fungal viruses in suppressing pathogenic virulence.

Overview: the place of fungi in genetic research

For more than 50 years the fungi have been major tools for classical genetical research because they have a combination of features unmatched in other eukaryotes.
• They are easy to grow in laboratory conditions and they complete the life cycle in a short time.
• They are amenable to biochemical studies, so they played a major part in development of the classical concept of 'one gene, one enzyme' (Fincham et al., 1979).
• They are haploid so they are easy to mutate and to select for mutants.
• They have a sexual stage for analysis of the segregation and recombination of genes, and all the products of meiosis can be retrieved in the haploid sexual spores.
• They produce asexual spores so that genetically uniform populations can be bulked up and maintained.

The application of molecular genetical approaches to fungi was relatively slow because of the need to develop protoplasting techniques, transformation systems and appropriate vectors. These are still major limitations for some fungi, although molecular approaches are now well developed for *Saccharomyces cerevisiae*, *Schizosaccharomyces* spp., *Neurospora crassa* and *Aspergillus nidulans*. These ascomycota are especially valuable because the molecular approaches can be combined with classical genetics (sexual crossing) for the mapping of genes on chromosomes. Two features make the fungi particularly valuable for molecular genetical studies: (i) their small genome for eukaryotic organisms (the genome of *S. cerevisiae* is only about three times larger than the genome of *Escherichia coli*); and (ii) their small amount of repetitive or redundant deoxyribonucleic acid (DNA) so that much of the fungal genome codes for something.

The recent development of 'reverse genetics' has opened up a new and valuable area. By this approach, if an enzyme of another gene product is known, or if a particular messenger ribonucleic acid (mRNA) is detectable at a particular stage of development, then the gene can be identified and subjected to targeted gene disruption. This is true for any fungus, even if nothing is known about its genetics and even if the fungus cannot be cultured in the laboratory. So, now it is possible to identify and manipulate the specific pathogenicity determinants of a fungus or to identify specific genes involved in developmental pathways, etc. These and other aspects of fungal genetics are discussed in the sections that follow.

Structure and organization of the fungal genome

The fungal genome has four components — chromosomal genes, mitochondrial genes, plasmids and mobile genetic elements, and fungal virus genes (which are truly resident genetic elements).

Chromosomes and chromosomal genes

Fungal nuclei were discussed in Chapter 2, but a brief resumé will be helpful here. In most fungi the nuclei are haploid, but the oomycota are diploid, a few fungi can alternate between haploid and diploid somatic phases, and some yeasts (e.g. *Candida albicans*) are permanently diploid. Some fungi also have polyploid series (e.g. *Allomyces* and *Phytophthora* spp., including *P. infestans*). This can be shown by staining the hyphae with a fluorochrome that intercalates in the DNA, then measuring the fluorescence of individual nuclei under a microscope. Fluorescence increases in a stepwise manner with each increase in the number of chromosome sets (e.g. Tooley & Therrian, 1991).

The chromosomes of fungi are small and highly condensed. They are difficult to count by conventional microscopy of stained cells because the nuclear membrane persists during most of the mitotic cycle. However, counts have now been obtained for several fungi by a combination of cytology, genetical linkage analysis and pulse-field electrophoresis of extracted chromosomes (Oliver, 1987; Sansome, 1987). As shown in Table 8.1, the haploid chromosome count of most fungi lies between six and 17.

The nuclear genome size of fungi is very small in comparison with other eukaryotes. For example, the genome of *Saccharomyces cerevisiae* is 1.35×10^4 kilobase pairs (kbp) and that of *Schizophyllum commune* (basidiomycota) is about 3.7×10^4 kbp. These values are only about three and eight times larger than that of *E. coli* (4×10^3 kbp) and they are much smaller than in the fruit fly *Drosophila* (16.5×10^4 kbp) or humans (290×10^4 kbp). Such a small genome distributed among several chromosomes makes it relatively easy to sequence the DNA of entire chromosomes. Some of the 17 chromosomes of *S. cerevisiae* have already been sequenced, and all will have been done within a few years. Part of the reason for the small genome size is that fungi have little multicopy (reiterated) DNA. It represents only 2–3% of the genome in *Aspergillus nidulans* and 7% in *S. commune*, where the reiterated DNA codes mainly for cell com-

Table 8.1 Haploid chromosome numbers in some representative fungi.

OOMYCOTA	
Phytophthora spp. (many)	9–10
Achlya spp.	3, 6, 8
Saprolegnia spp.	8–12
Pythium	Commonly 10 or 20
CHYTRIDIOMYCOTA	
Allomyces arbuscula	16
A. javanicus	14 (but variable in hybrids and polyploids)
ASCOMYCOTA	
Neurospora crassa	7
Saccharomyces cerevisiae	17
Aspergillus nidulans	8
BASIDIOMYCOTA	
Schizophyllum commune	11
Coprinus cinereus	13
Puccinia kraussianna	30–40

ponents that are needed in large amounts — ribosomal RNA, transfer RNA and chromosomal proteins. However, an abnormally large amount of the genome of the downy mildew pathogen *Bremia lactucae* (oomycota) is repetitive (65% of the total genome of 5×10^4 kbp); the reason for this is unknown.

Fungi transcribe a substantial amount of the nuclear DNA into mRNA—an estimated 33% in *S. commune* and 50–60% in *S. cerevisiae*. Compared with other eukaryotes, therefore, they have relatively little non-coding (redundant) DNA. Fungi resemble other eukaryotes in that their protein-encoding genes contain non-coding DNA sequences termed **introns** (Fig. 8.1). The introns are transcribed into mRNA but then need to be excised before the mRNA is translated into proteins. However, the introns of fungi are very short (often about 50–200 bp) compared with those of higher eukaryotes (often 10 kbp or more), and *S. cerevisiae* seems to be unusual because it has very few introns. We will see the significance of this when discussing the technological role of yeast for production of foreign proteins.

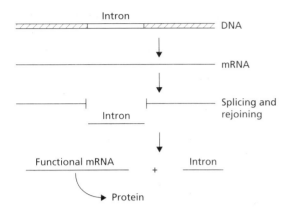

Fig. 8.1 Introns (non-coding regions) in DNA (coding regions are shown by cross-hatching).

Mitochondrial genes: normal functions and involvement in ageing

Mitochondria contain a small circular molecule of DNA. The size of the mitochondrial genome in fungi is similar to that in other eukaryotes (19–121 kbp); for example, 70 kbp in *S. cerevisiae*, and 50 kbp in *S. commune*. Any variations are due mainly to the amount of non-coding material, because all mitochondrial DNAs code for the same things: some components of the electron transport chain (including cytochrome c and adenosine triphosphatase (ATPase) subunits), some structural RNAs of the mitochondrial ribosomes and a range of mitochondrial transfer RNAs. Both the nuclear and the mitochondrial genes are needed to produce complete, functional mitochondria.

The mitochondrial DNA of fungi has received special attention in relation to ageing, because in several filamentous fungi (*Podospora, Neurospora, Aspergillus*) a single mutation in a single mitochondrion can lead to senescence of the whole colony, when the mutant gene causes the gradual displacement of wild-type mitochondrial DNA (reviewed by Esser, 1990; Bertrand, 1995). The strains of *Podospora* that exhibit the senescence phenotype can be maintained indefinitely as repeatedly subcultured young colonies, but they stop growing, become senescent and die after they have been grown continuously for about 25 days. It has long been known that a non-nuclear 'infective factor' is involved, because non-senescent strains (which never undergo senescence) acquire the ability to senesce when their hyphae anastomose with senescence-prone strains. Moreover, mitochondria were implicated because the onset of senescence could be postponed indefinitely by growing strains in the presence of sublethal doses of inhibitors of mitochondrial DNA synthesis or mitochondrial protein synthesis, but senescence occurred when the inhibitors were removed. More recent studies showed that DNA from strains that were undergoing senescence could be transformed into protoplasts of healthy strains, and the protoplast progeny senesced immediately. The cause of this seems to be a plasmid which normally exists as an integral part of the mitochondrial DNA of healthy strains or of juvenile (pre-senescent) cultures of senescent strains. But, as the senescent strains age, this DNA is excised from the mitochondrial genome, becomes a closed circular molecule and self-replicates, causing senescence. Precisely how it does this is still in doubt, but the plasmid shows DNA homology with an intron in one of the mitochondrion genes — the gene that codes for a subunit of cytochrome c oxidase, an enzyme essential for normal function of the respiratory electron transport chain. Senescent strains of *Podospora* lack cytochrome c oxidase activity, perhaps because the plasmid inserts in the mitochondrial DNA, leading to disruption of gene function or causing mitochondrial gene rearrangements. Dysfunction of cytochrome c oxidase or other components of electron transport would be lethal for *Podospora* because this fungus seems unable to grow anaerobically by fermenting sugars.

Plasmids and transposable elements

Plasmids usually are closed-circular molecules of DNA with the ability to replicate autonomously in a cell. However, they can also be linear DNA molecules if the ends are 'capped' (like chromosomes) to prevent their degradation by endonucleases. Plasmids or plasmid-like DNAs have been found in several fungi. The most notable example is the 'two-micron' plasmid of the yeast *S. cerevisiae*, so-called because of its 2 μm length as seen in electron micrographs. This plasmid is a closed-circular molecule of 6.3 kbp, and it is unusual because it is found in the nucleus, where it can be present in up to 100 copies. It has no known function, but it has major practical applications in the construction of 'vectors' for gene cloning in yeast, as discussed later.

Most other plasmids of fungi are found in the mitochondria. The best-characterized are the linear DNA plasmids of *Neurospora crassa* and *N. intermedia*; they show a degree of base sequence homology to the mitochondrial genome, suggesting that they are defective, excised segments of the mitochondrial genes. However, some other mitochondrial plasmids of *Neurospora* are closed-circular molecules with little or no homology to the mitochondrial genome. They have a variable 'unit' length of about 3–5 kbp (in different cases) and the units can join head-to-tail to form larger repeats. None of these fungal plasmids has any known function, so they are not like bacterial plasmids that code for antibiotic resistance, pathogenicity or the ability to degrade pesticides, etc.

Transposons (transposable elements) are short regions of DNA that remain in the chromosome but encode enzymes for their own replication. They produce RNA copies of themselves and they encode an enzyme, **reverse transcriptase**, that synthesizes new copies of the DNA from this RNA template, similar to the action of retroviruses such as human immunodeficiency virus (HIV). The new copies of DNA can then insert at various points in the same or other chromosomes, leading to alterations in gene expression. Transposons seem to be rare in filamentous fungi, but there are several types in *S. cerevisiae*. The best-studied of these are the chromosomal **Ty elements**, present in about 30 copies in yeast cells. Oliver (1987) described the known and possible roles of Ty elements (Fig. 8.2). In addition to a role in altering gene expression, they could have significant effects on chromosomal rearrangements when the 'delta sequences' on the ends of these elements combine with one another.

The mating-type genes of *S. cerevisiae* also are transposable elements, causing mating-type switching as discussed in Chapter 4. However, mating-type switching has not been found in *N. crassa*, so it is not a general feature of ascomycota.

Viruses and viral genes

Fungal viruses were first discovered in the 1960s, associated with 'La France' disease of the cultiv-ated mushroom *Agaricus bisporus*. In this disease the fruitbodies are distorted and the fruitbody yield is poor. Electron micrographs of both the hyphae and the fruitbodies showed the presence of many isometric virus-like particles (VLPs), assumed to be the cause of the problem, although a causal relationship still has not been fully demonstrated. VLPs were then discovered in other fungi, and by the 1980s they were known in over 150 species, including representatives of all the major fungal groups (Buck, 1986). With a few notable exceptions, however, the presence of VLPs was not associated with any obvious disorder, so most fungal viruses seem to be symptomless.

Studies on a range of fungi have shown that fungal viruses (or VLPs) have similar basic features.

• They are isometric particles, 25–50 nm diameter, with a genome of double-stranded RNA (dsRNA), a capsid composed of one major polypeptide, and they code for a dsRNA-dependent RNA polymerase for replication of the genome.

• The genome size is extremely variable. Even within a single fungus it ranges from about 3.5 to 10 kbp. In some cases this variation is due to internal deletions of a full-length molecule, but in other cases the genome is divided between different particles.

• In most fungi the VLPs are found infrequently in hyphal tips, but they can occur as large crystalline arrays in the cytoplasm of older hyphal regions (Fig. 8.3), often closely associated with sheets of endoplasmic reticulum that partly enclose the aggregates.

• The only natural mechanism of transmission of VLPs is via the cytoplasm during hyphal anastomosis. They can, however, enter the asexual spores, ensuring their transmission from one generation to another. They can also enter the sexual spores of some basidiomycota and in *Saccharomyces*, but this seems to be rare in the sexual spores of mycelial ascomycota.

Because the VLPs have no natural means of 'external' transmission, they can be considered as truly resident genetic elements of fungi, just like the mitochondrial and nuclear genes. This creates problems in determining their functions, because in La France disease and many other cases the evidence is merely correlative, attempting to relate the presence of VLPs (by electron microscopy)

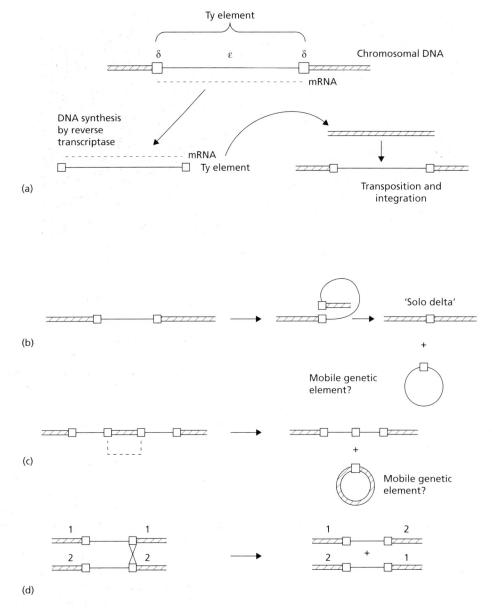

Fig. 8.2 Ty elements of yeast. (a) The Ty element is composed of an epsilon region which encodes its own replication, flanked by two delta sequences. The copies are then inserted elsewhere in the genome. (b–d) Possible roles of the Ty elements in causing chromosomal changes. (b) The delta regions of one Ty element can undergo homologous recombination, leaving a 'solo delta' in the chromosome. (c) Homologous recombination between two Ty elements on a chromosome can isolate an intervening segment of the chromosome. (d) Homologous recombination between Ty elements on different chromosomes can lead to crossing over, creating hybrid chromosomes. (Based on Oliver, 1987.)

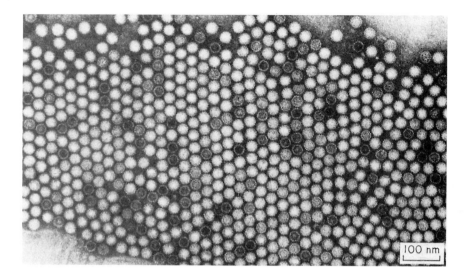

Fig. 8.3 Isometric VLPs extracted from hyphae of *Colletotrichum*. The particles have aggregated in a crystalline array *in vitro*. (From Rawlinson *et al.*, 1975.)

with changes in fungal behaviour. However, two major developments have changed this and opened the field to critical investigation. First was the discovery that virus-like dsRNA can be present in fungi even when VLPs are absent. In these cases it can be assumed that the virus has lost the ability —and the need—to produce a capsid polypeptide, because it never leaves the cytoplasm in any case. Second, protoplasting techniques and transformation systems have now been developed for a few fungi, so that dsRNA can be extracted, purified and introduced into protoplasts (the process termed **transfection**) or complementary DNA (cDNA) can be derived from dsRNA *in vitro* and then **transformed** into protoplasts. These approaches, discussed later in this chapter, have shown that the viral dsRNA of *Saccharomyces* and several other yeasts can cause the cells to produce **killer toxins** which act on other strains of the same species. Also, in the chestnut blight fungus *Cryphonectria parasitica*, the dsRNA causes a marked reduction in pathogenic virulence, creating **hypovirulent** strains that can be used in biocontrol of the disease. In recognition of this, a new name has been given to the dsRNA viruses that reduce fungal virulence—the **hypovirus** group.

Genetic variation in fungi

Mutation is the basis of all variation, but mutations are expressed and recombined in different ways depending on the biology of the organism. The peculiar organization of fungal hyphae sets them apart from most other organisms in this respect.

Non-sexual variation: the significance of haploidy

Fungi are the only major group of haploid eukaryotes; the others are diploid. To understand why this is so, we must consider the relative advantages of haploidy and diploidy. Haploid organisms express all their genes and thus perpetually expose them to selection pressure. Any mutation in a gene will either cause a loss of fitness, in which case the mutants are eliminated, or it will lead to an increase in fitness (e.g. antibiotic or fungicide resistance) in which case the mutant will flourish. This can be beneficial in the short term but the corresponding disadvantage is that haploid organisms cannot accumulate mutations that are not of immediate value: they cannot store variation. Diploid organisms have exactly the opposite features. Mutations often are recessive to the wildtype and so they are not immediately expressed; instead, they accumulate and can be recombined in various ways during sexual crossing, so that some of the progeny might have advantageous combinations of mutant genes.

The predominance of diploidy in the eukaryotic world shows that this is the favoured option. But, mycelial fungi differ from other organisms in having several nuclei in a common cytoplasm, and so they can shield any recessive mutations from selection pressure, provided that the mutation is only present in some nuclei. It will be complemented by the wild-type nuclei. Yet, the mycelial fungi can also expose their genes to selection pressure periodically — whenever they produce uninucleate spores or when hyphal branches develop from only one 'founder' nucleus. In other words, the hyphal growth habit enables fungi to be haploid and yet have at least some of the advantages of diploidy; the mycelial fungi would have little to gain by being diploid. However, this is not found in the oomycota, which are diploid, presumably because of their separate origin (see Chapter 1). The situation is different for yeasts because they grow as uninucleate cells, unable to shield mutations in a haploid genome. It is notable that several yeasts (e.g. *Candida* spp.) are permanently diploid, and even *Saccharomyces* grows as a diploid in nature, owing to its mating-type switching (see Chapter 4).

Non-sexual variation: heterokaryosis

The comments above raise the question of how commonly fungi exist with mixtures of genetically different nuclear types in the cytoplasm of the hyphae. This phenomenon is termed **heterokaryosis** (*hetero* = different; *karyos* = kernal or nucleus), so a fungus exhibiting this is termed a **heterokaryon**, in contrast to a **homokaryon** with one nuclear genotype.

How do heterokaryons arise?

Heterokaryons can arise in two ways.
1 When a mutation occurs in one of the nuclei in a hypha and the mutated nucleus proliferates along with the wild-type nuclei. This must happen frequently but a stable, functional heterokaryon will develop only if the genetically different nuclei proliferate in the apical cells so that all the newly formed hyphae contain both types.
2 When the hyphae of any two strains fuse at points of contact (anastomosis) so that their nuclei are present in the common cytoplasm. Again, the

nuclei would need to proliferate in the apical cells to form a stable heterokaryon.

Most experimental studies on heterokaryosis have involved the pairing of strains with defined mutations, such as a requirement for specific amino acids. These strains are termed amino acid **auxotrophs**, in contrast to the wild-type **prototroph**. For example, if a mutant requiring histidine (his⁻) is allowed to anastomose with a mutant requiring arginine (arg⁻) then the heterokaryon behaves as a prototroph. It can grow on a medium lacking both amino acids because the two nuclear types complement one another's deficiency. The most interesting feature in these cases is the finding that the ratio of nuclear types can vary within wide limits and is influenced by environmental conditions. So, at least in theory, a single heterokaryotic strain can change the frequency of its different genes in response to selection pressure. Table 8.2 illustrates this for a heterokaryon constructed from two parent homokaryons, one of which (termed A for simplicity) grew best on minimal medium, low in organic nutrients, and the other (B) grew best on apple-pulp medium. When the heterokaryon was grown on apple-pulp medium the proportion of B-type nuclei was very high, but as the amount of apple pulp was lowered so the proportion of A-type nuclei increased, dramatically so when the heterokaryon was grown on minimal medium. Other experiments of this type have shown that the nuclear ratio in a heterokaryon can vary by up to 1000:1 in each direction.

How do heterokaryons break down?

Heterokaryons can break down in two ways (Fig. 8.4): (i) when branches arise that contain only one nuclear type; and (ii) during the production of uninucleate spores.

A branch that, by chance, contains only one nuclear genotype can produce further branches and eventually give rise to a homokaryotic sector of the colony. If the homokaryon is favoured more than the heterokaryon in the prevailing environment then it will expand and become dominant; if it is not favoured it will be suppressed. Sectors are quite commonly seen when fungi are isolated from natural environments and grown on agar plates, suggesting that either nuclear genes or non-

Table 8.2 Effects of composition of the growth medium on the ratio of nuclear types (represented by A and B) in a heterokaryotic colony of *Penicillium cyclopium*. (Data based on Jinks, 1952.)

Per cent composition of medium		Percentage of nuclei in heterokaryon		Relative growth rates of homokaryons A and B
Minimal nutrients	Apple pulp	Type A	Type B	A : B
0	100	8.6	91.4	0.47 : 1
20	80	7.8	92.2	0.53 : 1
40	60	11.1	88.9	0.54 : 1
60	40	12.7	87.3	0.67 : 1
80	20	13.5	86.5	1 : 1
100	0	51.8	48.2	1.56 : 1

The heterokaryon was constructed from A and B homokaryons, and nuclear ratios were estimated by testing random samples of uninucleate spores produced by the heterokaryon. The ratio of type A nuclei in the hetrerokaryon rises dramatically when the fungus is grown on minimal nutrient medium which favours the growth of homokaryon A more than homokaryon B.

nuclear genes have segregated in culture. An example is shown in Fig. 8.4 where the sectors differ in hyphal pigmentation, and in other cases they can differ in branching frequency, sporulation, etc. However, many sectors would go unnoticed if they differed in pathogenicity, antibiotic production, etc.

Heterokaryons break down automatically during asexual sporulation if the spores are uninucleate. The nuclear ratios in Table 8.2 were obtained in this way — by analysing the uninucleate spores produced from the heterokaryon. In this respect we can regard the spore-bearing phialides of *Penicillium* or *Aspergillus* spp. as turnstiles, because only one nucleus enters each phialide and it divides to produce the nuclei for all the spores produced from that phialide (Fig. 8.4). This also is true for multinucleate spores that develop from phialides (e.g. *Fusarium*; Fig. 8.4) because a single nucleus enters the developing spore and then divides. But, some other fungi (e.g. *Neurospora*, *Monilinia*) produce conidia directly from multinucleate hyphal tips or buds, and these spores will be either homokaryotic or heterokaryotic, depending on whether the cells that produced them were homokaryotic or heterokaryotic.

Significance of heterokaryosis

Clearly, heterokaryosis is a potentially powerful phenomenon. It could enable fungi to accumulate

mutations, shielded from immediate selection pressure and yet exposed periodically in hyphal branches or via spores. It could also enable a growing colony to alter the nuclear (gene) ratio in response to the prevailing conditions as it grows. Thus, we might think of a colony as being a mosaic of different nuclear types and ratios, responsive to the environment at any one time and place.

However, caution is required because the degree of heterokaryosis in natural environments is largely unknown. Experiments on paired auxotrophic strains could be misleading because there would be strong selection pressure to maintain these 'forced heterokaryons' on nutrient-deficient media. Moreover, there are significant barriers to the creation of heterokaryons in nature, because many fungi have nuclear-encoded compatibility genes which lead to cytoplasmic death after anastomosis of incompatible strains. On this basis the natural populations can be shown to be composed of a mixture of **vegetative (somatic) compatibility groups** (VCGs), so that heterokaryosis will occur only when two strains of the same VCG are paired (Anagnostakis, 1992).

Non-sexual variation: parasexuality

Many common deuteromycota (*Aspergillus*, *Penicillium*, *Fusarium*, *Trichoderma*, etc.) seem to have abandoned sexual reproduction entirely. Others can be induced to form sexual stages in specific

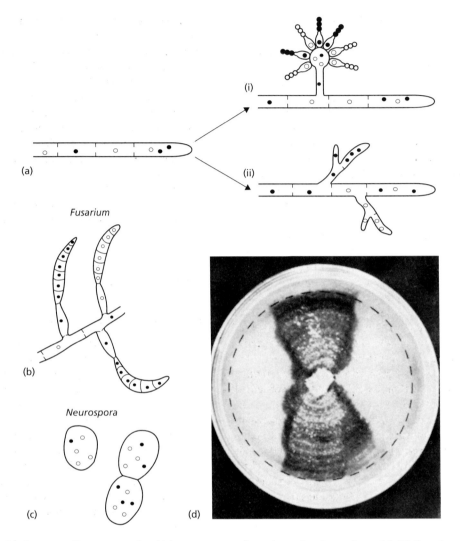

Fusarium

Neurospora

Fig. 8.4 (a) Diagram to illustrate ways in which a heterokaryon can revert to homokaryons: (i) during production of uninucleate spores; (ii) during hyphal branching when all the nuclei in a branch develop from one founder nucleus. (b) Multinucleate conidia in which all nuclei are derived from a single nucleus during spore development (homokaryotic). (c) Multinucleate conidia in which several nuclei enter the developing spore (homokaryotic or heterokaryotic). (d) Sectoring of a fungal colony on agar (colony margin shown by broken line). An originally melanized colony has produced non-melanized sectors and they have expanded because they are favoured more than the melanized hyphae in the prevailing environment. Light-coloured zones within the melanized part of the colony are tufts of aerial hyphae.

laboratory conditions but have not been found to produce them in nature, so they exist as essentially clonal, asexually reproducing populations. However, some of these fungi might have developed an alternative means of recombining their genes by a process termed **parasexuality**.

This was discovered by Pontecorvo in the 1960s, during studies on heterokaryosis in *Aspergillus nidulans* (which actually does have a sexual stage in culture). He had constructed a heterokaryon from two parental strains that differed at two gene loci (we will call them strains Ab and aB) and was analysing the homokaryotic spores produced by the heterokaryon. As expected, most of the spores had nuclei of the 'parental' types, either Ab or aB, but a significant number were found to be recombinants (AB or ab) and their frequency was too high to be explained by mutation. Evidently, the parental genes had recombined in the heterokaryon, although this cannot occur by heterokaryosis alone: the nuclei remain as distinct entities regardless of how they are mixed in the cytoplasm, like coloured marbles that are shaken in a bag. Further investigation led Pontecorvo to propose a parasexual cycle, involving three stages.

1 Diploidization. Occasionally, two haploid nuclei fuse to form a diploid nucleus. The mechanism is largely unknown, and this seems to be a relatively rare event, but once a diploid nucleus has been formed it can be very stable and divide to form further diploid nuclei, along with the normal haploid nuclei. Thus, the heterokaryon consists of a mixture of the two original haploid nuclear types as well as diploid fusion nuclei.

2 Mitotic chiasma formation. Chiasma formation is common in meiosis (see later) where two homologous chromosomes break and rejoin at the same point, but crossing between the two chromosomes so that the resulting chromosomes are hybrids of the parental types. It can also occur during mitosis but at a much lower frequency because the chromosomes do not pair in a regular arrangement so that they lie adjacent to one another. Nevertheless, the result will be the same when it does occur—the creation of chromosomes that are hybrids of the parental chromosomes.

3 Haploidization. Occasionally, non-disjunction of chromosomes occurs during division of a diploid nucleus, such that one of the daughter nuclei has $2n+1$ chromosomes and the other has $2n-1$ chromosomes. Such nuclei with incomplete multiples of the haploid number are termed **aneuploid** (as opposed to euploid nuclei, with n or complete multiples of n). They tend to be unstable and to lose further chromosomes during subsequent divisions. So, the $2n+1$ nucleus would revert to $2n$, whereas the $2n-1$ nucleus would progressively revert to n. Consistent with this, in *Aspergillus nidulans* ($n=8$) nuclei have been found with 17 ($2n+1$), 16 ($2n$), 15 ($2n-1$), 12, 11, 10 and nine chromosomes.

It must be emphasized that each of these events is relatively rare, and they do not constitute a regulated cycle like the sexual cycle involving meiosis. Nevertheless, the outcome would be the same — once a diploid nucleus has formed by the fusion of two haploid nuclei from different parents in a heterokaryon then the parental genes could be recombined. This could happen in two ways, equivalent to the events in meiosis (Fig. 8.5).

1 Chiasma formation would generate recombinant chromosomes.

2 Even if chiasma formation does not occur, the process of haploidization by random loss of chromosomes could produce haploid nuclei with chromosomes that came from the different parents. We can regard this as 'independent loss' of chromosomes, equivalent to the 'independent assortment' of chromosomes during meiosis.

Significance of parasexuality

Parasexuality has become a valuable tool for industrial mycologists to produce strains with desired combinations of properties. However, its significance in nature is largely unknown, and it will depend in any case on the frequency of heterokaryosis, discussed earlier. Assuming that it does occur, we can ask why the deuteromycota might have abandoned an efficient sexual mechanism of genetic recombination in favour of a more random and seemingly less efficient process. The answer might be that the parasexual events can occur at any time during normal, somatic growth and with no preconditions like those needed for the production of sexual stages. Although each stage of the parasexual process is relatively rare, there can be many millions of nuclei in a single colony, so that the chances of the parasexual cycle occurring within the colony as a whole may be quite high.

Sexual variation

Sex is the major mechanism for producing genetic recombinants. The pairing of parental chromosomes in meiosis leads to multiple crossing over (chiasma formation). Also, the independent assortment of homologous chromosomes from the two parents as they align during metaphase will mean that the individual daughter haploid nuclei

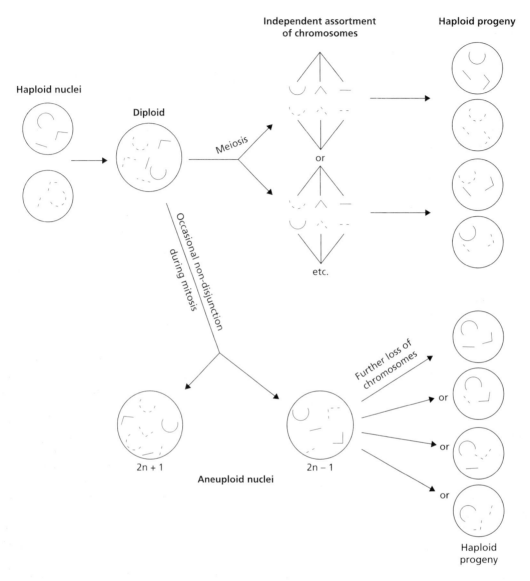

Fig. 8.5 Diagram to illustrate the parallel between independent assortment of chromosomes in meiosis and random loss of chromosomes during haploidization in the parasexual cycle. Only three different chromosomes are shown. It is seen that recombinant genotypes can be generated by haploidization through random loss of chromosomes in successive nuclear divisions, even if chiasmata are not formed.

might have, say, chromosomes 1 and 3 from one parent and chromosomes 2 and 4 from the other parent. With eight chromosomes (*A. nidulans*) this alone could generate 2^8 (i.e. 256) different chromosome combinations in the meiotic products. Of course, all this depends on an efficient outcrossing mechanism. As noted in Chapter 4, many fungi are heterothallic but a significant number are homothallic (e.g. most *Pythium* spp., about 10% of ascomycota and a few basidiomycota) and some normally heterothallic species exhibit **secondary homothallism** — the sexual spores are binucleate with one nucleus of each mating type (e.g. *Agaricus bisporus*). The common role of sexual spores as dormant spores creates a more immediate requirement than the generation of variation *per se*.

Tetrad analysis

Ascomycota have proved extremely valuable as genetic tools because all the daughter nuclei from each meiotic division are retained (in ascospores) within an ascus. The asci can be four-spored (e.g. *Saccharomyces*) or eight-spored if the nuclei undergo mitosis before the ascospores are formed. In any case, the spores can be dissected from a single ascus then germinated to see how the parental genes from a sexual cross have recombined in the progeny. This process is termed **tetrad analysis**. It is illustrated in Fig. 8.6 for the asci of *Neurospora*, *Sordaria*, etc., where the ascospores occur in a linear sequence.

Suppose that two haploid parents differ at two gene loci: parent Ab and parent aB (Fig. 8.6). If the two loci (A/a and B/b) are close together on a chromosome, then the chance of chiasma formation between these loci is low — the genes are said to be tightly linked, and almost all the progeny ascospores from a cross will be parental type (50% Ab and 50% aB). As the distance between gene loci increases, so the chance of chiasma formation between them increases, giving progressively more recombinant progeny, up to a maximum of 50% (25% Ab, 25% aB, 25% AB, 25% ab) because of multiple chiasma formation as the loci get further apart. If the A/a and B/b loci were on different chromosomes then we would also expect 50% recombination. So, in the early stages of tetrad analysis it is necessary to create many mutant strains and pair them in all combinations until any degree of linkage is detected (less than 50% recombination). Then, further mutations (at other loci) can be tested for linkage to the known genes, and eventually a linkage map can be developed, expressing the relative distances of gene loci from one another. In the same way, genes can be mapped in relation to the centromere (the point where spindle microtubules are attached for chromosome separation) if the ascospores are linearly arranged, because the chance of a chiasma forming between a gene and the centromere is related to the distance between these, and it governs whether the genes segregate in the first or second division of meiosis (Fig. 8.6).

For the best-studied fungi such as *Saccharomyces* and *Neurospora* there are banks of reference strains with defined genetic markers along the lengths of each chromosome, so that any new gene can be mapped in relation to these markers.

Applied molecular genetics of fungi

In this section we consider some examples of molecular approaches to the understanding of fungal behaviour or for direct, practical applications. The basic molecular methodologies for fungi are described in Bennett and Lasure (1985, 1991) and Peberdy *et al.* (1991).

Protoplasts, vectors and genetic transformation: the basic methodologies

Genetic transformation methods depend on:
1 production of protoplasts and subsequent regeneration of walled cells;
2 cloning of DNA and its introduction into protoplasts by means of a suitable vector;
3 stable integration and expression of the introduced DNA in the recipient cells.

Protoplasts can be produced from many fungi by incubating young colonies, with many thin-walled hyphal tips, in osmotically stabilized solutions (sorbitol, mannitol, etc.) containing mixtures of wall-degrading enzymes. The commercial enzymes often used include snail gut juice ('Helicase') or enzyme mixtures from *Streptomyces* (e.g. 'Novozym') or other microbes. A proportion of the protoplasts regenerate walls and give rise to viable hyphae when transferred to enzyme-free media.

The uptake of DNA by protoplasts is facilitated if the DNA is precipitated with calcium and added with polyethylene glycol. However, the efficiency of uptake and subsequent expression is relatively low in fungi compared with bacteria and cultured animal cells. Much of the introduced DNA seems to be digested by endonucleases, the DNA-degrading enzymes that serve a general housekeeping role in cells. For this reason, transformation usually requires that the DNA is introduced in high copy number, using a gene vector that can self-replicate in the fungus. Then, the introduced DNA can persist until it integrates into the host genome. Integration can be either homologous or non-homologous (random). In the former case the introduced DNA directly replaces a corresponding DNA sequence in the genome,

Assume that two allelic pairs, Aa and Bb, are on one chromosome

1 What is the distance between these allelic pairs?

The distance between Aa and Bb determines the likelihood of a cross-over event occurring between them. Therefore look for evidence of recombination.

	Progeny	Interpretation
Parent 1 AB × Parent 2 ab	AB 50% ab 50%	Aa and Bb loci are close together (i.e. closely linked); there is a negligible chance of crossing-over between them.

	Progeny	Interpretation
Parent 1 AB × Parent 2 ab	AB 25% Ab 25% aB 25% ab 25%	Aa and Bb loci are widely spaced; there is maximum likelihood of crossing-over between them (50% of parental types and 50% recombinant types in the progeny is the maximum one might expect because of multiple cross-over events).

Thus, in the second case above:

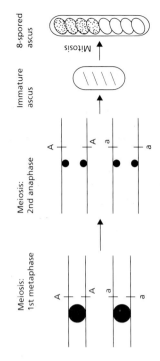

1st metaphase of meiosis (homologous chromosomes pair)

2nd anaphase of meiosis (chromatids separate)

Note that one can determine the distance between genes only over relatively short lengths of the chromosome; the frequency of crossing-over is such that we soon approach the maximum percentage of recombinant nuclear types as the distance between gene loci is increased.

2 Is the allelic pair Aa situated close to the centromere?

By the same reasoning as before, the chance of a cross-over event occurring between Aa and the centromere depends on the distance between these.

Thus, if 'A' codes for black spores and 'a' for white spores:

Negligible chance of crossing-over:

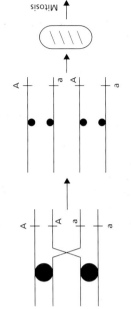

Meiosis: 1st metaphase | Meiosis: 2nd anaphase | Immature ascus | 8-spored ascus

Mitosis

Significant chance of crossing-over:

Mitosis

Note that centromere-linkage can be detected easily if spores are arranged in a linear sequence within the ascus. The confined space within the ascus ensures that chromosome sets do not overlap but remain in the order in which they separate during successive divisions. Centromere-linkage is important to establish, because the centromere is a useful reference point in chromosome mapping.

Fig. 8.6 Simplified scheme to show the principles of chromosome mapping (tetrad analysis) by meiosis in ascomycota.

which is of value for targeting specific genes in gene-disruption experiments (see later).

The availability of vectors is the major limitation for fungi, because the best vectors are based on plasmids that are associated with the nucleus, and only the 'two-micron' plasmid of *S. cerevisiae* is of this type. We will use this example to illustrate how a vector is created. However, it works only for *Saccharomyces*, not for *A. nidulans, Neurospora crassa* or other mycelial fungi, suggesting that the regulation of gene function is different in yeast than in these fungi. Artificial cloning vectors have been constructed for *A. nidulans* and *N. crassa*, but they are much less efficient that the two-micron plasmid.

The Saccharomyces *vectors*

Figure 8.7 shows an example of a gene vector based on the two-micron plasmid, and its features are outlined below. Initially the two-micron plasmid is modified substantially *in vitro*, by using specific **endonucleases** (restriction enzymes) to excise parts of the DNA that are not required and to create breaks where other DNA can be inserted. Different endonucleases recognize different nucleotide sequences and cut the DNA wherever these sequences occur. For example, the enzyme *Bam*H1 recognizes the sequence GGATCC (with CCTAGG on the complementary strand of DNA) and cuts it between GG, leaving the two DNA strands as follows:

–G⎯⎯GATCC–
–CCTAG⎯⎯G–

The resulting breaks are termed 'sticky ends' and any other piece of DNA can insert at these positions if that DNA also has been treated with *Bam*H1.

The vector shown in Fig. 8.7 contains the yeast **origin of replication** so that the vector will multiply in yeast cells. It also contains parts of a plasmid from *E. coli*, including the bacterial origin of replication so that it can be introduced into *E. coli* and will multiply there, producing a high copy number for treatment of yeast protoplasts. There are several further features. The vector contains a marker gene, leu+ which codes for ability to synthesize leucine, so that the plasmid can be introduced into a leucine-auxotrophic strain of yeast and only the cells that contain the plasmid will grow on a leucine-free medium. Similarly, the bacterial part of the plasmid has a gene coding for resistance to ampicillin, so only plasmid-containing cells of the bacterium will grow on media containing the antibiotic. There is a single site with the nucleotide sequence GGATCC (for *Bam*H1) so that the plasmid can be cut at this point *in vitro* and any desired piece of DNA can be inserted there. Moreover, this site is located within the gene that codes for resistance to tetracycline (an antibacterial antibiotic), so the insertion of any DNA will disrupt this gene. This allows the presence of the DNA insert to be monitored during

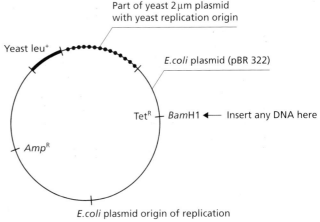

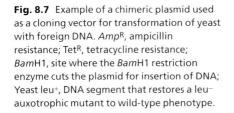

Fig. 8.7 Example of a chimeric plasmid used as a cloning vector for transformation of yeast with foreign DNA. *Amp*R, ampicillin resistance; *Tet*R, tetracycline resistance; *Bam*H1, site where the *Bam*H1 restriction enzyme cuts the plasmid for insertion of DNA; Yeast leu+, DNA segment that restores a leu− auxotrophic mutant to wild-type phenotype.

replication of the plasmid in *E. coli*, because the cells will be ampicillin resistant but tetracycline sensitive (testable by replica-plating of colonies).

In short, this vector—a chimeric plasmid—incorporates all the necessary features for it to: (i) be detected in both *E. coli* and yeast; (ii) replicate in both types of cell; and (iii) be monitored for the successful insertion of foreign DNA. The vector is bulked up in *E. coli*, then used to transform yeast protoplasts where it will replicate, and sooner or later the DNA insert will integrate into a yeast chromosome.

Production of heterologous proteins in *Saccharomyces cerevisiae*

The availability of efficient vectors has enabled *S. cerevisiae* to be used as a 'factory' for the products of many foreign (heterologous) genes. One of the most significant products is a vaccine against hepatitis B virus; it consists of one of the viral surface antigens, HBsAg, produced by transforming yeast cells with the antigen gene. It was the first genetically engineered product approved for use in humans. Many other proteins have been produced experimentally from yeast, including cellulases, amylases, interferon, epidermal growth factor and β-endomorphin. However, there have also been problems in using *Saccharomyces* for heterologous protein production. In particular, yeast has a relatively poor ability to remove introns from foreign genes (its own introns are few and small) so it is most efficient when transformed with cDNA, derived *in vitro* from the mRNA of a protein (the introns are spliced out during the processing of mRNA). Yeast also fails to recognize the promoter regions of the genes of other fungi, and it does not always faithfully glycosylate foreign proteins. This can be important because several bioactive proteins, including pharmaceuticals, are glycoproteins that depend on the sugar chains for their activity.

These difficulties have served to demonstrate that *S. cerevisiae* is genetically quite different from the mycelial fungi because *A. nidulans*, for example, can recognize the promoter sequences of the genes of other fungi and also can excise their introns. It may be possible to use *A. nidulans* or the fission yeast *Schizosaccharomyces* (see Chapter 1, Fig. 1.2) as an alternative to *Saccharomyces* for het-

erologous protein production. But, in general, it is now thought that the best approach is to use cell lines related to the natural producer organism—mammalian cell lines for mammalian gene products, and so on.

Molecular approaches to population structure

Fungal species are dynamic entities. Their populations fragment by geographical isolation or the development of somatic compatibility barriers, then the fragments diverge by genetic drift or in response to local selection pressure. This is particularly true for the clonal deuteromycota, because strains of different VCGs are isolated permanently from one another. Many sexual species also have VCGs but the mating-type genes override the somatic incompatibility genes so that strains of different VCGs can mate.

In general, fungi have too few morphological features for identification of population subunits, so biochemical and molecular tools must be used for this. One approach is to compare the electrophoretic banding patterns of proteins on gels (reviewed by Barrett, 1987), using either total protein extracts or different forms (isozymes) of particular enzymes such as pectic enzymes, visualized on the gels by colour reactions with the enzyme substrate. These **zymograms** (Fig. 8.8) reflect random mutations in the DNA encoding the enzyme, although not at the enzyme active site which is highly conserved. About 30% of the amino acid changes resulting from mutations will affect the net charge on the protein and thus alter its electrophoretic mobility. Since these changes are random, they tend to accumulate over time and thus reflect the history of a population. As one practical example, MacNish *et al.* (1993) used a combination of pectic zymogram and VCG typing to identify different subgroups of the soil-borne fungus *Rhizoctonia solani* that causes bare-patch (stunting) disease of wheat in Australia. All fungal isolates from within each patch were of an identical VCG–zymogram group, but different patches can be caused by different VCG–zymogram groups. Figure 8.9 shows how this approach can be used to understand the biology of the fungus. The patches can merge, but never overlap because the subgroups seem unable to grow into one another's

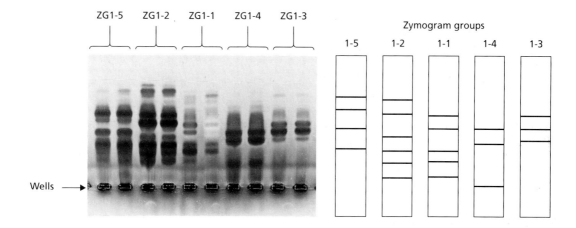

ZG1-5 ZG1-2 ZG1-1 ZG1-4 ZG1-3

Zymogram groups

1-5   1-2   1-1   1-4   1-3

Wells →

Fig. 8.8 Five distinctive pectic zymogram groups (ZGs) of *Rhizoctonia solani* strains that cause bare patch disease of wheat in Australia. Protein extracts were run on acrylamide gels containing pectin then stained to develop the bands of pectic isozymes. (Courtesy of M. Sweetingham; from MacNish & Sweetingham, 1993.)

territory. The patches (and therefore the fungal populations) are also dynamic—they can expand, contract or even disappear in different cropping seasons.

The population structure of a fungus can also be analysed by the use of restriction enzymes on DNA extracted from the cells. Any one enzyme (e.g. *Bam*H1) will cut the DNA at specific 'target' points (GGATCC in this case) giving fragments of different lengths that band on gels according to their size, and can be visualized by adding ethidium bromide, which intercalates in the DNA and

fluoresces under ultraviolet illumination. The banding patterns are termed **restriction fragment length polymorphisms** (RFLPs), and they are like fingerprints, reflecting the accumulations of point mutations that generate or delete the target nucleotide sequence (e.g. GGATCC) or chromosomal rearrangements that changed the relative positions of these sequences. Different enzymes give different RFLPs by cutting the DNA at different sites. They also give different numbers of fragments, because some recognize four-base sequences which are more common than, say, eight-base sequences. Also, these enzymes can be used on mitochondrial DNA, which is more highly conserved than the total DNA, so different levels of sensitivity can be selected to analyse both minor and major changes within a fungal population. Kohn (1995) described a good example of this approach for comparing the inter- and intracontinental clonal subgroups of the plant pathogen

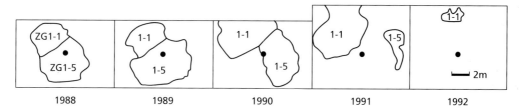

ZG1-1 ZG1-5 1-1 1-5 1-1 1-5 1-1 1-5 1-1 2m

1988   1989   1990   1991   1992

Fig. 8.9 Mapped positions of patches of stunted cereal plants (bare-patch disease) caused by *Rhizoctonia solani* in a field trial site in Australia. Two adjacent disease patches were caused by different pectic zymogram groups of the pathogen (ZG1–1 and ZG1–5) which never

invaded the territory of the other. The patches expanded or contracted in successive cropping seasons. The central dot is a fixed reference point. (Derived from MacNish *et al.*, 1993.)

Sclerotinia sclerotiorum in both wild and agricultural plant communities.

Finally, the **polymerase chain reaction** (PCR) has been used not only to analyse populations but also to develop diagnostic probes for specific plant pathogens. The simplest and most common approach involves the random amplification of DNA from a crude DNA extract by adding a random primer composed, say, of 10 nucleotides, which anneals to a complementary nucleotide sequence on the extracted DNA. Then, a DNA polymerase enzyme extends along the DNA, reading the sequence of bases along from the primer, and this DNA is amplified during 25–40 successive rounds of synthesis. The technique is known as **RAPD**, pronounced 'rapids' (random amplified polymorphic DNA). Alternatively, the DNA that codes for particular proteins can be targeted with primers based on knowledge of the partial amino acid sequence of the protein. In any case, selected fragments of the amplified DNA can be suitably tagged and used as diagnostic probes that will bind to equivalent DNA sequences in extracts of a sample. Probes of this type are available commercially for detecting several individual plant pathogens (Fox, 1994). For analysis of population structure, both RFLP analysis and RAPD have been used to distinguish the different pathogenic strains of *Ophiostoma* spp. that cause Dutch elm disease, helping to trace the origin and progress of the recent epidemics (Pipe *et al.*, 1995; see Chapter 9).

Identification of genes for plant pathogenicity and differentiation

We saw in Chapter 4 how differentiation-specific mRNAs were identified in *Schizophyllum commune* by comparing the mRNA profiles of cultures grown in conditions where fruitbodies were or were not produced. The specific mRNAs can then be used as templates to produce cDNA *in vitro*. This cDNA, produced from labelled nucleotides, becomes a probe for binding to complementary sequences of extracted chromosomal DNA. In this way a specific gene can be identified even if nothing is known about the basic genetics of a fungus. A purified protein also can lead to the gene. One approach is to obtain a partial amino acid sequence, then deduce the nucleotide

sequence and use this as a primer for PCR. Alternatively, the protein can be used to raise an antibody, which then can be used to probe for clones of *E. coli* that produce the protein after transformation with DNA segments of the fungus (Turner, 1991). These techniques of 'reverse genetics' have made it possible to perform **targeted gene disruption**. For the fungal gene of interest, a cDNA is produced *in vitro* and disrupted to make it nonfunctional. Then it is transformed into the fungus, to replace the original gene by homologous recombination. The following example shows the power of this technique.

Role of pre-formed inhibitors in plant resistance to pathogens

The take-all fungus of cereals, *Gaeumannomyces graminis* (Fig. 8.10), grows on the surface of cereal roots by dark 'runner hyphae' then penetrates the root cortex and enters the vascular system, causing phloem breakdown and blockage of the xylem. If enough roots are killed in this way by spread over the root system during the growing season then the plants die prematurely, with much-reduced grain yield. *G. graminis* has two main pathogenic forms — variety *tritici* (GGT) which attacks wheat roots but not oat roots, and variety *avenae* (GGA) which attacks both wheat and oats. This difference was explained in the 1960s when oat roots were found to contain pre-formed inhibitors. The most potent of these is **avenacin A** (Fig. 8.11), a saponin which combines with sterols in the fungal membrane, creating ion-permeable pores. In laboratory culture both GGT and GGA grow readily in aqueous extracts of wheat roots, but GGT is totally suppressed by aqueous extracts of oat roots whereas GGA is unaffected by them. The reason was suggested to be that GGA detoxifies avenacin, cleaving the terminal sugars from the molecule (Fig. 8.11), by producing a glycosidase enzyme termed **avenacinase**. This enzyme would thus be a key pathogenicity determinant, allowing GGA to extend its host range to oats.

This has now been proved beyond doubt by targeted disruption of the avenacinase gene (Osbourn *et al.*, 1994). GGA was transformed with a marked, disrupted cDNA of the avenacinase gene. When the marked cDNA had been inserted at the site of the gene by homologous recombina-

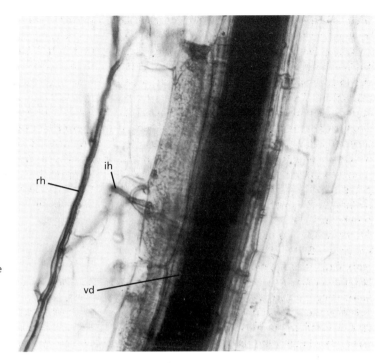

Fig. 8.10 Wheat root infected by the take-all fungus *Gaeumannomyces graminis*, showing growth on the root surface by melanized runner hyphae (rh), invasion of the root cortex by infection hyphae (ih), and vascular discolouration (vd) and blockage when the fungus enters the xylem vessels.

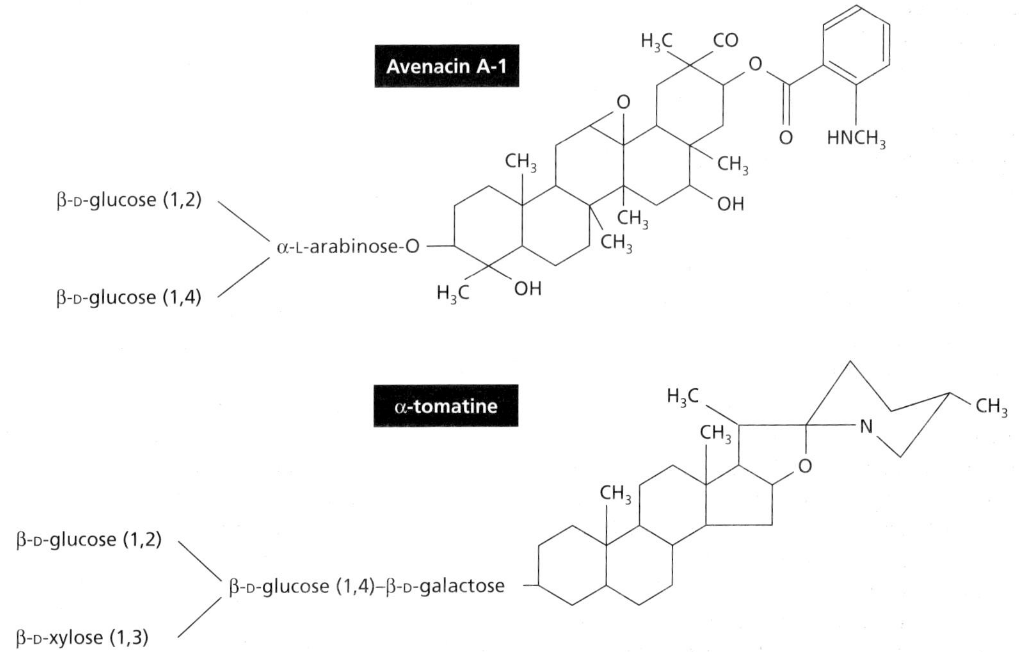

Avenacin A-1

β-D-glucose (1,2)
β-D-glucose (1,4)
α-L-arabinose-O

α-tomatine

β-D-glucose (1,2)
β-D-xylose (1,3)
β-D-glucose (1,4)–β-D-galactose

Fig. 8.11 Structure of two saponins that are pre-formed resistance chemicals in plants: avenacin in roots of oats (*Avena sativa*) and α-tomatine in tomato. Pathogens can overcome this resistance by producing an enzyme that cleaves some of the sugar residues from the saponin.

tion, the fungus lost its pathogenicity to oats but retained its normal pathogenicity to wheat. But, when the marked cDNA had been inserted elsewhere in the genome (non-homologous recombination), the fungus was still fully pathogenic to oats. Extending from this work, Osbourn *et al.* examined other host–pathogen systems where saponins have been implicated as plant-resistance factors. In particular, tomatoes are known to contain the saponin α-**tomatine**, and a pathogen of tomatoes, *Septoria lycopersici*, is known to detoxify this by cleaving a single sugar from the molecule (Fig. 8.11). The enzyme responsible (**tomatinase**) was found to be very similar to avenacinase: it was recognized by an anti-avenacinase antibody, and cDNA of avenacinase hybridized with DNA components of *S. lycopersici*, presumably by recognizing the gene for tomatinase. This might be explained by the fact that all the saponin-detoxifying enzymes are β-glycosidases, with perhaps some common structure that would be reflected in the DNA sequences. Of interest, however, the ability of avenacinase (from GGA) to detoxify tomatine was only 2% of its ability to detoxify avenacin, and tomatinase (from *S. lycopersici*) had negligible ability to detoxify avenacin. So, it seems that these pathogenic fungi have evolved saponin-detoxifying enzymes with quite specific activity against the saponins of their hosts.

Roles of viral dsRNA

The general features of fungal viruses and viral dsRNA were given earlier in this chapter. Here we consider two cases in which the dsRNA has a major effect on fungal behaviour.

Yeast killer systems

Individual species of at least eight genera of yeasts (*Saccharomyces*, *Candida*, *Kluyveromyces*, etc.) have been found to contain **killer strains.** These strains secrete a protein that kills other strains of the same species but not of unrelated species. The toxin binds to a receptor on the wall of susceptible cells, then passes to the membrane where it causes leakage of hydrogen ions (H+), and therefore the cells die owing to loss of transmembrane potential and disruption of amino acid uptake, potassium ion (K+) balance, etc. The toxins are stable only at low pH and are thought to give a significant advantage to killer strains over non-killer strains of the same species in acidic environments.

In *Kluyveromyces lactis* the toxin is encoded by a linear DNA plasmid, but in all other cases it is encoded by dsRNA. Both the killer and non-killer strains can contain VLPs so this feature alone does not correlate with killer activity. However, these particles are found to be of two types: the dsRNA in the 'L type' encodes the virus coat protein, whereas the dsRNA in the 'M type' codes for the toxin. The M-dsRNA is thus a satellite of L, dependent on it for the coat protein, but L can occur alone with no effect on the cells. The molecular biology of the killer system in *S. cerevisiae* has been studied intensively and has shown why the toxin-producers are not affected by their own toxin. They produce it as a large precursor protein which undergoes changes during passage through the secretory system to produce a **protoxin**. This protoxin is finally cleaved at the cell membrane, to release the active toxin, but leaves part of the molecule in the membrane. This residual part seems to interact with a toxin receptor, making the cell resistant to active toxin in the external environment.

Similar dsRNA killer systems are found in the yeast phase of *Ustilago maydis* (basidiomycota) which causes smut disease of maize. A killer system might also occur in the take-all fungus, *G. graminis*, because some dsRNA-containing strains can markedly inhibit the growth of other strains at low pH. However, the mechanism in this case remains unknown. In terms of evolution it is difficult to understand how individual strains of a fungal species could benefit by killing other, essentially identical strains of the same species. Perhaps the killer dsRNA is merely a 'selfish gene', ensuring its own perpetuation by killing any cells that compete with the host cell in which it resides.

Hypovirulence of plant pathogens

Virus-associated dsRNA has been shown conclusively to reduce the virulence of *Cryphonectria parasitica* (ascomycota), an aggressive pathogen that causes chestnut blight. The fungus infects through wounds in the bark of chestnut trees (*Castanea* spp.) then spreads in the cambium, progressively girdling the stem and killing the plant above the infection point (Fig. 8.12). The disease was a significant problem in Europe earlier this century, but in Italy in the 1950s some heavily infested sweet chestnut plantations (*Castanea sativa*) started to recover spontaneously. The lesions stopped spreading round the trunks, and strains of *C. parasitica* isolated from them showed abnormal features. They grew slowly and erratically on agar,

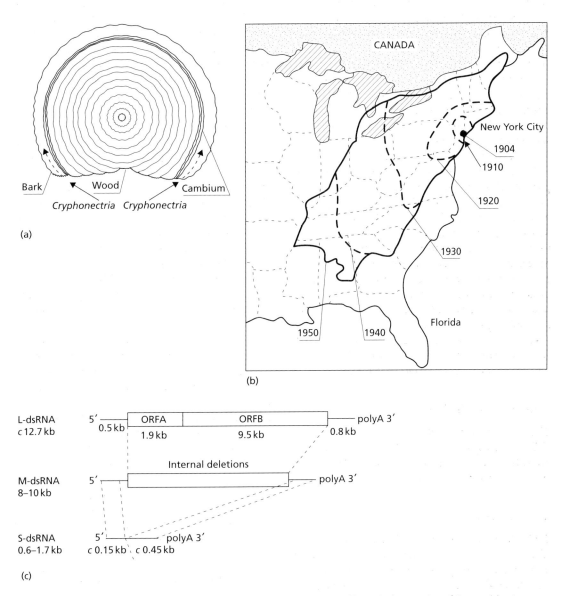

(a)

(b)

L-dsRNA
c 12.7 kb

M-dsRNA
8–10 kb

S-dsRNA
0.6–1.7 kb

(c)

Fig. 8.12 Chestnut blight caused by *Cryphonectria parasitica*. (a) Cross-section of a tree trunk showing how *C. parasitica* enters wounds from air-borne spores and then progressively girdles the trunk by growing in the cambium. (b) Recorded spread of the disease over the eastern part of the USA after it was first recorded in the New York Zoological Garden in 1904; in the next 50 years it destroyed billions of mature chestnut trees.

(c) dsRNA of hypovirulent strains of *C. parasitica*. Large (L) dsRNA is the full-length molecule comprising two conserved end regions and a central coding region of two open reading frames (ORFA and ORFB) which confer hypovirulence. Medium (M) and small (S) dsRNA are also commonly found in the hyphae. They are internally deleted copies of the L-dsRNA. (From Nuss, 1992.)

were white rather than the normal orange colour, produced significantly fewer conidia and showed only low virulence when wound-inoculated into trees. Moreover, this low virulence (**hypovirulence**) was transmitted to other strains during hyphal anastomosis on agar, so it was coined **transmissible hypovirulence**. None of the hypovirulent strains contained 'conventional' VLPs, but (fortuitously in the light of more recent evidence) all the hypovirulent strains contained dsRNA whereas fully virulent strains had no dsRNA. Electron micrographs have now shown that this dsRNA is contained in rounded or club-like membrane-bound vesicles in the cytoplasm (Newhouse *et al.*, 1983) and that, unlike many VLPs, these can occur in significant amounts in the apical cells of *C. parasitica.*

The spontaneous disease decline in Italy led French workers to develop a highly successful biological control programme. Hypovirulent strains of *C. parasitica* were cultured in the laboratory and inoculated at the expanding margins of cankers in the field. Within a short time the cankers stopped growing, and only hypovirulent strains could be recovered where once the virulent strain had been. This transmissible change of phenotype was always accompanied by the transmission of dsRNA. Chestnut blight was also a major problem in the USA, where the fungus had almost destroyed the native American chestnut, *Castanea dentata*, over most of its range. However, when hypovirulent strains were introduced into the USA from Europe they gave only partial and local disease control. The reason was that the pathogen population in the USA consists of numerous VCGs, limiting the natural transfer of dsRNA because of cytoplasmic death during anastomosis. In one early study, about 0.5 ha of natural chestnut forest was found to contain at least 35 VCGs of the pathogen. More recently the VCGs were found to be in a continuous state of flux: samplings of identical trees over several years showed that some of the predominant VCGs declined, while new ones arose (Anagnostakis, 1992). The European population of *C. parasitica* is much more uniform than in the USA, accounting for the success of biocontrol in Europe.

Progress in understanding the role of dsRNA in *Cryphonectria* was significantly delayed by the lack of a suitable transformation system for this fungus. However, when a system was eventually developed it led to rapid progress, reviewed by Nuss (1992). The dsRNA of *C. parasitica* was found to be extremely variable, with lengths falling into three broad size ranges, termed S (small), M (medium) and large (L, about 12.7 kbp). However, all these types had the same terminal regions—a polyA tail at the 3' end and a 28-nucleotide conserved sequence at the 5' end. They differed only (or mainly) in the degree of internal deletion of the molecule (Fig. 8.12). The large form seems to be the full-length molecule, and the smaller forms seem to be defective (presumably functionless) derivatives of it which accumulate in the hyphae. This high degree of variability of RNA genomes is not unusual because there is no effective proof-reading system for RNA genomes, to ensure that they are replicated faithfully, in contrast to DNA genomes.

With the development of a transformation system, Nuss and his colleagues were able to produce a cDNA from the full-length dsRNA and transform it into virulent strains. It caused the virulent strains to become hypovirulent and to exhibit all the features typical of hypovirulent strains — slow growth, pale coloration and reduced sporulation (the hypovirulence-associated traits). Thus, dsRNA was shown unequivocally to be the cause of hypovirulence. The cDNA could also be used for molecular analysis of the dsRNA, and it was shown to consist of two open reading frames (ORFs)—ORFA of 622 codons (nucleotide triplets) and ORFB of 3165 codons. When ORFB was deleted from the cDNA this cDNA did not reduce the virulence when transformed into *C. parasitica*. In similar tests, the deletion of ORFA led to loss of the hypovirulence-associated traits (slow growth, low sporulation, etc.). Thus, it seems that ORFB is necessary for the expression of hypovirulence, whereas ORFA encodes many of the associated traits which are potentially disadvantageous in a biocontrol strain, reducing its environmental fitness. It may be possible, therefore, to manipulate the cDNA *in vitro* so that it has only the most desirable traits for biocontrol.

Perhaps the most significant point, however, was that cDNA, when transformed into *Cryphonectria* (causing the fungus to be hypovirulent), became stably integrated in the chromosomal genome so that it was replicated along with the other chromosomal genes during nuclear division. Thus, the cDNA was maintained throughout the life cycle, even entering the sexual spores, whereas dsRNA seldom enters the sexual spores of *Cryphonectria* or other ascomycota. This would mean that biocontrol strains could be produced with permanent, stable hypovirulence, subject to proof-reading like the rest of the chromosomal genes. Moreover, it overcomes the problem of transmission between VCGs, because *Cryphonectria* has only two mating types (the mating-type genes override the VCG genes) and so a single biocontrol strain could mate with 50% of the pathogen population in each generation, transmitting hypovirulence.

Recent developments

The question arising from the stable integration of 'hypovirulence' cDNA is, can there still be cytoplasmically transmitted hypovirulence? The answer seems to be 'yes' because the chromosomally integrated cDNA is transcribed into dsRNA rather than single-stranded mRNA (as in the rest of the genome) and this dsRNA accumulates in the cytoplasm where, presumably, it acts as a template for production of mRNA. This still leaves unanswered the question of how dsRNA suppresses virulence and other phenotypic traits. The preliminary evidence suggests that it does so by down-regulating some of the normal chromosomal genes, including the gene for production of laccase, an enzyme involved in lignin breakdown (see Chapter 10).

Most recently, the hypovirulence cDNA (derived from dsRNA of *C. parasitica*) has been transformed into other canker-forming *Cryphonectria* spp. and into a less closely related *Endothia* spp., but it failed to convert these to hypovirulence. However, when the same cDNA was used as a template to produce RNA and this was transfected into the other fungi, it gave rise to full-length dsRNA in their cytoplasm and caused both a marked reduction of virulence and altered growth rate and pigmentation (Chen *et al.*, 1994). The success achieved with RNA but not cDNA seems to be explained by the fact that *C. parasitica* produces RNA from the cDNA but then splices this RNA to delete a 73-base sequence before the

RNA can act as a template for dsRNA production. The other fungi may lack some of the processing system for this. In any case, this work on the hypovirus system of *Cryphonectria* has raised many issues of fundamental interest in fungal genetics, as well as holding the prospect of developing entirely new approaches to plant disease control.

Conclusion: the genetic fluidity of fungi

The molecular genetical studies of the past 10 years have revealed a genetic fluidity in fungi that could never have been imagined. Transposons and other mobile elements can switch the mating types of fungi and cause chromosomal rearrangements. Deletions of mitochondrial genes can accumulate as either symptomless plasmids or as disruptive elements leading to cellular senescence. dsRNA genomes are resident elements in many fungi, often defective because of internal deletions but sometimes conferring major effects such as toxin production or loss of pathogenic virulence. We shall also see in Chapter 12 that plant-pathogenic fungi can have entirely dispensable chromosomes involved in pathogenicity but not in normal growth of the fungi. Many aspects of the genetic fluidity of fungi remain to be resolved, and probably many more remain to be discovered.

References

Anagnostakis, S.L. (1992) Diversity within populations of fungal pathogens on perennial parts of perennial plants. In: *The Fungal Community: its Organization and Role in the Ecosystem* (eds G.C. Carroll & D.T. Wicklow), pp. 183–92. Marcel Dekker, New York.

Barrett, J. (1987) Molecular variation and evolution. In: *Evolutionary Biology of the Fungi* (eds A.D.M. Rayner, C.M. Brasier & D. Moore), pp. 83–95. Cambridge University Press, Cambridge.

Bennett, J.W. & Lasure, L.L. (1985) *Gene Manipulations in Fungi.* Academic Press, San Diego.

Bennett, J.W. & Lasure, L.L. (1991) *More Gene Manipulations in Fungi.* Academic Press, San Diego.

Bertrand, H. (1995) Senescence is coupled to induction of an oxidative phosphorylation stress response by mitochondrial DNA mutations in *Neurospora. Canadian Journal of Botany,* **73**, S198–S204.

Buck, K.W. (1986) *Fungal Virology.* CRC Press, Boca Raton.

Chen, B., Choi, G.H. & Nuss, D.L. (1994) Attenuation of fungal virulence by synthetic hypovirus transcripts. *Science*, **264**, 1762–4.

Esser, K. (1990) Molecular aspects of ageing: facts and perspectives. In: *Frontiers in Mycology* (ed. D.L. Hawksworth), pp. 3–25. CAB International, Wallingford.

Fincham, J.R.S., Day, P.R. & Radford, A. (1979) *Fungal Genetics*, 4th edn. Blackwell Scientific Publishers, Oxford.

Fox, R.T.V. (1994) *Principles of Diagnostic Techniques in Plant Pathology*. CAB International, Wallingford.

Jinks, J.L. (1952) Heterokaryosis: a system of adaptation in wild fungi. *Proceedings of the Royal Society of London, Series B*, **140**, 83–99.

Kohn, L.M. (1995) The clonal dynamic in wild and agricultural plant-pathogen populations. *Canadian Journal of Botany*, **73**, S1231–S40.

MacNish, G.C. & Sweetingham, M.W. (1993) Evidence of stability of pectic zymogram groups within *Rhizoctonia solani* AG–8. *Mycological Research*, **97**, 1056–8.

MacNish, G.C., McLernon, C.K. & Wood, D.A. (1993) The use of zymogram and anastomosis techniques to follow the expansion and demise of two coalescing bare patches caused by *Rhizoctonia solani* AG8. *Australian Journal of Agricultural Research*, **44**, 1161–73.

Newhouse, J.R., Hoch, H.C. & MacDonald, W.L. (1983) The ultrastructure of *Endothia parasitica*. Comparison of a virulent with a hypovirulent isolate. *Canadian Journal of Botany*, **61**, 389–99.

Nuss, D.L. (1992) Biological control of chestnut blight: an example of virus-mediated attenuation of fungal pathogenesis. *Microbiological Reviews*, **56**, 561–76.

Oliver, S.G. (1987) Chromosome organisation and genome evolution in yeast. In: *Evolutionary Biology of the Fungi* (eds A.D.M. Ryner, C.M. Brasier & D. Moore), pp. 33–52. Cambridge University Press, Cambridge.

Osbourn, A., Bowyer, P., Bryan, G., Lunness, P., Clarke, B. & Daniels, M. (1994) Detoxification of plant saponins by fungi. In: *Advances in Molecular Genetics of Plant–Microbe Interactions* (eds M. Daniels, J.A. Downie & A.E. Osbourn), pp. 215–21. Kluwer Academic, Dordrecht.

Peberdy, J.F., Caten, C.E., Ogden, J.E. & Bennett, J.W. (1991) *Applied Molecular Genetics of Fungi*. Cambridge University Press, Cambridge.

Pipe, N.D., Buck, K.W. & Brasier, C.M. (1995) Molecular relationships between *Ophiostoma ulmi* and the NAN and EAN races of *O. novo-ulmi* determined by RAPD markers. *Mycological Research*, **99**, 653–8.

Rawlinson, C.J., Carpenter, J.M. & Muthyalu, G. (1975) Double-stranded RNA virus in *Colletotrichum lindemuthianum*. *Transactions of the British Mycological Society*, **65**, 305–8.

Sansome, E. (1987) Fungal chromosomes as observed with the light microscope. In: *Evolutionary Biology of the Fungi* (eds A.D.M. Rayner, C.M. Brasier & D. Moore), pp. 97–113. Cambridge University Press, Cambridge.

Tooley, P.W. & Therrien, C.D. (1991) Variation in ploidy in *Phytophthora infestans*. In: *Phytophthora* (eds J.A. Lucas, R.C. Shattock, D.S. Shaw & L.R. Cooke), pp. 204–17. Cambridge University Press, Cambridge.

Turner, G. (1991) Strategies for cloning genes from filamentous fungi. In: *Applied Molecular Genetics of Fungi* (eds J.F. Peberdy, C.E. Caten, J.E. Ogden & J.W. Bennett), pp. 29–43. Cambridge University Press, Cambridge.

Chapter 9

Spores, spore dormancy and spore dispersal

Fungi produce an astonishing variety of spores, differing in shape, size, motility, surface properties, etc. This is illustrated in Fig. 9.1 for the spores of aquatic fungi which have some extremely bizarre shapes. Such diversity is in stark contrast to the uniformity of somatic hyphae and yeasts, indicating that the diversity is functional — it is needed to ensure that the spores are dispersed to appropriate sites. In this chapter we will discuss several examples of this fine-tuning, and we will see that the spore tells us much about the biology of a fungus.

General features of spores

Because of their extreme diversity we can only define fungal spores in a general way: as microscopic propagules that lack an embryo and are specialized for dispersal or dormant survival.

The spores produced by a sexual process usually function in dormant survival (oospores, zygospores, ascospores) whereas asexual spores usually serve for dispersal. However, many basidiomycota do not produce asexual spores, or produce them only rarely, and instead the basidiospores are their main dispersal agents. Some fungi have an additional spore type: the chlamydospore. It is a thick-walled, melanized cell that develops from an existing hyphal compartment (or sometimes from a spore compartment) in conditions of nutrient stress. The properties of these different spore types vary considerably but, in general, the spores of fungi differ from somatic cells in the following ways.

1 The wall is often thicker, with additional layers or additional pigments such as melanins.

2 The cytoplasm is dense and some of its components (e.g. endoplasmic reticulum and mitochondria) are poorly developed.

3 There is a relatively low water content, low respiration rate and low rates of protein and nucleic acid synthesis.

4 There is a high content of energy storage materials such as lipids, glycogen or trehalose.

Dormancy and germination

Almost all spores are dormant, in the sense that their rate of metabolism is low. But, they can be assigned to two broad categories in terms of their ability to germinate. The sexual spores often show **constitutive dormancy**. They do not germinate readily when placed in conditions that are suitable for normal, somatic growth (appropriate nutrients, temperature, moisture, pH, etc.). Instead, some of them require a period of ageing (postmaturation) before they will germinate, and others require a specific activation trigger such as a heat shock or chemical treatment. Other spores show **exogenously imposed dormancy** — they remain dormant if the environment is unsuitable for growth, but otherwise they germinate readily. Given this difference, all spores germinate in essentially the same way. The cell becomes hydrated, there is a marked increase in respiratory activity, followed by a progressive increase in the rates of protein and nucleic acid synthesis. An outgrowth (the germ-tube) is then formed, and it either develops into a hypha or, in the case of some sexual spores, it produces an asexual sporing stage. The germination process usually takes 3–8 h, but zoospore cysts can germinate much faster (20–60 min) and some sexual spores can take longer (12–15 h).

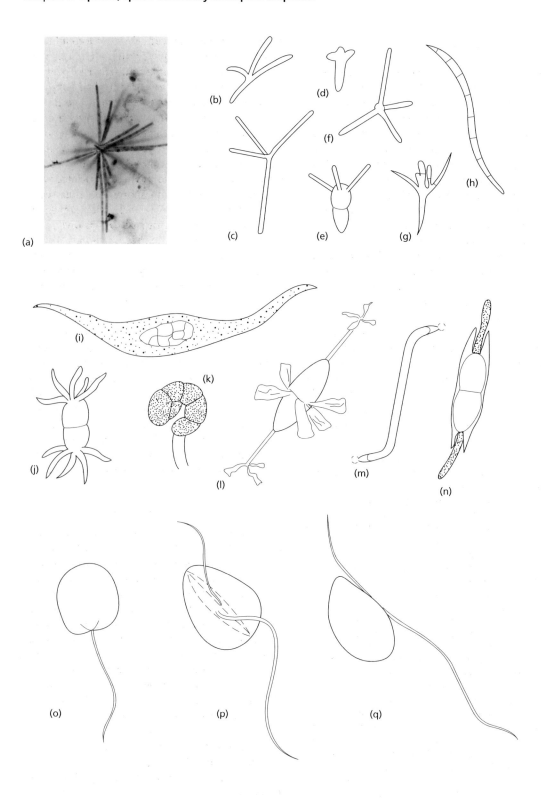

Constitutive dormancy

Constitutive dormancy has been linked to several factors but is still poorly understood. The oospores of many *Pythium* and *Phytophthora* spp. seem to need a post-maturation phase before they can germinate. During this phase the initially thick oospore wall (about 2 μm in *Pythium ultimum*) becomes progressively thinner (about 0.5 μm in *P. ultimum*) by digestion of its inner layers. This process is hastened by keeping the spores in nutrient-poor conditions at normal temperature and moisture levels. Then, after some weeks, the spores will germinate in response to nutrients or another environmental trigger. For example, *Pythium* oospores germinate in response to common nutrients (sugars and amino acids) or volatile metabolites (e.g. acetaldehyde) released from germinating seeds.

Ascospores can eventually become germination-competent by ageing, but can be triggered to germinate at any time by specific triggers — in some cases a heat shock (e.g. 60°C for 20–30 min) a cold shock (–3°C) or exposure to lipid-soluble chemicals such as alcohols and furfuraldehyde. The heat-activated ascospores of *Neurospora tetrasperma* have been studied most thoroughly in this respect. Their dormancy cannot be explained in terms of a general permeability barrier, because they are permeable to radiolabelled oxygen, glucose and water. Instead, the dormancy is linked to an inability to use their major storage reserve, trehalose, which is not metabolized during dormancy, but is metabolized immediately after activation. The enzyme **trehalase**, which cleaves trehalose to glucose, is found to be associated with the walls of the dormant spores and thus separated from its substrate. So, activation somehow causes the enzyme to enter the cell, as one of the earliest detectable events in germination.

Constitutive dormancy of some other spores has been linked to endogenous inhibitors. For example, uredospores of the cereal rust *Puccinia graminis* contain methyl-*cis*-ferulate, and those of bean rust, *Uromyces phaseoli*, contain methyl-*cis*-3,4-dimethoxycinnamate. Prolonged washing of spores can remove these inhibitors, and this perhaps occurs when the spores are bathed in a water film on a plant surface. At first sight it seems surprising that a spore adapted for dispersal should have an endogenous inhibitor. However, this might prevent the spores from germinating in a sporing pustule (they do not require exogenous nutrients for germination) and ensure that they germinate only after they have been dispersed.

Ecology and constitutive dormancy

The behaviour of constitutively dormant spores often has clear ecological relevance. For example, a characteristic assemblage of fungi grow on the dung of herbivorous animals (see Chapter 10). Their spores are ingested with the herbage, are activated during passage through the gut, and are deposited in the dung where they germinate to initiate a new phase of growth. The spores of many of these **coprophilous** (dung-loving) fungi can be activated in the laboratory by treatment at 37°C in acidic conditions, simulating the gut environment. Examples include the ascospores of *Sordaria* and *Ascobolus*, sporangiospores of several zygomycota and basidiospores of *Coprinus* and *Bolbitius*.

Heating to 60°C activates the spores of many thermophilic fungi of composts (see Chapter 10). It also activates the ascospores of **pyrophilous** (fire-loving) fungi such as *Neurospora tetrasperma* which

---

Fig. 9.1 Spores of some aquatic fungi. (a–h) Freshwater species; (i–n) estuarine or marine species; (o–q) zoospores. Approximate spore lengths, excluding flagella, are shown in parentheses below. (a) Conidium of *Dendrospora* (150–200 μm); (b) conidium of *Alatospora* (30–40 μm); (c) conidium of *Tetrachaetum* (70–80 μm); (d) conidium of *Heliscus* (30 μm); (e) conidium of *Clavariopsis* (40 μm); (f) conidium of *Lemonniera* (60–70 μm); (g) conidium of *Tetracladium* (30–40 μm); (h) conidium of *Anguillospora* (150 μm); (i) ascospore of *Pleospora* with mucilaginous appendages (stippled) (400 μm); (j) ascospore of *Halosphaeria* with chitinous wall appendages (25 μm); (k) conidium of *Zalerion* (25 μm); (l) ascospore of *Corollospora* with membranous appendages (70 μm); (m) ascospore of *Lulworthia* with terminal mucilaginous 'pouches' (60 μm); (n) ascospore of *Ceriosporiopsis* with mucilaginous appendages (stippled) (40 μm); (o) zoospore of chytridiomycota (5–8 μm) with single posterior whiplash flagellum; (p) zoospore of oomycota (15–20 μm) with anteriorly directed tinsel-type flagellum and posteriorly directed whiplash flagellum, both arising from a longitudinal groove in the spore body; (q) zoospore of plasmodiophorid (4–5 μm) with short anterior and long posterior whiplash flagella.

grow on burnt ground or charred plant remains. Most of these fungi are saprotrophs of no economic importance, but one of them, *Rhizina undulata* (ascomycota), causes the 'group-dying' disease of coniferous trees in Britain and elsewhere. It infects trees replanted into clear-felled forests, and the foci of infection correspond to the sites where the trash from the felled trees was stacked and burned. The ascospores are heat activated around or beneath the fires, then the fungus grows as a saprotroph on the stumps and dead roots and produces mycelial cords that infect the newly planted trees. Hence, the name, group dying. Once the cause had been recognized, the problem was easily solved by abandoning the practice of burning. However, this is not possible in regions where lightning-induced fires are a periodic, natural occurrence. Some of the plants in these areas have become adapted to fire — their seeds remain dormant for years until they are heat activated. Some of the mycorrhizal fungi are similarly adapted, an example being the ascospores of the mycorrhizal *Muciturbo* spp. in Australian eucalyptus forests.

Role of dormancy in mycorrhizal successions

Many forest trees in the cool temperate and boreal regions (pine, birch, oak, beech, chestnut, etc.) form ectomycorrhizal associations with basidiomycota which produce a sheath around the individual feeder roots and extend as mycelia into the soil (Plate 9.1, facing p. 210). Some of these fungi are relatively host-specific and tend to become dominant on the root systems as the trees age (e.g. *Suillus luteus*, 'slippery jack', on pines) whereas others are generalists and infect a wide range of trees, especially in nurseries (e.g. *Thelephora terrestris*, the 'earth fan'). Nevertheless, it is common to find several mycorrhizal fungi on the roots of a single tree, and studies on their patterns of establishment suggest an important role for basidiospore dormancy.

The morphology of ectomycorrhizas can be sufficiently distinct to enable them to be categorized into types and even identified to species level of the fungus. On this basis, a succession of mycorrhizal fungi has been found to occur on young birch trees (*Betula* spp.), reviewed by Deacon and

Fleming (1992). Birch seedlings raised from seeds in a glasshouse were colonized by a few common mycorrhizal fungi that established from air-borne or water-borne spores, especially *T. terrestris*, *Inocybe* and *Hebeloma* spp. These also are among the pioneer mycorrhizal fungi in tree nurseries. But, when the seedlings were transplanted to a previously treeless field site, a range of further mycorrhizal fungi appeared over the years. As shown in Fig. 9.2, the pioneer fungi could always be found on root tips at the spreading periphery of the root system, but in the older part of the root zone, near the tree bases, they were displaced progressively by the later colonizers (characteristic of older trees). This pattern was explained by two types of experiment. First, basidiospores were collected from fruitbodies of both 'pioneer' and 'later' mycorrhizal fungi, then added to soil in pots, and non-mycorrhizal birch seedlings were planted into these soils. Only the pioneer fungi formed mycorrhizas in these conditions. This was related to basidiospore germination, because only the pioneer fungi have basidiospores that germinate readily; the spores of the 'later' fungi germinate extremely poorly — often less than 0.1% germination in any conditions that have been tested. Second, birch seedlings were raised aseptically (without mycorrhizas) then planted beneath older trees in a field site so that they would be infected by the fungi already established there. Some of the seedlings were planted directly into the undisturbed soil, but others were planted where the soil had been removed as a core and then replaced immediately, using the tool for making the potting holes on golf courses. All the seedlings in the cored positions became infected by pioneer fungi, presumably from spores in the soil. In contrast, all the seedlings planted in undisturbed positions were infected by the 'later' fungi, presumably from mycelial networks which needed to be attached to a food base (i.e. the living parent tree roots) in order to infect.

So, the pattern of mycorrhizal establishment on birch in previously treeless sites can be summarized as shown in Fig. 9.3. The pioneer fungi infect young seedlings in nurseries or in the field, from basidiospores that land on the soil surface and are washed into the root zone. They probably have annual cycles of infection from basidiospores as the root system grows and expands into new soil

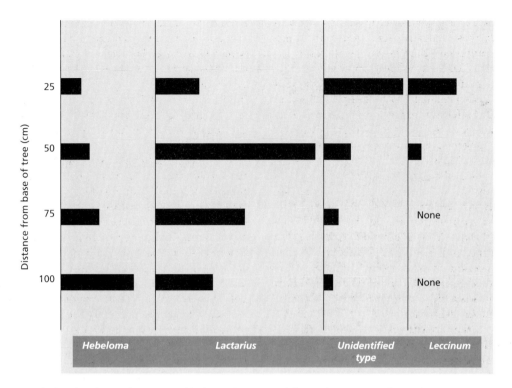

Fig. 9.2 Relative abundance of ectomycorrhizal root tips of distinctive types at different distances from the stem of a young (5–7 year) birch tree in a field site. Mycorrhizas of *Hebeloma* spp. predominate at the root periphery, whereas mycorrhizas of *Lactarius* spp. dominate the mid-root zone, and mycorrhizas of *Leccinum* spp. and an unknown fungus dominate in the older root zone. (Adapted from Deacon *et al.*, 1983.)

zones. The later fungi cannot establish initially because their spores germinate poorly. But, they germinate eventually, especially in the older parts of the root zone where soil conditions might favour them, and then they become dominant by spreading as mycelial networks to infect further root tips. This leads to a spatial distribution of mycorrhizal types as shown in Fig. 9.2— pioneer fungi at the expanding root margin, and waves of colonization by later mycorrhizal types behind the margin. In natural woodlands and forestry plantations, we can expect that seedlings will be infected directly by the 'later' fungi; but in new sites they will initially be infected by pioneer fungi. This has practical consequences, because only the pioneer fungi are suitable for mycorrhizal inoculation programmes, commonly used in land-reclamation sites (Marx & Cordell, 1989). The fungus often used for this is the puffball, *Pisolithus tinctorius*, a pioneer fungus that tolerates high levels of toxic minerals and the low water-retention properties of mine-spoil sites. The inoculum used for these programmes often is basidiospores from field-collected fruitbodies.

Exogenously imposed dormancy

In laboratory conditions, most asexual spores germinate readily at suitable temperature, moisture, pH and oxygen levels. Some germinate even in distilled water, although most require at least a sugar source, and a few have multiple nutrient requirements. However, in nature all these spores can be held in a dormant state for long periods by the phenomenon termed **fungistasis** (or **mycostasis**). This is very common in soil (Lockwood, 1977) but also occurs on leaf surfaces (Blakeman, 1981).

Fungistasis is a microbially induced phenomenon. Spores often fail to germinate in topsoil, where the microbial content is high, but they germinate in sterilized soil or in subsoils of low microbial activity. The germination in sterilized soil can

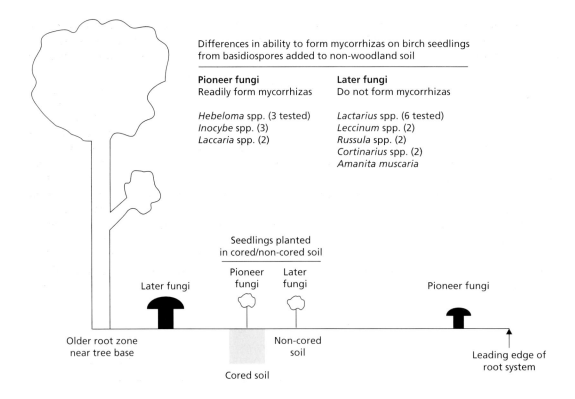

Differences in ability to form mycorrhizas on birch seedlings from basidiospores added to non-woodland soil

Pioneer fungi	Later fungi
Readily form mycorrhizas	Do not form mycorrhizas
Hebeloma spp. (3 tested)	*Lactarius* spp. (6 tested)
Inocybe spp. (3)	*Leccinum* spp. (2)
Laccaria spp. (2)	*Russula* spp. (2)
	Cortinarius spp. (2)
	Amanita muscaria

Seedlings planted in cored/non-cored soil

Pioneer fungi   Later fungi

Later fungi

Pioneer fungi

Older root zone near tree base

Non-cored soil

Cored soil

Leading edge of root system

Fig. 9.3 Diagram of the factors that cause the succession of mycorrhizal types on birch seedlings planted into previously treeless sites. (The information on infection from basidiospores is from Fox, 1986.)

be prevented if the soil has been recolonized by micro-organisms, and even single micro-organisms of various types will restore the suppression. This suggests that fungistasis is caused by nutrient competition or by general microbial metabolites (or both), but not by specific antibiotics or other inhibitors from particular micro-organisms. A long history of research on this topic has suggested that volatile germination inhibitors such as ethylene ($H_2C{=}CH_2$), allyl alcohol ($H_2C{=}CHCH_2OH$) and ammonia can play some role in some soils. But, the strongest body of evidence implicates **nutrient deprivation** as a key component of fungistasis.

According to the nutrient-deprivation hypothesis advanced by Lockwood and his colleagues (Lockwood, 1977), even spores that can germinate in distilled water are inhibited in soil because, during hydration, they leak nutrients into their immediate environment and these nutrients are rapidly and continuously metabolized by other organisms. Thus, the spores are held in a dormant state by some form of feedback inhibition which signals that the environment is unsuitable. A simple experimental system was developed to test this (Fig. 9.4). Spores were placed on sterile membranes overlying sterile sand or sterile glass beads, then sterile water was percolated slowly through the sand or beads so that any nutrients released from the spores were continuously removed. As shown in Table 9.1, except for the special case of activated ascospores of *Neurospora*, which germinated in all conditions, the spores did not germinate in the 'nutrient-leaching' system, but they germinated if the flow of water was stopped for 24 h or when a flow of glucose solution was used in place of water. By using very slow rates of water percolation it was possible to simulate the fungistatic effects of natural soils. The spores of different fungi have different fungistatic sensitivities (related to spore size, spore nutrient reserves and their typical speed of germination) but, with few exceptions, there is remarkably good agreement between their sensitivity to nutrient leaching in the model system and their sensitivity to soil-imposed fungistasis.

Ecological implications of fungistasis

Fungistasis causes spores to remain quiescent in soil or other natural environments until nutrients become available. Thus, saprotrophs can lie in wait for organic matter, and root pathogens or mycorrhizal fungi can wait for a root to pass nearby. In many cases these responses are non-specific because the spores of root pathogens, for example, seem to respond to exudates from host or non-host plants. But, this has some advantages in terms of disease control, because these fungi are often quite specific in their parasitic abilities (see Chapter 12), and so in many cases the spores will

germinate but fail to establish an infection and the sporeling dies. This phenomenon, termed **germination-lysis,** might largely explain the success of conventional crop rotations in reducing the damage caused by soil-borne pathogenic fungi.

There are very few proven cases of host-specific triggering of germination of fungi, and they involve sclerotia rather than spores. The best example is *Sclerotium cepivorum* (basidiomycota) which causes 'white rot' of onions, garlic and closely related *Allium* spp. (Coley-Smith, 1987). The sclerotia are produced abundantly in infected onion bulbs and they can survive for up to 20 years in soil until they are triggered to germinate by the host. The germination triggers in this case are volatile sulphur-containing compounds (alkyl thiols and alkyl sulphides) such as diallyl disulphide. However, the host seems to release the non-

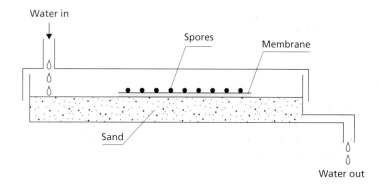

Fig. 9.4 Nutrient-leaching system used to study soil fungistasis.

Table 9.1 Relationship between germination of fungal spores incubated on natural soil and on a nutrient-leaching system (see Fig. 9.4) designed to simulate the continuous removal of spore nutrients by soil micro-organisms. (Data from Hsu & Lockwood, 1973.)

Fungus and spore type	Percentage germination		Leaching system	
	Distilled water	**Natural soil**	**Water flowing**	**Flow stopped for 24 h**
CONIDIA				
Verticillium albo-atrum	60	9	8	—
Thielaviopsis basicola	89	4	5	89
Fusarium culmorum	94	20	9	91
Curvularia lunata	95	16	13	91
Cochliobolus sativus	97	21	19	91
Alternaria tenuis	95	54	71	—
ACTIVATED ASCOSPORES				
Neurospora crassa	98	87	84	—

—, no data.

volatile precursors of these compounds (alkyl sulphoxides and alkylcysteine sulphoxides) which then are metabolized to volatile triggers by common soil bacteria. Knowledge of this system has suggested some novel approaches for control of white-rot disease, but unfortunately with only partial success to date. One approach would be to breed crop cultivars that are non-stimulatory, but this has met with the problem that some of the germination-triggering compounds are also the flavour and odour components of onions. (Cynics might argue, with some justification, that this has not been a barrier to the marketing of some apple cultivars.) A more promising approach would be to use chemical triggers to stimulate germination in the absence of the crop. Artificial onion oil, which is used commercially as a flavour component of processed foods (including cheese-and-onion potato crisps), contains large amounts of diallyl disulphide. When artificial onion oil is applied to soil it triggers the germination of sclerotia, which then die by germination-lysis. Up to 95% of the sclerotia can be eradicated by this means, but even the remaining few per cent can be sufficient to cause serious crop damage, and so the use of germination stimulants will need to be combined with other treatments such as fungicides for effective disease control.

Spore dispersal

Spore dispersal has been described in detail by Ingold (1971) and Gregory (1973), so here we will deal with selected aspects, focusing on how the spores or spore-bearing structures of fungi are precisely tailored for dispersal. In doing so, we cover many topics of practical and environmental significance.

'Self-dispersal' by coprophilous fungi

Coprophilous fungi grow on the dung of herbivores and help to recycle the vast amounts of plant material that are deposited annually in the grazing food chain (see Chapter 10). The spore-dispersal mechanisms of these fungi are highly attuned to their lifestyle, because the main requirement is to project the spores from the dung onto the surrounding vegetation, where they will be ingested and pass through the gut to repeat the cycle.

In many cases this is achieved by ballistic mechanisms.

Pilobolus (zygomycota) is one of the most striking examples (Fig. 9.5; Plate 9.2, facing p. 210). Each sporing structure consists of a large black sporangium, mounted on a swollen vesicle which is part of the sporangiophore. At maturity the sporangiophore develops a high turgor pressure, the wall that encloses both the sporangium and the vesicle breaks down locally by enzymic means, and the vesicle then suddenly ruptures, squirting its contents forwards and propelling the sporangium for 2 m or more. Mucilage released from the base of the sporangium during this process serves to stick the sporangium to any plant surface on which it lands, then the spores are released from the sporangium and can be spread by water or other agencies. As a further adaptation for dispersal, the sporangiophore is phototropic, ensuring that the sporangium is shot free from any crevices in the dung. The light signal is perceived by a band of orange carotenoid pigment at the base of the vesicle, and the vesicle itself seems to act as a lens that focuses light on the pigment. A unilateral light signal is thereby translated into differential growth of the sporangiophore stalk, aligning the sporangium towards the light source. *Pilobolus* thus exhibits three special types of adaptation that also are found, to different degrees, in several other coprophilous fungi (Fig. 9.5).

1 The spore-bearing structure is **phototropic**, an adaptation also seen in the tips of the asci of *Ascobolus* and *Sordaria*.

2 There is an **explosive discharge** mechanism. This also is seen in the asci of *Ascobolus* and *Sordaria* because the asci act as guns, shooting spores up to 1–2 cm into the air. A different discharge mechanism is found in *Basidiobolus ranarum* (zygomycota) which grows on the faeces of lizards and frogs, because in this case the sporangium is mounted on a subsporangial vesicle, as in *Pilobolus*, but the vesicle ruptures at its base, squirting the sap backwards and propelling the sporangium forwards. Yet another variation is found in *Sphaerobolus stellatus* (basidiomycota) which produces basidiospores in a large ball-like structure within a cup-shaped fruitbody; at maturity, the inner layer of the cup separates from the outer layer and suddenly inverts, springing the spore mass into the air, like a trampoline.

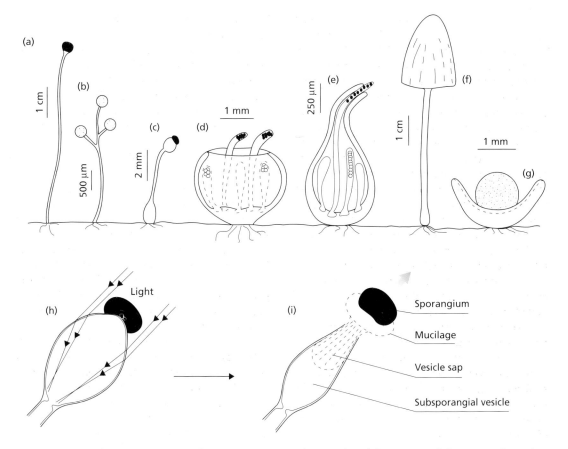

Fig. 9.5 Diagrammatic representation of some coprophilous fungi. (a) *Pilaira anomala* (zygomycota); the sporangiophore elongates to several centimetres at maturity and the spores 'flop' onto the surrounding vegetation. (b) *Mucor racemosus* (zygomycota). (c) *Pilobolus* spp. (zygomycota; see also (h) and (i)). (d) *Ascobolus* spp. (acomycota); the tips of the mature asci project from the apothecium and are phototropic. (e) *Sordaria* spp. (ascomycota); the perithecium neck is phototropic and the mature asci elongate up the neck to discharge the ascospores. (f) *Coprinus* spp. (basidiomycota). (g) *Sphaerobolus* spp. (basidiomycota); the large spore mass is shot from the cup-shaped fruitbody when the layers of this separate and the inner layer suddenly inverts. (h,i) *Pilobolus*, showing how the terminal vesicle of the sporangiophore acts as a lens to focus light and orientate the sporangiophore, and the mechanism of discharge of the sporangium.

3 There is a **large projectile**, based on the ballistic principle that large (heavy) objects travel further than small objects if released at the same initial velocity. *S. stellatus* has a spore mass of about 1 mm diameter, allowing it to be thrown 2 m vertically.

However, not all coprophilous fungi employ ballistic mechanisms. For example, in *Pilaira* (zygomycota; Fig. 9.5) the sporangiophore merely extends several centimetres at maturity so that the sporangium 'flops' onto the surrounding vegetation.

Insect-dispersed fungi

Dispersal by arthropod vectors can be extremely efficient because a fungus can exploit the searching behaviour of the vector to reach a new site. There are many of these fungus–vector associations, ranging from cases where the association is almost incidental to cases of highly evolved mutualism (Cooke, 1977). Here we consider two examples: (i) dispersal of the Dutch elm disease pathogen; and (ii) the association between a woodrotting fungus and its vector, a wood wasp.

The spread of Dutch elm disease

Ophiostoma ulmi and the related species *O. novo-ulmi* (ascomycota) cause Dutch elm disease (Fig. 9.6). They are vascular wilt pathogens which enter through wounds, grow in a yeast-like phase in the xylem and kill the host tree by vascular blockage (see Chapter 12). They are transmitted from one tree to another by bark beetles (e.g. *Scolytus* spp.) which carry the spores on their body surface. The disease cycle starts when young, contaminated beetles feed on the young bark of the elm twigs in early spring and cause incidental damage to the xylem, thereby introducing the fungus into the tree. The fungus then spreads in the xylem, killing the whole tree or some of its major branches, and the bark of the newly killed trees is then used by the female beetles for egg laying. The female tunnels into the inner bark and eats out a channel, depositing eggs along its length — the 'brood gallery' (Fig. 9.6). The eggs hatch and the young larvae eat out a series of radiating channels before they pupate for overwintering. Meanwhile, the fungus that killed the tree grows from the xylem into the bark and sporulates in the beetle tunnels. In this way, the young adult beetles that emerge from the pupae in the following spring become contaminated with spores; they leave the bark and fly in search of new trees, repeating the disease cycle.

The fungus–vector relationship clearly benefits both partners, because the fungus is dispersed to its host plant, owing to the feeding preferences of the beetle, while the beetle is ensured of a fresh supply of breeding sites in the bark of newly killed trees—it will not lay eggs in older dead bark. The association works, even though the beetle seems to have no special adaptations for transmitting the spores. The fungus, however, is adapted for insect dispersal in at least two ways (Fig. 9.6). Its sporing structures project into the brood galleries as aggregates of conidiophores termed **coremia**, and the conidia are borne in a mucilaginous head so that they adhere to the body of the beetle. The fungus also can produce a sexual stage in the bark — a cleistothecium with a long neck, and within this the asci break down to release the ascospores in a mucilaginous matrix which is extruded from the neck. This sexual stage can ensure the generation of recombinant strains, and again the beetle is involved in this because adult beetles feed on the bark late in the season, introducing new strains of the pathogen, different from the strain that killed the tree.

As testimony to the efficiency of the fungus–vector association, new strains of *C. novo-ulmi* have swept across Britain and much of continental Europe, decimating the elm population, since they were introduced on shipments of imported elm logs in the 1960s. A similar epidemic spread across the USA earlier this century, apparently introduced in the same way. In all these cases the problem arose because the logs were not de-barked, to remove any beetle vectors, in contravention of quarantine regulations. Molecular characterization of *Ophiostoma* by the methods discussed in Chapter 8 has been used to trace the origins and histories of these epidemics (Mitchell & Brasier, 1994; Brasier, 1995). This has revealed that *O. ulmi* has been present for many years in Britain and much of continental Europe, but as a heterogeneous population of non-aggressive strains comprising several vegetative compatibility (VC) groups This population was in balance with the tree host, causing relatively little damage. The recent epidemics have been caused by two aggressive subgroups, one imported from North America (termed NAN) and one of Eurasian origin (EAN). These aggressive forms are sexually incompatible with the original non-aggressive population, so they have been separated as a new species, *O. novo-ulmi*. At the advancing margins of the disease in Europe, the pathogen population is almost genetically pure and exists as a single VC 'super-group'. We could expect this from the fungus–vector relationship, because the most aggressive strain will kill most of the trees at the advancing front, and the beetle population will proliferate in these trees, carrying the strain to new trees. However, behind the disease fronts the incidence of this super-group declines to only some 20–30% of the fungal population, suggesting that the population is returning to a more stable form. One of the reasons may be that *Ophiostoma*, like *Cryphonectria* discussed in Chapter 8, can contain virulence-suppressing double stranded ribonucleic acid (dsRNA). A diversity of VC groups can act as a barrier to the transmission of hypovirus genes — perhaps a natural fungal defence against these extrachromosomal elements.

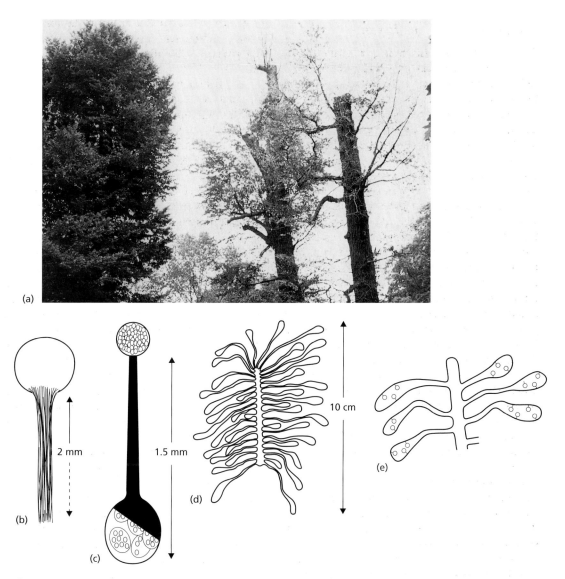

Fig. 9.6 Dutch elm disease caused by *Ophiostoma novo-ulmi*. (a) Dead and dying elm trees (right) 2–3 months after symptoms first appeared. The healthy tree (left) was killed later in the same season. The truncated appearance of the trees is due to the normal practice of lopping by municipal authorities because elms tend to shed major branches. (b) Asexual spores (conidia) borne in a large mucilaginous head on a coremium. (c) Sexual stage: a long-necked cleistothecium in which the asci break down at maturity and the ascospores are extruded in mucilage from the neck. (d,e) Beetle-brood galleries in the bark of a newly killed tree. The female lays eggs along the central channel and the emerging larvae eat out the radiating channels; the fungal sporing stages are produced in these channels.

Fungi and wood wasps

Wood wasps of the genus *Sirex* lay their eggs in the wood of moribund trees by means of an auger-like ovipositor (Fig. 9.7), and the larvae then eat the rotting wood. In most cases these wasps cause little economic damage because the trees are already dead or dying, but the species *Sirex noctilio*

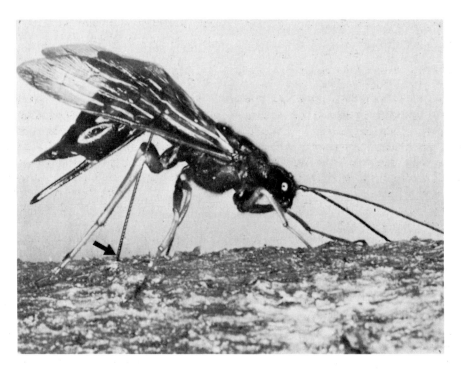

Fig. 9.7 The wood wasp *Sirex noctilio*, in the process of boring egg tunnels and introducing fungal spores through the bark of a tree. Photograph by courtesy of M. P. Coutts, J.E. Dolezal and the University of Tasmania. (See Madden & Coutts, 1979.)

injects its eggs into healthy trees and it simultaneously introduces spores of the wood-rotting fungus *Amylostereum areolatum* (basidiomycota). The wood is killed locally by a toxic secretion of the wasp and by growth of the fungus which progressively dries the wood; this has led to significant economic losses in Australia and New Zealand (Madden & Coutts, 1979). This fungus–vector relationship is quite highly evolved, because the wasp carries conidia of the fungus in special pouches, termed **mycangia**, at the base of its ovipositor. The wasp is also behaviourally adapted, because it usually makes two or three tunnels from a single bore-hole in the trunk; the first one or two of these contain the eggs and the last one contains the fungal spores. If the wasp is disturbed during this process it will return to the same bore-hole to complete the inoculation. The positioning of the tunnels is such that the fungus colonizes the wood and causes a brown rot (see Chapter 10) which reaches the egg tunnels by the

time the eggs hatch. Then, the larvae feed on the rotten wood and perhaps also on the fungal mycelium.

There are several other insects that have mutualistic associations with the fungi they transmit, so that the fungus will provide the insect with a food source. In some of these cases the fungus grows in a yeast phase in the mycangia, utilizing oily secretions of the insect as its food source. Also, in some cases the fungus, once inoculated into a substrate, produces hyphae with balloon-like, detachable swellings so that the insects feed on these without causing damage to the hyphae themselves.

Dispersal of aquatic fungi: appendaged spores

Fungi that grow as saprotrophs in aquatic environments often have spores with conspicuous appendages (see Fig. 9.1). The most common type is the tetraradiate (four-armed) spore. Examples of this are found in the conidia of deuteromycota that grow on fallen tree leaves in fast-flowing streams, in the basidiospores of two marine basidiomycota, in the sporangia of *Erynia conica* (zygomycota) which parasitizes freshwater insects, and even in the somatic cells of the yeast

Vanrija aquatica which grows in mountain tarns. In contrast, the wood-rotting ascomycota of estuarine and marine environments often have ascospores with flakes of wall material or mucilaginous appendages. These various types of appendage must be functionally significant in aquatic environments, and the fact that they occur in a range of unrelated fungi is clear evidence of convergent evolution (Webster, 1980).

The appendages could have several different functions rather than a single common role. The yeast *V. aquatica* forms them in response to nutrient-poor conditions in laboratory culture, suggesting that they might increase the surface area for nutrient absorption. The tetraradiate shape of conidia is probably significant for dispersal or attachment to substrates. These conidia of freshwater deuteromycota have been shown to sediment slowly in water, at about 0.1 mm s^{-1}, although differences in sedimentation rates are unlikely to be important in the turbulent, fast-flowing streams where these spores are found naturally. Perhaps more important is the role of spore shape in entrapment, because appendaged spores accumulate better than non-appendaged spores on objects suspended in fast-flowing streams. In addition, the spores settle like a tripod on a surface and they respond rapidly to contact by releasing mucilage from the ends of the arms in contact with a surface, but not from the fourth (free) arm. Then they germinate from the contact sites, so that the fungus can establish itself from three points, which perhaps increases the efficiency of resource capture. Yet another role was suggested by the recent discovery that some freshwater deuteromycota occur on the attached leaves of trees that grow near streams. They produce the spores on leaves above the water, and the spores are easily dislodged from the conidiophores by raindrops owing to the large surface area for interaction with the surface tension of water drops. These different possibilities are not mutually exclusive. On the contrary, the tetraradiate design might represent an ideal solution to a multiplicity of needs in the habitats of these fungi.

The mucilaginous appendages of the marine ascomycota have been shown to function in attachment to surfaces, because the ascospores become attached to wood veneers in estuarine environments (Moss, 1986). These fungi have important roles as decomposers of woody materials in saltwater habitats.

Dispersal by zoospores

Zoospores are naked, wall-less cells that swim by means of flagella. They are the characteristic dispersal spores of chytridiomycota, oomycota and plasmodiophorids (see Chapter 1). However, the role of swimming for spore dispersal is not as clear as it might seem. In appropriate conditions zoospores can swim for 10 h or more, at rates of at least 100 μm s^{-1}, so they could move as far as 3–4 m. In practice, however, they make frequent random turns, and in still water their rate of dispersion is found to be little more than the rate of diffusion of a small molecule like hydrogen chloride. So, their swimming activity must serve other roles. First, it keeps zoospores in suspension so that they can be carried in moving water, and this is how zoospores are dispersed in nature. Second, the swimming is linked to sensory receptors, so that zoospores can swim towards attractants such as nutrients or oxygen (positive **chemotaxis**) or avoid unsuitable chemical environments (negative chemotaxis). In addition, the zoospores of some fungi have been shown to respond to pH, to electrical or ionic fields (**electrotaxis**) and to accumulate by autoaggregation. This extreme responsiveness of zoospores enables them to act as site-selection agents. They can accumulate in large numbers near the root tips or at wound sites on roots, around single stomata on a leaf surface, or at specific sites on the surface of a nematode host, etc. Then, the zoospores go through a characteristic sequence: they round off, secrete an adhesive, encyst by producing a cell wall and show orientated germination.

This series of events, from zoospore taxis through to germination, has been termed the homing sequence (Fig. 9.8). It can take as little as 20–30 min in the plant-pathogenic oomycota (*Pythium* and *Phytophthora*), so these fungi can infect a host very rapidly from zoospores. During the whole swimming phase, and throughout the homing sequence, zoospores seem to depend on stored nutrient reserves. The cells take up organic nutrients only after they have germinated.

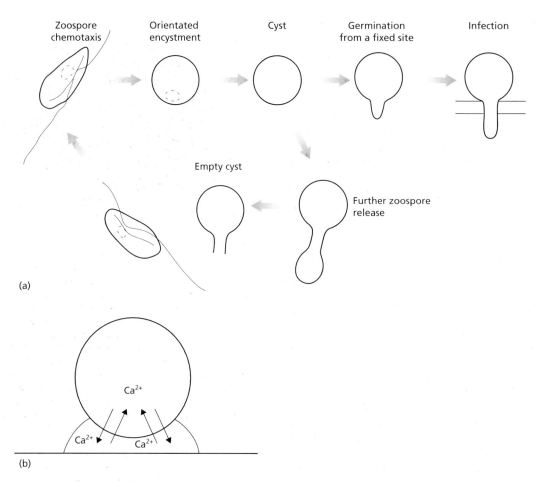

Fig. 9.8 (a) Stages in infection of a plant from a zoospore of the oomycota (e.g. *Pythium, Phytophthora*). The zoospore locates the host by chemotaxis, then settles, encysts and adheres with precise orientation (the position of the water-expulsion vacuole which opens into the ventral groove is shown). The cyst germinates from a fixed site, leading rapidly to infection. The zoospore has a default option if it does not find a suitable host: it encysts before the spore reserves are exhausted, then the cyst germinates to release a further zoospore. (b) Model of autosignalling for germination: calcium ions (Ca^{2+}) are released from the spore in the early stages of encystment, become trapped in the adhesive between the cyst and the host surface and then are reabsorbed to trigger germination. (Based on Deacon & Donaldson, 1993.)

Structure and organization of zoospores

Two representative types of zoospore are shown in Fig. 9.9. Those of chytridiomycota are small and, except for the rumen chytrids, have a single posterior flagellum of the **whiplash** (smooth) type. In contrast, the zoospores of oomycota typically are larger and kidney-shaped, with two flagella inserted in a ventral groove. The longer flagellum is whiplash type and trails behind the swimming spore; the other projects forwards and is **tinsel-**type, with short glycoprotein hairs along its length. In addition to these types, the plasmodiophorids have small zoospores with two whiplash flagella: a short one directed forwards and a longer one directed backwards. These differences in number and arrangement of the flagella reflect the quite different evolutionary lineages of these organisms. Nevertheless, all flagella have the same basic structure consisting of a central shaft, the **axoneme,** which is composed of microtubules and associated motor proteins, and this is surrounded by a membrane which is continuous with the cell membrane. At the base of the axoneme is a complex and variable rooting structure, and a **kinetosome** which is a modified centriole. This basal apparatus is often linked to the nucleus by microtubular arrays.

The arrangement of internal organelles differs markedly in the zoospore types. The zoospores of chytridiomycota have a relatively simple structure. The nucleus is located towards the rear of the spore and is surmounted by a conspicuous nuclear cap which is an aggregation of ribosomes (Fig. 9.9). There is often a single large mitochondrion near the base of the flagellum, but a hydrogenosome (see Chapter 6) in the same position in the anaerobic rumen chytrids; its location suggests that it is involved in providing energy for flagellar beating. The chytridiomycota also have a microbody–lipid globule complex beside the nucleus near the base of the flagellum. Its function is not fully known but it is suggested to be involved in encystment and adhesion. The zoospores of chytridiomycota do not seem to have an osmoregulatory apparatus, despite the fact that they are wall-less cells. Further details of these zoospores and their functional properties can be found in Powell (1994).

The organization of the zoospores of oomycota is markedly different (Fig. 9.9). Again, the nucleus is located near the site of flagellar attachment, connected to this by a complex system of microtubules. There is also a conspicuous water-expulsion vacuole which discharges regularly into the ventral groove, like a contractile vacuole in some protists. The rest of the zoospore contains various organelles, such as mitochondria, lipid bodies and vacuoles, and immediately beneath the spore membrane there are many **peripheral vesicles**. These are of three types, distinguished by the binding of their proteinaceous contents to specific monoclonal antibodies or lectins (Fig. 9.10). The large peripheral vesicles which are seen beneath most of the cell periphery contain a glycoprotein that is thought to serve as a protein store after encystment. These vesicles migrate towards the centre of the cells when zoospores encyst. The smaller 'dorsal' vesicles contain a different glycoprotein which is released by exocytosis at an early stage of encystment, and it accumulates on the cell surface as a cyst coat. The third class of vesicles, the ventral vesicles, are found around the

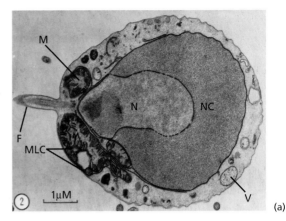

Fig. 9.9 (a) Electron micrograph of a longitudinal section of a posteriorly uniflagellate zoospore of *Blastocladiella emersonii* (chytridiomycota). The spore is surrounded only by a plasma membrane which is continuous with the flagellar membrane. Only part of the flagellum (F) is seen in the section. The zoospore contains a conspicuous nucleus (N) and a nuclear cap (NC) composed of ribosomes. Surrounding the base of the flagellum is a large mitochondrion (M), and a microbody-lipid globule complex (MLC) is also seen in this region. (Courtesy of M.S. Fuller; from Reichle & Fuller, 1967.) (b,c *overleaf*) Two parallel sections through a zoospore of *Phytophthora palmivora* (oomycota). One section passes through the water-expulsion vacuole (WEV) which opens into the ventral groove of the zoospore. The WEV consists of a central vacuole (CT) which discharges periodically, and surrounding vacuoles (SU) which fuse to form a new central vacuole. The other section passes through the region of flagellar insertion in the groove, and shows the nucleus (N) extending to the base of the flagella. Beneath the zoospore plasma membrane are sheets of peripheral membraneous cisternae (PC), large peripheral vesicles (PV) and small peripheral vesicles (labelled SPV). The cells also contain mitochondria (M), Golgi dictyosomes (D) and fingerprint vesicles (FV) which contain β-1,3-glucans that probably serve as carbohydrate reserve. The cells were prepared by freeze-substitution (see Chapter 2) so the lipids have been lost by extraction with organic solvent (regions marked*). (Courtesy of M.S. Fuller; from Cho & Fuller, 1989.)

zoospore ventral groove. They also contain a protein which is released during encystment, but it is deposited locally between the cyst and the surface on which encystment has occurred, and it is thought to act as an adhesive. A newly encysted zoospore has no cyst wall, only an amorphous gly-

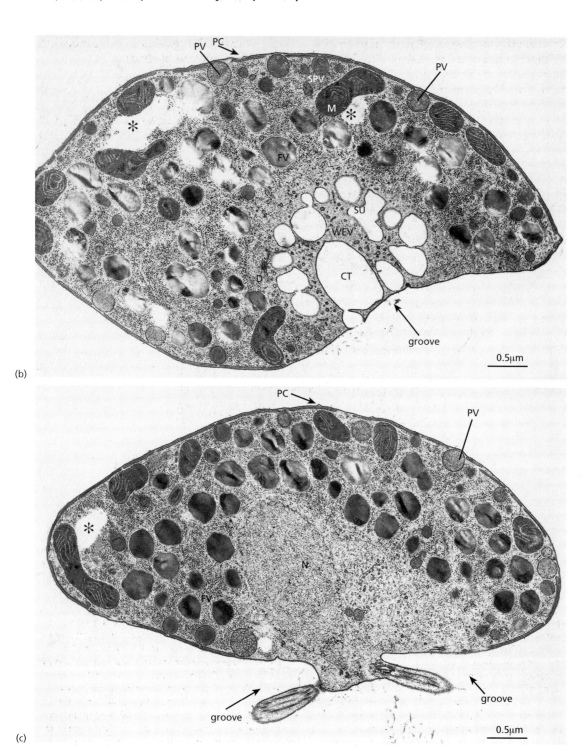

(b)

(c)

Fig. 9.9 *Continued.*

coprotein coat deposited by the dorsal vesicles (Fig. 9.10). But, a true wall is synthesized beneath the cyst coat in the first few minutes of encystment. It is derived from the peripheral membrane cisternae beneath the plasma membrane, and at this stage the water-expulsion vacuole disappears because it no longer needs to serve an osmoregu-latory function. Details of all these features can be found in Hardham *et al.* (1994).

The zoospore homing sequence

The events in the homing sequence (see Fig. 9.8) are of critical importance for infection from

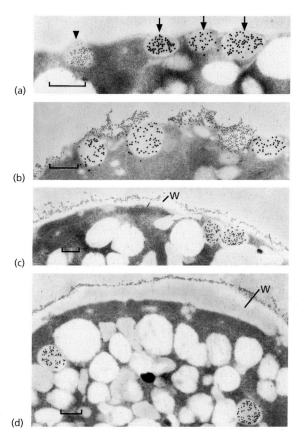

Fig. 9.10 Electron micrographs of zoospores of *Phytophthora cinnamomi* subjected to immunogold labelling with monoclonal antibodies (mAbs) specific to the contents of different peripheral vesicles. The micrographs were taken at different stages of zoospore encystment. (a) Outer region of a motile zoospore, showing two classes of peripheral vesicles: the contents of large peripheral vesicles bind to a a specific mAb that has been complexed to large (18 nm) gold particles (arrows), whereas the contents of small dorsal vesicles bind to a different mAb complexed to 10 nm gold particles (arrowhead). (b) Zoospore labelled at 1 min after addition of pectin to induce encystment. Most of the small dorsal vesicles have released their contents onto the cell surface by exocytosis; the large peripheral vesicles are still present below the cell surface. (c) Spore labelled at 5 min after induction of encystment. The spore has started to synthesize a cell wall (W), the surface of which is coated with material from the small dorsal vesicles. (d) Spore labelled at 10 min after induction of encystment. The cell wall is now thicker and coated with the former contents of the small dorsal vesicles; the large peripheral vesicles have migrated towards the centre of the cyst; their contents are thought to be a protein store for use during cyst germination and germ-tube growth. Bars = 0.5 μm. (Photographs courtesy of F. Gubler and A. Hardham; (a) from Hardham, 1995; (b–d) from Gubler & Hardham, 1988.)

zoospores of plant-pathogenic fungi, and have been studied in detail for *Pythium* and *Phytophthora* spp. (see Deacon & Donaldson, 1993).

Zoospore taxis can be studied using capillaries filled with attractant substances. Often the zoospores show chemotaxis to some individual sugars and amino acids, or to volatile compounds such as ethanol and aldehydes which might be released as fermentation products of roots in moist soil conditions. But, the responses usually are stronger to mixtures of compounds, like the mixtures in seed and root exudates. Most of these fungi show taxis to the roots of host and non-host plants, but a few interesting examples of host-specific taxis have been reported. The host-specific pathogen *Phytophthora sojae* shows chemotaxis *in vitro* to the flavonoids daidzein and genistein of its soybean host, and *Aphanomyces cochlioides* shows chemotaxis to the flavonoid 'cochliophilin A' from spinach plants. It should not be assumed that these compounds are the only factors involved in attraction to the host roots, but the findings are notable because they parallel the behaviour of *Rhizobium* spp., which show *in vitro* chemotaxis to the specific flavonoids of their hosts.

Zoospore encystment seems to occur by recognition of a host surface component, but often can be triggered by pectin and other polyuronates (e.g. alginate) *in vitro*. There can also be a degree of host-specificity at this stage, because the *Pythium* spp. that parasitize the grass family (Gramineae) show significantly more encystment on grass than on dicotyledonous roots.

Zoospores always **orientate precisely** during encystment, with the ventral groove located next to the host. The zoospores of *Pythium* and *Phytophthora* also have a pre-determined point of germ-tube outgrowth, close to the point where the flagella are attached in the ventral groove. So, the precise orientation of encystment on a host surface serves to locate this germination site next to the host, for rapid infection. It also ensures that the adhesive (released from ventral vesicles) will be deposited next to the host surface. Receptors on the flagella probably are involved in this process, because monoclonal antibodies that bind to the two flagella are found to cause rapid encystment *in vitro*.

Cyst germination can be induced by a few individual amino acids and sugars *in vitro*, but it is

suggested to be an autonomous process on a host, triggered by events that occur in the earliest stages of encystment. In particular, zoospores seem to release a large amount of calcium, from intracellular stores, in the first few minutes of encystment, and they also rapidly absorb or reabsorb calcium. Zoospores that are encysted artificially by mechanical agitation *in vitro* will only germinate if they are brought rapidly into contact with a surface such as a glass slide, on which they adhere. If they are kept away from a surface then they need to be supplied with high exogenous levels of calcium for germination. So, it is suggested that, in normal conditions, the calcium released during encystment becomes trapped in the adhesive (against a host surface) and is reabsorbed to trigger germination (see Fig. 9.8). If the zoospore encysts away from a surface (e.g. by agitation *in vitro*) then the released calcium diffuses away from the spore and a high level of exogenous calcium must then be supplied to trigger germination. This has parallels with the model of fungistasis discussed earlier, where spores were suggested to remain dormant if the small amounts of nutrients that they continuously leak into their vicinity are removed by continuous leaching. The germination of zoospore cysts is remarkable because it occurs so rapidly—often within 20 or 30 min of the zoospore settling on a surface. This is the fastest known rate of spore germination among fungi, and it illustrates how zoospores are ideally adapted to locate their hosts or other substrata and then show a rapid response sequence to initiate a new phase of growth.

Zoospores as vectors of plant viruses

About 20 plant viruses are currently known to be transmitted by fungal zoospores (Table 9.2), and in many cases this is their main or only means of transmission (Hiruki & Teakle, 1987). It testifies to the efficiency of zoospores in locating a host. However, the vectors belong to just three genera, *Olpidium* (chytridiomycota), *Polymyxa* and *Spongospora* (plasmodiophorids, similar to *Plasmodiophora brassicae* in Chapter 1, Fig. 1.4). These fungi are common and usually symptomless parasites of roots, and their significant feature as virus vectors is that the zoospores encyst and then germinate to release a naked protoplast into the plant. The fungi (e.g. oomycota) that produce walled hyphae from

Table 9.2 Plant viruses that are transmitted by fungal zoospores. (Based on data in Hiruki & Teakle, 1987; Brunt & Richards, 1989.)

Virus type and features	Examples	Host	Vector
Furoviruses			
Straight tubular particles, 250–300 + 100–150 × 20 nm	Soil-borne wheat mosaic	Wheat, barley	*Polymyxa graminis*
	Beet necrotic yellow vein	Sugar beet, spinach	*P. betae*
Single-stranded RNA	Potato mop top	*Solanum* spp.	*Spongospora subterranea*
Genome divided between more than one particle	Peanut clump	Peanut	*P. graminis*
	Oat golden stripe	*Avena* spp.	*P. graminis*
Always vectored by fungi	Broad bean necrosis	*Vicia faba*	*P. graminis*
Barley yellow mosaic type			
Filamentous particles, 350–700 × 13 nm	Barley yellow mosaic	*Hordeum*	*P. graminis*
	Wheat yellow mosaic	*Triticum*	*P. graminis*
Always vectored by fungi	Wheat spindle streak	*Triticum*	*P. graminis*
	Oat mosaic	*Avena*	*P. graminis*
	Rice necrosis mosaic	*Oryza*	*P. graminis*
Tobacco stunt type			
Straight tubular particles, 200–375 × 22 nm	Tobacco stunt	*Nicotiana* spp.	*Olpidium brassicae*
	Lettuce big vein	Lettuce	*O. brassicae*
Double-stranded RNA			
Characteristically vectored by fungi			
Tobacco necrosis type			
Isometric particles, 26–30 nm	Tobacco necrosis	Tulip, *Solanum*, *Vicia*, many other plants	*O. brassicae*
Single-stranded RNA, full genome in one particle	Cucumber necrosis	Cucumber	*Olpidium* spp.
Various means of transmission, not only by fungi	Melon necrotic spot	Melon, cucumber	*Olpidium radicale*

zoospore cysts seem unable to act as virus vectors.

There are different degrees of specialization in these virus–vector relationships. *Olpidium* spp. usually transmit isometric viruses such as cucumber necrosis virus and tobacco necrosis virus which have additional (and perhaps more important) modes of transmission. The zoospores acquire these viruses when they swim in virus-contaminated soil, and this can be blocked experimentally by adding virus-specific antiserum, indicating that the binding is mediated by a zoospore surface component. The viruses can also be removed easily by mild chemical treatments that probably remove the zoospore surface component. In any case, the virus presumably remains on the plasma membrane when the zoospore encysts, and will later be carried on the membrane of the protoplast that enters the host.

In contrast to this, the zoospores of *Polymyxa* and *Spongospora* cannot acquire viruses by swimming in virus suspensions, but the fungus acquires the virus inside the co-infected plant tissues. Many of these viruses are of a distinct type, termed **furoviruses** (fungally transmitted rod-shaped viruses), and they have no other natural means of transmission. They include some of the most economically important soil-borne viruses, such as beet necrotic yellow vein virus, soil-borne wheat mosaic virus and potato mop-top virus. Other fungally transmitted viruses, such as the filamentous types (Table 9.2), show some affinities to furoviruses but are excluded from this group by their particle shape. Many details of these virus–vector relationships remain unclear, but there is no evidence that the viruses multiply within the fungal vectors. They are, however, carried internally in the resting spores produced by the fungi at the end of their parasitic phase.

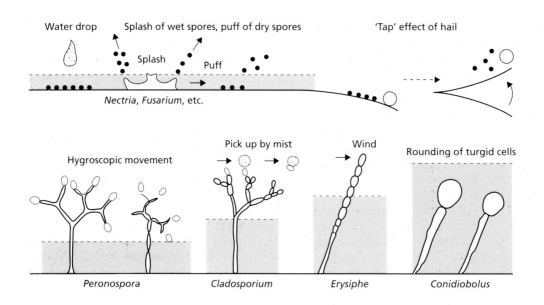

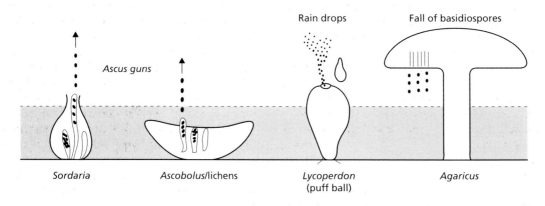

Fig. 9.11 Various mechanisms of spore liberation through a boundary layer of still air (shown by shading).

These spores can persist in soil for 20 years or more, and only a small proportion of them can be induced to germinate at any one time. This means that, once established in a field site, the furoviruses are almost impossible to eradicate. Further details can be found in Adams (1991).

Dispersal of air-borne spores

Most terrestrial fungi produce air-borne spores that are dispersed either by wind or by rainsplash.

These are the spores of most significance in plant pathology and for allergies and fungal infections of humans, as will be seen in the sections below. In terms of the mechanisms of aerial dispersal it is customary to distinguish three phases: (i) spore liberation (take-off); (ii) spore dispersal (flight); and (iii) spore deposition (landing).

Take-off

The essential feature of spore liberation is that the spore needs to break free from the **boundary layer** of still air that surrounds all surfaces. Above this layer the air becomes progressively more turbulent in local eddies, until there is net movement of the air mass which can carry the spore to a new site. The depth of the boundary layer can vary from a fraction of 1 mm on a leaf surface on a windy day to 1 m or more on a forest floor on a perfectly calm day. So, the fungi that grow in these different types of environment require different strategies for getting their spores air borne. Some of these strategies are shown in Fig. 9.11. Often they involve adaptations of the spore-bearing structures rather than of the spores themselves.

Fungi of leaf surfaces sometimes produce chains of spores from a basal cell so that the mature spores are pushed upwards through the boundary layer as the chain extends (e.g. *Erysiphe graminis* and other powdery mildew pathogens). The spores then are removed by wind or, sometimes more effectively, by mist-laden air (*E. graminis*). Other spore types are flung off the supporting structures by sudden hygroscopic movements during drying (e.g. *Phytophthora infestans* and downy mildew fungi). Fungi that grow on more rigidly supported surfaces can release the spores by active processes. The asci of many ascomycota function as small guns, shooting ascospores to a 1- or 2-cm distance. Spores can also pop from a supporting structure when an enclosing wall layer suddenly ruptures, like two balloons that pop apart if compressed and suddenly released. Some other fungi are characteristically dispersed by rainsplash. In these cases the spores often are linear or curved (e.g. *Fusarium*) and are produced in mucilage on a pad of tissue (an acervulus) or in a splash cup so that raindrops are caused to fragment on impact and rebound as many tiny droplets which can then be carried by wind. In puffballs (basidiomycota) the mature basidiospores are enclosed in a papery fruitbody with an apical pore so that raindrops 'puff' the spores into the air.

Toadstools display a different strategy from all those above, made necessary by the deep layer of still air on a woodland floor or in grass turf where the fairy-ring fungi grow on the 'thatch' of dead tissue at soil level. The toadstool projects into turbulent air and drops the spores from the gills or pores (Fig. 9.11). Much of the variation in shape and size of toadstools is related to this strategy — those with thick, rigid stipes (e.g. *Boletus*, *Amanita*) and the large woody brackets of tree-rotting fungi (e.g. *Ganoderma applanatum*) have very closely spaced and deep gills or pores, just wide enough for spores to be popped from the basidia and then to fall vertically into turbulent air. Toadstools with thin, bendable stipes (e.g. *Marasmius oreades*, the common fairy-ring fungus of turf) have widely spaced, shallow gills to ensure that the spores fall free.

This is more than just a catalogue of examples because it demonstrates how fungi have an **integrated lifestyle**. The only reason for producing a fruitbody is to disperse the microscopic spores, and the only reason for producing a massive fruitbody from the microscopic hyphae is to overcome the constraints to spore dispersal imposed by a boundary layer. Furthermore, the only reason why this constraint exists for the toadstool-forming basidiomycota is because the hyphae exploit a particular resource, as degraders of wood, leaf-litter, etc., or as mycorrhizal fungi on tree roots. The somatic hyphae, therefore, must pool resources to build a massive fruitbody, and this, in turn, means that these fungi must exploit a substrate resource that is large enough to support a mass of hyphae. Given these points, we can understand why many basidiomycota do not have asexual spores — the supporting structures would be too small to release spores through the boundary layer. Instead, these fungi produce large numbers of sexual spores for dispersal (the normal function of asexual sporulation). The dikaryotic system serves this role by ensuring that a single fusion of compatible monokaryons leads to conjugate nuclear division and, ultimately, to the production of millions of basidia that produce basidiospores (see Fig. 1.10).

Flight

The fate of spores in the air is determined largely by forces over which fungi have no control, but at least two features of spores are significant for long-distance dispersal: (i) resistance to desiccation which can be conferred by hydrophobins in the walls; and (ii) resistance to ultraviolet rays, conferred by wall pigmentation. Thus, the hyaline, thin-walled conidia of *E. graminis* or the wind-borne sporangia of *Phytophthora infestans* (potato blight) are known to remain viable for only a short time on bright, cloudless days, whereas the pigmented uredospores of rust fungi (e.g. *Puccinia graminis*) and conidia of *Cladosporium* can remain viable for days or even weeks in air.

The use of spore-trapping devices on the outside of aircraft has shown clear evidence of intercontinental spread of fungi. Figure 9.12 shows an example, where spores were carried across the North Sea from England to Denmark on the westerly winds. From knowledge of wind speeds on the days preceding the flight it was possible to distinguish between spores released on different days in England and also to distinguish between spores released in daytime (e.g. *Cladosporium*) and those released at night (the pink yeast *Sporobolomyces*, and various ascospores). Such long-distance dispersal is not particularly significant for the general needs of a fungus, because of the substantial dilution effect. It can, however, be significant for plant disease epidemiology, as illustrated by black stem rust of wheat (*Puccinia graminis*) in North America. The uredospores of this fungus have been shown to travel over 3000 km, in stages from the extreme south of the USA to the northern states and even Canada, because wheat is grown at different times of the year in these different climatic zones. When early studies in the northern USA showed that *P. graminis* overwinters on its alternate host, barberry (see Chapter 1, Fig. 1.10), a barberry eradication programme was initiated to try to reduce the infection of wheat. But, the progressive northwards spread of air-borne spores rendered this ineffective. In a different context, the rust fungi are notorious for their ability to evolve new pathogenic races to overcome resistance genes that have been bred into crop plants (see Chapter 12). These pathogenic races can spread similarly between different cropping regions, negating the efforts of the breeding programmes.

Landing

Spores are removed from the air in three major ways: by sedimentation, impaction and wash-out. In each case the shape, size and surface properties of spores have major effects.

Sedimentation

Spores settle out of the air in calm conditions, and the heavier (larger) spores settle faster than lighter spores. The sedimentation rates have been measured in closed cylinders and, except for unusually shaped spores for which correction factors are needed, the rates are found to agree closely with Stokes's Law for perfect spheres of unit density (1.0). The relevant equation is:

$$V_t = 0.0121r^2$$

where V_t is the terminal velocity (in centimetres

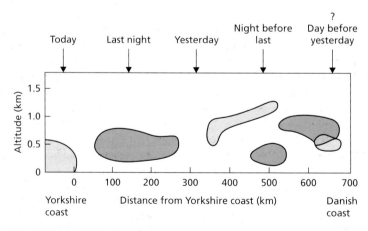

Fig. 9.12 Positions of peak concentrations of *Cladosporium* spp. spores (light shade) and damp air spore types (dark shade) at various altitudes over the North Sea down-wind of the English coast, 16 July 1964. Interpreted as windborne remnants of alternating day and night spore clouds liberated from the English land surface. (From Lacey, 1988, based on work by P.H. Gregory and J.L. Monteith.)

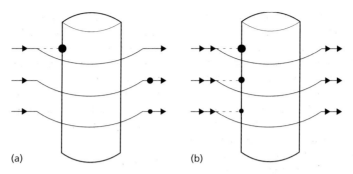

(a)   (b)

Fig. 9.13 Illustration of the relationship between spore size, wind speed and impaction of spores onto a cylinder. (a) At low wind speeds only the largest (heaviest) spores impact. (b) At higher wind speeds progressively smaller (lighter) spores impact.

per second) and r is the radius (in micrometres). For example, spores of the cereal smut pathogen *Tilletia caries* (17-μm diameter) had a measured V_t of 1.4 cm s^{-1}, whereas puffball spores (*Bovista plumbea*, 5.6 μm diameter) had a V_t of 0.24 cm s^{-1}. Differences of this order probably have relatively little effect in outdoor environments, so that spores of all types would remain air borne or settle according to the prevailing conditions. However, small differences in sedimentation rate can be significant in buildings and in the respiratory tract, discussed later.

Impaction

Impaction is one of the major mechanisms by which large spores are removed from the air, and it has special significance for plant pathogens. As shown in Fig. 9.13, when spore-laden air moves towards an object (or vice versa) the air is deflected around the object and tends to carry spores with it. But, the momentum (mass × velocity) of a spore will tend to carry it along its existing path for at least some distance. Three points arise from this:
1 at any given air speed the larger (heavier) spores have more chance of impacting than the smaller spores;
2 as the air speed increases, so progressively smaller spores can impact;
3 as the size of the receiving object increases, so the deflection of air is greater and this reduces the chances of impaction.

These points are directly relevant to spores in nature. All the fungi that infect leaves or that grow on leaf surfaces have large spores, with sufficient momentum to impact at normal wind speeds (up to 5 m s^{-1}). Examples include the common phyllosphere fungi *Cladosporium herbarum* (spores 8–15 μm diameter) and *Alternaria* spp. (about 30 μm diameter), and the leaf-infecting pathogens *E. graminis* (30 μm diameter) and *P. graminis* (40 μm diameter). In contrast, the typical soil fungi such as *Penicillium* and *Aspergillus* spp. have spores of about 4 μm diameter, too small to impact but large enough to sediment out of the air in calm conditions.

The effect of the receiving object on efficiency of spore impaction was demonstrated by Carter (1965). He placed the young branches of apricot trees in wind tunnels and exposed them to air containing spore clusters of the pathogen *Eutypa armeniacae* (ascomycota). As shown in Fig. 9.14, at all wind speeds the spore clusters impacted best on the narrow leaf petioles (about 1–2 mm diameter) and less well on the broader leaf blades or woody stem, and the efficiency of impaction increased as the wind speed increased. It might be considered that the impaction efficiencies were quite low in all cases, but the results should be seen in relation to an apricot orchard where the spores that do not impact on one shoot system would impact on another. So, Carter estimated that most of the spore clusters would be removed from the air as it travelled through an orchard at 2 m s^{-1}, which was the speed measured in field conditions. This example raises some further points, because the experiments were carried out with spore clusters, each composed of eight ascospores held together by mucilage. This is how *Eutypa* discharges its spores in nature, to increase the size for efficiency of impaction and so that the spores stick to the

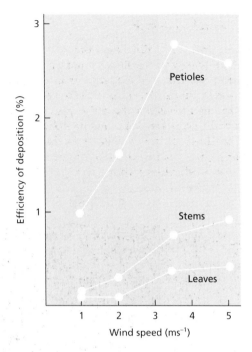

Fig. 9.14 Efficiency of impaction of ascospore tetrads of *Eutypa armeniacae* on leaf blades, petioles and branch stems of apricot shoots at different wind speeds. Most impaction occurs on the narrower objects (leaf petioles) and in all cases the efficiency of impaction is increased as the wind speed increases. (From Carter, 1965.)

plant by mucilage. *Eutypa* is a wound pathogen, which commonly infects grapevines and apricots through natural wounds or pruning wounds. So, after impaction, the fungus relies on secondary spread by rain or irrigation water, which disperses the mucilage and carries the individual spores down the plant to wound sites.

Wash-out

Rain is the most effective mechanism of removing spores from air, because even light, steady rain will remove almost all suspended particles. However, the spore surface properties then come into play. Wettable spores become incorporated in the raindrops and finally come to rest where the water does—by spread across a wettable surface or dripping from a non-wettable one. Non-wettable spores remain on the surface of the drop, so a raindrop landing on a non-wettable surface such as a leaf cuticle will roll off it but leave the spores

behind as a trail. This is easily demonstrated in a laboratory by rolling a water drop over a *Penicillium* colony, when the drop becomes coated with the non-wettable spores. If the spore-laden drop is then rolled across an empty plastic Petri dish, it loses its surface coating of spores which are seen as a trail across the plastic. Non-wettable spores that are left on a surface in this way can become air borne again when another water drop lands on the surface. As shown in Fig. 9.11, the water drop 'splodges' sideways when it contacts a flat surface, causing a local puff of air that disturbs the boundary layer and can get spores air borne again. Hailstones hitting a leaf surface can also get spores air borne again, by 'tapping' the surface, causing it to reverberate (see Fig. 9.11).

Spores in the respiratory tract

The respiratory tract of humans and other animals is a natural spore-trapping device, as people with allergies know only too well. The respiratory tract is also the route of entry of opportunistic fungal pathogens, discussed in Chapter 13. The mechanisms of spore deposition in these cases are similar to those discussed earlier—impaction, sedimentation and **boundary layer exchange**.

During each intake of breath the air speed is fastest in the nose, trachea and main bronchi (about $100\,cm\,s^{-1}$), and it diminishes with successive branching of the bronchioles. Therefore, impaction will only occur in the upper respiratory tract. The hairs in the nose are narrow and covered with mucilage, making them highly efficient for intercepting the larger fungal spores and pollen grains by impaction. Some of these air-borne particles cause rhinitis and other typical symptoms of hay-fever. All other particles are too small to impact, so they are carried deep into the lungs and reach the terminal bronchioles and alveoli. This applies to most particles of $5\,\mu m$ diameter or less. Many of them are expelled again, but some are large enough ($2\text{–}4\,\mu m$) to settle onto the mucosal membranes by sedimentation during the brief period (usually less than 1 s) when the air in the alveoli is static between inhalation and exhalation. Particles even smaller than this can be trapped by boundary layer exchange — a process involving small eddies that carry small particles into the boundary layer. The air-borne spores of potentially pathogenic fungi such as *Aspergillus fumigatus*, *Blastomyces* and *Histoplasma* spp. settle in the alveoli by sedimentation, as do the spores of some *Aspergillus* and *Penicillium* spp. (e.g. *A. clavatus*) that cause acute allergic alveolitis. Once a spore has been deposited in the alveoli it persists until it is engulfed by a phagocyte. In contrast, the upper region of the respiratory tract is lined with ciliated epithelium which continuously sweeps mucus upwards and removes any particles deposited there.

Air-sampling devices

The importance of air-borne spores in relation to crop pathology, human ailments and air quality in general has led to the design of air-sampling devices for monitoring of spore loads. We end this chapter by considering the main types of device and the principles on which they operate.

The simplest type of device is the **rotorod sampler**, used mainly as an experimental tool. It consists of a thin U-shaped metal rod, attached to a

(a)

(b)

(c)

(d)
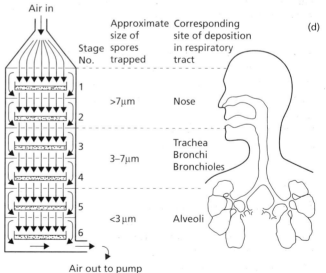

Fig. 9.15 Air-sampling devices. (a) Burkard spore trap in assembled form, ready for use. Air is drawn through the narrow slit orifice (arrowed) which is protected by a rain shield above. (b) The sampler has been opened to show the slowly central rotating reel which is covered with adhesive tape (arrowed) to catch spores drawn in through the slit orifice. (c) Anderson sampler mounted in a field site with a wind vane. (d) Diagrammatic representation of the Anderson sampler, showing how the impaction of spores of successively smaller sizes on agar plates down the cylinder simulates the deposition of spores of different sizes in the human respiratory tract.

spindle. The upright arms revolve rapidly (about 2000 rev min^{-1}) by a battery-driven electric motor. The arms are covered with transparent adhesive tape so that spores impact and are retained on them. Then the tapes are removed and examined microscopically. This apparatus is highly efficient at collecting relatively large particles, in the range 10–20 µm diameter, including pollen grains and the spores of most leaf fungi. It is light and portable, so it can be used to 'home-in' on the source of a particular type of spore, by making successive samplings in a small area. For example, it was used in a field site in southern England to find the source of spores of *Pithomyces chartarum*, a toxigenic fungus that causes facial eczema of sheep (see Chapter 10).

The **Burkard spore trap** (Fig. 9.15a,b) is a continuous monitoring device that works on the same principle as the rotorod sampler. It is used commonly in crop epidemiology and for monitoring allergen levels, including the pollen counts announced on radio and television. It consists of a drum with a narrow slit orifice, through which air is sucked at high speed by a motor. Inside the drum is a reel covered with transparent sticky tape, and this rotates slowly past the air-intake orifice on a daily or weekly cycle. Thus, the spores present in the air at any one time impact on a small zone of the tape, giving a continuous daily or weekly record. Again, the method only detects the larger particles, but these include the common allergens and leaf-infecting pathogens.

Perhaps the most ingenious air-sampling device is the **Anderson sampler** (Fig. 9.15c,d) which is claimed, with some justification, to simulate the human respiratory tract. It consists of a stack of perforated metal plates which fit together to form an air-tight cylinder, with space for open agar-filled Petri dishes to be inserted between them. The metal plates have the same number of holes, but these holes become progressively smaller down the stack. Air is drawn in at the top, and down through the apparatus by a motor-driven suction pump at the base.

The air striking the first agar dish is travelling at low speed, so large particles impact on this. But, the air speed increases progressively as it is forced to pass through successively smaller perforations, so that progressively smaller particles impact on the plates lower down the column. After it has run

for an appropriate time, depending on the spore load, the apparatus is dismantled and the agar dishes are incubated to identify the organisms that grow. The top plates bear colonies of the common 'impacting' fungi, some of which cause hay-fever or similar conditions. The middle plates bear colonies of fungi such as *Aspergillus* and *Penicillium* that can be opportunistic pathogens or cause occupational allergies. The lowest plates bear colonies of actinomycetes such as *Thermoactinomyces vulgaris* and *Faenia rectivirgula*, which cause the debilitating 'farmer's lung' allergic condition. These have extremely small spores (1–2 µm) and would not be detected by normal impaction methods. Thus, the Anderson sampler achieves a size separation of particles, similar to the respiratory tract, but it achieves this by impaction, not by the combination of impaction, sedimentation and boundary layer exchange which occurs in the respiratory system.

References

Adams, M.J. (1991) Transmission of plant viruses by fungi. *Annals of Applied Biology*, **118**, 479–92.

Blakeman, J.P. (1981) *Microbial Ecology of the Phylloplane*. Academic Press, London.

Brasier, C.M. (1995) Episodic selection as a force in fungal microevolution, with special reference to clonal speciation and hybrid introgression. *Canadian Journal of Botany*, **73**, S1213–21.

Brunt, A.A. & Richards, K.E. (1989) Biology and molecular biology of furoviruses. *Advances in Virus Research*, **36**, 1–32.

Carter, M.V. (1965) Ascospore deposition of *Eutypa armeniacae*. *Australian Journal of Agricultural Research*, **16**, 825–36.

Cho, C.W. & Fuller, M.F. (1989) Ultrastructural organization of freeze-substituted zoospores of *Phytophthora palmivora*. *Canadian Journal of Botany*, **67**, 1493–9.

Coley-Smith, J. (1987) Alternative methods of controlling white rot disease of *Allium*. In: *Innovative Approaches to Plant Disease Control* (ed. I. Chet), pp. 161–77. Wiley, New York.

Cooke, R.C. (1977) *The Biology of Symbiotic Fungi*. Wiley, Chichester.

Deacon, J.W. & Donaldson, S.P. (1993) Molecular recognition in the homing responses of zoosporic fungi, with special reference to *Pythium* and *Phytophthora*. *Mycological Research*, **97**, 1153–71.

Deacon, J.W. & Fleming, L.V. (1992) Interactions between ectomycorrhizal fungi. In: *Mycorrhizal Functioning* (ed. M.F. Allen), pp. 249–300. Macmillan Press, New York.

Deacon, J.W., Donaldson, S.J. & Last, F.T. (1983)

Sequences and interactions of mycorrhizal fungi on birch. *Plant and Soil*, **71**, 257–62.

Fox, F.M. (1986) Groupings of ectomycorrhizal fungi on birch and pine, based on establishment of mycorrhizas on seedlings from spores in unsterile soils. *Transactions of the British Mycological Society*, **87**, 371–80.

Gregory, P.H. (1973) *Microbiology of the Atmosphere*, 2nd edn. Leonard Hill, Aylesbury.

Gubler, F. & Hardham, A.R. (1988) Secretion of adhesive material during encystment of *Phytophthora cinnamomi* zoospores, characterized by immunogold labelling with monoclonal antibodies to components of peripheral vesicles. *Journal of Cell Science*, **90**, 225–35.

Hardham, A. R. (1995) Polarity of vesicle distribution in oomycete zoospores: development of polarity and importance for infection. *Canadian Journal of Botany*, **73** (Suppl.), S400–407.

Hardham, A.R., Cahill, D.M., Cope, M., Gabor, B.K., Gubler, F. & Hyde, G.J. (1994) Cell surface antigens of *Phytophthora* spores: biological and taxonomic characterization. *Protoplasma*, **181**, 213–32.

Hiruki, C. & Teakle, D.S. (1987) Soil-borne viruses of plants. In: *Current Topics in Vector Research*, Vol. 3 (ed. K.F. Harris), pp. 177–215. Springer-Verlag, New York.

Hsu, S.C. & Lockwood, J.L. (1973) Soil fungistasis: behavior of nutrient-independent spores and sclerotia in a model system. *Phytopathology*, **63**, 334–7.

Ingold, C.T. (1971) *Fungal Spores and their Dispersal*. Clarendon Press, Oxford.

Lacey, J. (1988) Aerial dispersal and the development of microbial communities. In: *Micro-organisms in Action: Concepts and Applications in Microbial Ecology* (eds J.M. Lynch & J.E. Hobbie), pp. 207–37. Blackwell Scientific Publications, Oxford.

Lockwood, J.L. (1977) Fungistasis in soils. *Biological Reviews*, **52**, 1–43.

Madden, J.L. & Coutts, M.P. (1979) The role of fungi in the biology and ecology of woodwasps (Hymenoptera: Siricidae) In: *Insect–Fungus Symbiosis* (ed. L.R. Batra), p. 165. Allanheld, Osmun & Co., New Jersey.

Marx, D.H. & Cordell, C.E. (1989) The use of specific ectomycorrhizas to improve artificial forestation practices. In: *Biotechnology of Fungi for Improving Plant Growth* (eds J.M. Whipps & R.D. Lumsden), pp. 1–25. Cambridge University Press, Cambridge.

Mitchell, A.G. & Brasier, C.M. (1994) Contrasting structure of European and North American populations of *Ophiostoma ulmi*. *Mycological Research*, **98**, 576–82.

Moss, S. (1986) *The Biology of Marine Fungi*. Cambridge University Press, Cambridge.

Powell, M.J. (1994) Production and modifications of extracellular structures during development of chytridiomycetes. *Protoplasma*, **181**, 123–41.

Reichle, R.E. & Fuller, M.F. (1967) The fine structure of *Blastocladiella emersonii* zoospores. *American Journal of Botany*, **54**, 81–92.

Webster, J. (1980) *Introduction to Fungi*, 2nd edn. Cambridge University Press, Cambridge.

Chapter 10
Fungal decomposer communities

Previous chapters have dealt with the nutrient requirements of fungi (see Chapter 5) and environmental conditions for fungal growth (see Chapter 7). Now we turn to the decomposer activities of fungi in natural environments, where a single food base (a substratum or resource) often contains several different food sources (substrates), where environmental conditions seldom are constant, and where the activities of one organism will be influenced by others. We must integrate all these factors in order to understand the activities of fungi in nature.

This complexity can be tackled in essentially two ways. First, by development of concepts and models, following the approaches that ecologists have used for plant and animal communities. This is more relevant to ecological theory than to our present purposes, but is discussed briefly here and at length in Carroll and Wicklow (1992). The second way is more empirical: to identify (i) the activities of individual fungi in decomposition processes; (ii) how these activities change with time and in different conditions; and (iii) how fungi interact in clearly defined habitats and experimental model systems. In this way we can build an understanding of how fungi behave in complex communities. We begin this process in this chapter by discussing the major groupings of decomposer fungi and their roles in some well-defined types of organic matter. We continue in Chapter 11, where we consider fungal interactions.

Behavioural groupings of decomposer fungi

The niche concept

It is a basic ecological tenet that no two organisms have exactly the same niche. In other words, for every organism there is a particular set of conditions that favour it more than other organisms. However, it is almost impossible to define a niche in precise terms because it includes a combination of factors. For a fungus this combination will include the type of substrate utilized (food source), environmental factors such as temperature, pH, etc., and any influences of other organisms.

We can identify at least the limits of the niche of a particular fungus by pure culture studies — the temperatures at which it can grow, the substrates it can utilize, etc. But, in nature there are many fungi and other micro-organisms with overlapping ranges, so the niche of a fungus may be only a small part of its potential range. A simple demonstration of this was shown in Fig. 7.1 (see Chapter 7), where the take-all fungus was seen to grow best in natural conditions at temperatures lower than its optimum in pure culture, due to interaction with soil bacteria which were favoured at higher temperatures.

Life-history strategies

Several workers have tried to define the activities of fungi in terms of **life-history strategies** following the ecological theories developed by animal and plant ecologists. We can illustrate this by considering two extremes.

1 Some fungi can use only simple monomeric nutrients (sugars and amino acids, etc.) which almost all other organisms can use. So, these fungi must have evolved a strategy for rapid exploitation of these substrates: they produce many dispersal spores which maximize the chances of finding these substrates; they grow rapidly to exploit the substrate; and after a short time they

produce further dispersal spores or resting stages. Several species of *Mucor* and other zygomycota fall into this category.

2 At the other extreme, some fungi have the enzymic machinery to degrade very complex polymers that few other fungi can utilize. So they can grow more slowly (although not necessarily), have an extended phase of growth on the substrate during which they build up a substantial mycelial network, and then use the mycelial resources to produce a few, large fruiting bodies. Examples of this include many of the wood-rotting basidiomycota.

Fungi and other organisms of the first type are said to be **r-selected** (ruderals) whereas those of the second type are said to be **K-selected**. In truth, however, there is an r–K continuum, because different organisms exhibit different degrees of adaptation for exploiting a resource. This scheme has been extended by adding a third component (**stress-tolerance**) so that, in theory, any organism might be defined in terms of its degree of adaptation to ruderal behaviour (termed competition for fungi), K-selected behaviour (termed combat for fungi) and stress-tolerance. One way of visualizing this is shown in Fig. 10.1. It serves an important function in emphasizing that every fungus has a particular combination of features that fit it for a particular niche. But, it remains a theoretical approach, almost impossible to translate into practice, partly because the three axes in Fig. 10.1 cannot be quantified, and partly because 'stress' is a meaningless concept—a mesophilic fungus, for example, is stressed by high temperatures whereas a thermophilic fungus is stressed by normal temperatures, so which of these fungi is the more stress-tolerant? Andrews (1992) has discussed the application of these and other general ecological concepts to fungi.

Substrate groupings

Garrett (1951) proposed a scheme for categorizing the decomposer soil fungi into four broad groups according to their patterns of substrate utilization. This scheme has been followed by many mycologists and expanded to fungi in other types of habitat (e.g. Hudson, 1968).

1 Saprophytic (saprotrophic) sugar fungi (e.g. *Mucor*), characterized by their dependence on simple organic compounds, and their adaptations rapidly to exploit these.

2 Cellulolytic fungi (and equivalent types), able to compete for, and degrade, the common polymers and with a life history to match this.

3 Lignin-degrading fungi (and equivalent types), able to degrade the most complex polymers that most other fungi cannot use.

4 Secondary saprophytic (saprotrophic) sugar fungi, which co-exist with the polymer-degrading fungi and gain sugars or other simple organic nutrients from the activities of other fungi.

Clearly, this scheme is too simple to accommodate all the activities of decomposer fungi, but it emphasizes the primary requirement of any fungus — an organic substrate that it can utilize, just as light is the primary requirement of a plant.

Behavioural groupings

For purposes of this chapter we can combine the features in the schemes above, to recognize **five behavioural groupings** of decomposer fungi. These groupings take into account three factors:

1 the typical substrates utilized;

2 the typical environmental conditions for growth;

3 the interactions with other organisms.

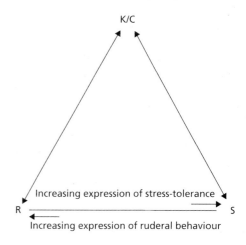

Fig. 10.1 A method of defining the life-history strategy of a fungus according to the degree of expression of the three 'primary' strategies: ruderal behaviour (R), K-selected or combative behaviour (C) and stress-tolerance (S).

The behavioural groups are summarized in Table 10.1, and discussed below with particular reference to decomposition of plant material. The fungi in these groupings often occur in an overlapping sequence (Fig. 10.2).

Group 1: pathogens and weak parasites

Pathogens and weak parasites colonize living plant tissues or tissues that are beginning to senesce. They gain an initial advantage over purely saprotrophic fungi because they tolerate or overcome the host defence factors (host stress-tolerance) or other stresses (see below). They can remain active for a time after the tissues have died, usually by exploiting the more readily utilizable nutrients, so they are truly a part of the decomposer sequence. However, they are soon replaced by the more specialized saprotrophs when the 'stress' factors that favoured them are removed.

The fungi that commonly grow on the surfaces of living leaves or fruits are good examples of this behavioural category. They include *Alternaria* spp., *Cladosporium herbarum*, *Aureobasidium pullulans*, *Epicoccum purpurascens* and *Botrytis cinerea* (all deuteromycota), and pigmented yeasts such as *Rhodotorula* (ascomycota) and *Sporobolomyces roseus* (basidiomycota). They utilize the simple sugars and other soluble materials that leak from the living plant tissues or that are exuded in aphid 'honeydew'. They also colonize senescing leaves and exploit the more readily available nutrients. Consistent with this, they seldom grow well on 'crystalline' cellulose (filter paper, cotton fabric) in culture, but many of them can grow on soluble cellulose (see Chapter 5). Some of them are highly tolerant of alternating wetting and drying (see Chapter 7) and have pigmented walls for protection against ultraviolet damage. They can remain active on fallen leaves, but are soon replaced by other fungi if the leaves are permanently moist.

Table 10.1 Characteristics of the behavioural groupings of decomposer fungi.

Group	Features
1 Pathogens and weak parasites	(i) Grow initially by tolerating host resistance factors or other special conditions (periodic drought stress, etc.) (ii) Generally utilize simple soluble substrates or storage compounds but not structural polymers (iii) Generally poor competitors for dead organic matter
2 Pioneer saprotrophic fungi	(i) Generally utilize simple soluble substrates or storage compounds but not structural polymers (ii) Good competitors, with fast growth, etc., and short life cycles (iii) Cannot defend a resource against subsequent invaders
3 Polymer-degrading fungi	(i) Degrade the main structural polymers (cellulose, hemicelluloses, chitin, etc.) (ii) Have extended growth phase, defending a resource by antibiosis or sequestering mineral nutrients, etc. (iii) Substrate-specialized, and sometimes tolerant of stress factors (extremes of temperature, pH, etc.)
4 Degraders of recalcitrant compounds	(i) Specialized to degrade recalcitrant organic materials (lignin, etc.) and thereby gain access to polymers (cellulose, etc.) complexed with them (ii) Long growth phase, and defend a resource by antagonism or mutual inhibition (deadlock) (iii) Can gain access to mineral nutrients (nitrogen, etc.) that previous colonizers have exploited
5 Secondary opportunists	(i) Nutritionally opportunistic: grow on dead remains of other fungi, insect exoskeletons, etc., or parasitize other fungi, or grow commensally with polymer-degraders (ii) Tolerant of metabolic by-products of other fungi (iii) Often antagonistic

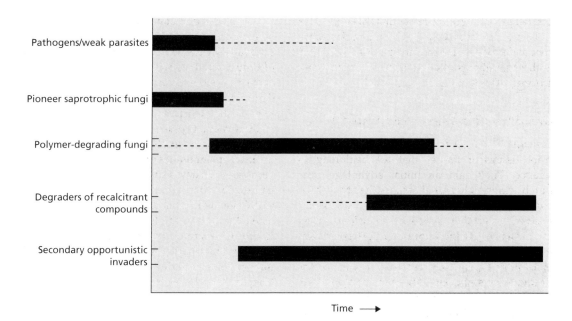

Fig. 10.2 Illustration of the overlapping phases of activity of different types of fungi during decomposition of a material. Main stages of activity are shown, but they might extend in some conditions (broken line).

Some major pathogens have an extended saprotrophic phase in the tissues they colonized while the host was alive. For example, the take-all fungus (*Gaeumannomyces graminis*) and the eyespot fungus (*Pseudocercosporella herpotrichoides*) can persist for 1 or 2 years, respectively, when cereal stem bases are buried in soil; this is how they survive to infect a subsequent crop. They seem to grow slowly but continuously, exploiting cellulose as a food source and deterring other fungi by prior possession. Yet, they have almost no ability to compete for uncolonized crop residues.

Group 2: pioneer saprotrophic fungi

Pioneer saprotrophic fungi colonize fresh plant or animal remains in the absence of 'group 1'. They grow rapidly and utilize the simple, soluble nutrients but usually cannot degrade the more complex structural polymers (cellulose, etc.). In laboratory culture they are found to be intolerant of the antibiotics or growth metabolites of other fungi, and they do not themselves produce antibiotics. So, these pioneer fungi typically have a short exploitative phase, rapidly followed by the production of asexual dispersal spores or sexual resting spores. Thus, this behavioural grouping is characterized by:

1 utilization of simple monomeric substrates;
2 tolerance or intolerance of environmental stress such as temperature, pH, etc.;
3 high competitive ability but poor ability to defend a resource.

Species of *Mucor*, *Rhizopus* and several other zygomycota are typical of this grouping. They are particularly common as primary colonizers of faeces or other animal remains, and also in the rhizosphere (root zone) of plants where they utilize the soluble root exudates. Several *Pythium* spp. also fall in this category because they rapidly colonize fresh plant residues in soil and then convert to resting spores (oospores) within a few days, but they are best known as pathogens of living root tips and other fleshy plant tissues (see Chapter 12).

Group 3: polymer-degrading fungi

Polymer-degrading fungi may colonize at the same time as fungi of 'group 2' but have a more extended growth phase, utilizing major structural polymers such as cellulose, hemicelluloses or chitin, or other polymers such as lipids, proteins, starch, etc. Once established, these fungi tend to defend the resource against potential invaders, either by sequestering a critically limiting nutrient such as nitrogen or by producing inhibitory metabolites (see Chapter 11). They are responsible for the major phase of decomposition of the most abundant natural polymers—cellulose in plant cell walls, chitin in insect exoskeletons, starch or other reserve carbohydrates in seeds and tubers, etc. This behavioural grouping is characterized in general by:

1 utilization of polymeric substrates;
2 tolerance or intolerance of physical environmental stress;
3 ability to defend a resource against potential invaders.

The large number of fungi in this category differ in the types of substrate they utilize, their different environmental requirements and their growth in different phases of a decomposition sequence — one fungus might replace another when a critical nutrient such as nitrogen becomes limiting (see Chapter 11). The following examples illustrate this diversity.

When cereal straw and other cellulose-rich materials are buried in soil they are colonized by fungi such as *Fusarium*, *Chaetomium*, *Stachybotrys*, *Humicola* or *Trichoderma* spp. The *Fusarium* spp. tend to be favoured by low water potentials (drought 'stress') because they can grow at almost undiminished rates when the water potential is lowered to −5 MPa in culture. In contrast, *Tricho-derma* spp. can be favoured by soil acidity (pH 'stress') because they grow at about pH 3.0 in culture. A completely different spectrum of cellulolytic fungi is found on plant remains in estuarine waters (*Lulworthia* spp., *Halosphaeria hamata* and *Zalerion varium*) or in freshwater streams (some of the fungi with tetraradiate spores, such as *Tetracladium*, *Lemonniera* and *Alatospora*). Chitinous materials are colonized by different fungi again — commonly by *Mortierella* spp. (zygomycota) in soil or by *Chytridium confervae* (chytridiomycota) in freshwater habitats. Starchy materials such as stored grains or germinated seeds are often colonized by *Penicillium* and *Aspergillus* spp. (see Chapter 7).

The polymer-degrading fungi also exploit different microhabitats, so they can co-exist in a resource. The best evidence for this has come from microscopical examination of transparent cellulose film ('Cellophane') buried in soil (Tribe, 1960). In wet soils the film is colonized by cellulolytic chytrids (e.g. *Rhizophlyctis rosea*; see Chapter 1, Fig. 1.6) and chytrid-like organisms (e.g. *Hyphochytrium catenoides*). They degrade the cellulose locally by forming finely branched rhizoids between the layers of the film (Fig. 10.3). Other parts of the film can be colonized by mycelial fungi such as *Humicola grisea*, *Fusarium* spp. and *Rhizoctonia*. The *Rhizoctonia* and *Fusarium* colonies form loose networks over the surface of the film, whereas *Humicola* is seen as localized, compact colonies that 'root' into the cellulose film and produce fans of hyphae within it, like the fans of *Pythium graminicola* shown in Fig. 10.3. Such communities of fungi, compatible with one another because they occupy different microsites, may be common in natural materials.

Group 4: fungi that degrade the more recalcitrant (persistent) polymers

These types of fungi, such as lignin degrades, often predominate in the later stages of decomposition. Several of them are basidiomycota, including *Mycena galopus* in the leaf-litter of woodlands, *Marasmius oreades* and *Entoloma* spp. which cause fairy-rings by growing in the 'thatch' of dead leaf sheaths in old grasslands, and the wood-decay fungi discussed later. The keratin-degrading fungi, related to the dermatophytic pathogens of humans and other animals (see Chapter 13), are found similarly in the later stages of decomposition of hair or animals' hooves. They might occur earlier, but they become conspicuous only after an initial phase of exploitation by zygomycota and *Penicillium* spp., which utilize the more readily available proteins and lipids.

The ecological success of the fungi in this grouping stems from their specialized ability to degrade complex polymers that most other fungi cannot degrade. However, it does not necessarily follow that they use these complex polymers as their main energy source. Indeed, *Mycena galopus* cannot degrade lignin in culture unless it is also supplied with cellulose or hemicelluloses as more readily utilizable substrates. The chief attribute of these fungi could be that they degrade or modify the recalcitrant polymers and so gain access to

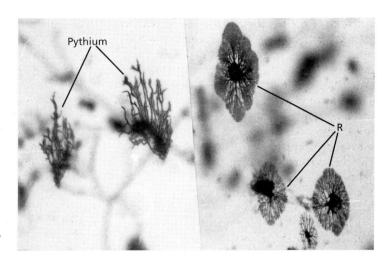

Fig. 10.3 Composite photograph of a piece of transparent cellulose film retrieved from soil. Four colonies of *Rhizophlyctis rosea* (R) (chytridiomycota) have produced finely branched rhizoidal systems in the cellulose film. At a different plane of focus *Pythium graminicola* (one of the few markedly cellulolytic *Pythium* spp.) has grown from inoculum placed on the surface of the film and produced fan-like hyphal branches between the layers of the film. (From Deacon, 1979.)

other substrates that are chemically or physically complexed with the resistant polymers. Much of the cellulose in plant cell walls is intimately associated with lignin, either covalently bonded to it or encrusted by it, and this 'lignocellulose' is largely unavailable to fungi that cannot modify the lignin. In one of the best-studied lignin-degrading fungi, *Phanaerochaete chrysosporium*, the production of lignin-degrading enzymes is markedly stimulated by a critical shortage of nitrogen. This could explain why these fungi occur in the later stages of decomposition of materials — they might need to wait until other fungi have sequestered most of the available nitrogen.

Group 5: secondary (opportunistic) invaders

Secondary (opportunistic) invaders are found at all stages of decomposition of natural materials. Some of these opportunists might grow in close association with polymer-degraders, using some of the breakdown products released by enzyme action. As discussed later, *Thermomyces lanuginosus* seems to behave this way in composts. Other secondary invaders are known to parasitize other fungi in culture (e.g. *Pythium oligandrum*; see Chapter 11) or grow on dead hyphal remains; yet others (e.g. *Mortierella* spp.) might grow on the faecal pellets or shed exoskeletons of microarthropods which are abundant in decomposing materials in nature. The range of potential behaviour

patterns of these fungi makes it difficult to generalize, but their features commonly include:
1 nutritional opportunism, because they can scavenge low levels of nutrients of various types;
2 the ability to tolerate the general metabolic products of other fungi.

The fungal community of composts

Composts are made from any type of decomposable organic matter, stacked into a heap with adequate nitrogen, moisture and aeration. They are used commercially for production of mushrooms — usually a mixture of cereal straw and animal manure — and for the processing of horticultural and urban wastes. In addition, composts made from tree bark are becoming increasingly important as soil conditioners and as peat substitutes in horticulture.

Irrespective of the material used for composting, the decomposition process and most of the fungi involved are essentially similar because the main selective factor is the high temperature reached during composting. This is shown in Fig. 10.4 for an experimental compost made from wheat straw and inorganic nitrogen. Within a few days the temperature can rise to 70–80°C because of the metabolic heat generated by micro-organisms that grow on the more readily utilizable, soluble nutrients. Bacteria are primarily responsible for this, and they include the thermophilic actinomycetes that cause farmer's lung

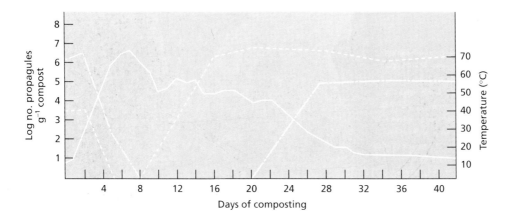

Fig. 10.4 Changes in temperature (——) and populations of mesophilic fungi (— · —) and thermophilic fungi (---) in a wheat straw compost. (Based on Chang & Hudson, 1967.)

disease (see Chapter 9). Fungi play only a minor role at this stage because they are slower to initiate growth, and the maximum temperature for growth of any fungus is about 62–65°C. After 'peak-heating' the compost starts to cool because most of the micro-organisms are killed in the centre of the compost, but some *Bacillus* spp. can survive as spores, and other bacteria can recolonize from the surface, giving rise to second and third smaller peaks. Then there is a long period of gradually declining temperature for 20–30 days before the compost reaches ambient temperature. Bacterial populations remain high throughout this time, but fungi are considered to play the most important role as they recolonize after peak-heating and degrade much of the organic matter during the prolonged high-temperature phase.

Chang and Hudson (1967) recognized several behavioural groups of fungi, occurring at different stages of the composting process. They are modified here for ease of presentation, and are shown in Fig. 10.5.

1 A mixture of **weak parasites and pioneer saprotrophic fungi** are found in the first few days. Many of them were present on the original material, including the leaf-surface fungi (*Cladosporium*, etc.) and *Fusarium* spp., which often grow on cereal straw as it starts to dry in field conditions. But, a few thermophilic fungi also grow in the first few days, notably *Mucor hiemalis* and *M. miehei*

(pioneer saprotrophic fungi), with maximum growth temperatures of 57–60°C. All these fungi are inactivated or killed by the peak temperature and they do not reappear.

2 A range of **cellulolytic** ascomycota and deuteromycota colonize after peak-heating and grow over the next 10–20 days. The high temperature might activate the dormant ascospores of some of these fungi (see Chapter 9). Although their incidence can differ between different types of compost, these cellulolytic fungi commonly include *Chaetomium thermophile, Humicola insolens, Thermoascus aurantiacus, Scytalidium thermophilum* and *Aspergillus fumigatus* (deuteromycota and ascomycota). It would be reasonable to assume that their temperature ranges for growth determine how quickly they start to colonize after peak-heating (see Chapter 7, Fig. 7.2). For example, *C. thermophile* (maximum temperature about 55°C, optimum about 50°C) might start earlier than *H. insolens* (maximum about 55°C, optimum about 37°C), and both these fungi are known to grow before *A. fumigatus* (maximum 52–55°C, optimum 37–40°C), which is a relatively late colonizer, when the temperature falls to about 40°C. During the prolonged phase of warm temperatures after peak-heating a compost can lose up to 50% of its initial dry weight, including nearly two-thirds of the main plant wall components such as cellulose and hemicelluloses (mixed polymers of arabinose, xylose, mannose and glucose). *Thermomyces lanuginosus* (deuteromycota) is also common and active during the high-temperature phase (maximum about 60°C, optimum about 45°C); it is one of the most common fungi in all types of

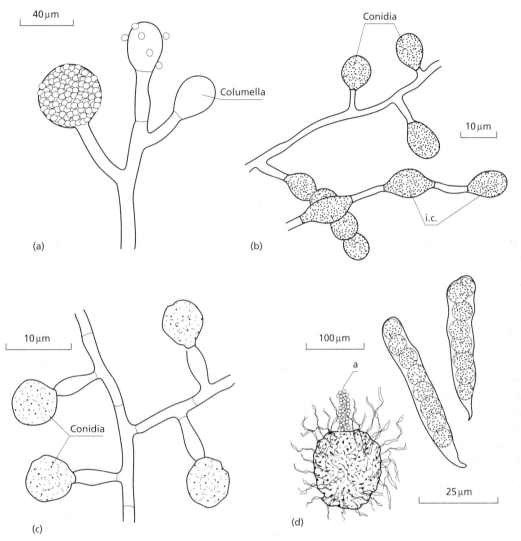

Fig. 10.5 Thermophilic fungi. (a) *Mucor pusillus* sporangiospore with three attached sporangia. One sporangium contains spores; the others have released the spores, leaving a columella which is a bulbous central part of the sporangium. (b) *Humicola insolens*, with melanized conidia formed either in intercalary positions (i.c.) or terminally on short hyphal branches.

(c) *Thermomyces lanuginosus*, with single conidia borne on short inflated hyphal branches, similar to *Humicola*. (d) *Chaetomium thermophile*, showing a perithecium covered in sterile hyphae and extruding ascospores (a) from an apical ostiole; two individual asci also are shown. (Based on Cooney & Emerson, 1964.)

compost. However, it is non-cellulolytic, so it seems to grow by utilizing some of the sugars generated by the enzymes of other fungi and perhaps also by using their mycelial breakdown products (see Chapter 11). In terms of the behavioural groupings mentioned earlier, it can be considered as a **secondary (opportunistic) invader**.

3 As the temperature falls below 35–40°C, the thermophilic fungi start to decline, but *A. fumigatus* remains active. The compost is then colonized progressively by mesophilic fungi, including some deuteromycota (e.g. *Fusarium*, *Doratomyces*) and basidiomycota such as *Coprinus cinereus* which has a maximum growth temperature of about 40°C, but its optimum is much lower than this. *Coprinus* spp. represent the **degraders of recalcitrant polymers.** They utilize the lignocellulose and they are highly antagonistic to many other fungi, damaging the hyphae on contact by a process termed **hyphal interference** (see Chapter 11). The commercial mushroom, *Agaricus bisporus*, could be introduced into the compost once this has cooled to below 30°C. It cannot be added earlier because of its temperature requirements, but it must be added before *Coprinus* becomes established or it will be antagonized. In practice, this problem is overcome (and the whole process accelerated) by pasteurizing a compost soon after peak-heating so that most of the resident fungi are killed and then adding inoculum of *A. bisporus* as explained in Chapter 4.

Role of oxygen

Composting, like most decomposition processes, is highly dependent on oxygen. In practice this is achieved by stacking mushroom composts and municipal composts in long heaps (windrows) which are turned regularly to ensure adequate aeration. We can compare this with the production of silage as winter feedstuff for farm animals — a strictly anaerobic process. Silage is produced from fresh grass or legume crops which are mown and left for a short time so that they dry partially, then chopped and 'bruised' to release the plant sap and compacted into an airtight container (a silo), which may be nothing more sophisticated than a polythene-lined box. A wide range of aerobic fungi and bacteria grow initially on the readily available nutrients, but they rapidly deplete the oxygen and their growth then slows or stops. Only the lactic acid bacteria (*Streptococcus*, *Lactobacillus*, *Pediococcus* and *Leuconostoc*) continue to grow at an appreciable rate because they are obligately fermentative (see Chapter 6) — they ferment sugars in the presence or absence of air. They generate lactic acid as the main fermentation product, and this rapidly lowers the pH to about 4.0. At this point their own growth is inhibited, and the combination of low pH and anaerobic conditions prevents all other organisms from growing. The silage is therefore kept in a stable form. However, many yeasts and mycelial fungi will spoil it rapidly if

the surfaces are exposed to air — it is a readily available substrate made unavailable simply by the anaerobic conditions.

Role of nitrogen

Nitrogen and other mineral nutrients are essential for composting and for decomposition in general. Moreover, the level of nitrogen must be quite high if decomposition is to proceed at a high rate. This is expressed in terms of the carbon:nitrogen (C:N) ratio of a material — the ratio of elemental nitrogen to elemental carbon, assuming that both the carbon substrate and the nitrogen are available in utilizable forms. In order to appreciate this, we must note two points.

1 Fungal hyphae have a C:N ratio of approximately 10:1, although it can vary depending on their age and other factors.

2 During their growth, fungi convert approximately one-third of the substrate carbon into cellular material (a substrate conversion efficiency of about 33% — see Chapter 3). The other two-thirds is respired to carbon dioxide.

These values are only approximate, but it follows that a material of C:N ratio about 30:1 is a 'balanced' substrate — it can be degraded rapidly and completely, as follows:

10 of the carbon 'units' (whatever the unit might be) are incorporated into fungal biomass;

20 units of carbon are released as carbon dioxide;

one unit of nitrogen is incorporated into fungal biomass.

Consider, now, a material of C:N ratio 100:1 (wheat straw is roughly 80:1; sawdust ranges from 350:1 to 1250:1; newsprint has essentially no nitrogen). A fungus starts to grow on this material, but the available nitrogen is depleted long before all the organic carbon has been used. Viewed in simplistic terms, decomposition will stop at the point where:

10 carbon units are combined with one nitrogen unit in the mycelium;

20 carbon units have been released as carbon dioxide;

70 carbon units are left in the residual substrate.

In effect, the fungus is starved because there is a carbon substrate available but no further nitrogen available for growth, enzyme production, etc.

Several things can happen as this stage is approached. Some fungi seem to be well adapted to recycle their cellular nitrogen, perhaps by controlled autolysis of the older hyphae, or they respond to low-nitrogen environments by preferentially allocating their nitrogen to essential metabolic processes. Wood-decay fungi, such as *Coriolus versicolor*, seem to be highly adept at this, as discussed later. Alternatively, nitrogen can be recruited from the external environment, which commonly happens when wheat straw is added to soil. But, in a 'closed' system such as a compost, some of the existing cells are likely to die of starvation and the nitrogen present in these cells can then be reused either by the remaining cells of the same species or by other species. Whatever happens, the rate of decomposition is nitrogen-limited because at least some of the initial nitrogen will be locked up in spores and other resting stages of the early colonizers. The same considerations apply to any other essential mineral nutrient that is present in limited amounts.

Can nitrogen-depletion drive fungal successions?

A succession of activities of fungi might be driven by the availability of different organic substrates—the early colonizers exhaust the substrates they can use, then they are replaced by other fungi that use further substrates, and so on. A succession could also occur in response to changes in environmental conditions (temperature, pH, etc.), as happens in composts. But, a third possibility is that the availability of nitrogen drives a succession: the early colonizers use the available nitrogen and might not be efficient in recycling their mycelial nitrogen reserves, so they starve and other fungi replace them, using their lytic products as a nitrogen source. In fact, fungi that enter late in a sequence—for example, *Coprinus* spp. in composts —would need to gain access to nitrogen or other mineral nutrients that previous fungi have used.

Some of these points have been addressed in laboratory experiments (Fig. 10.6) where flasks

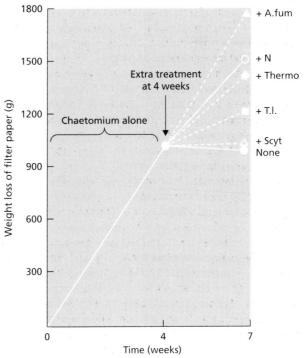

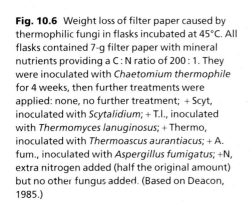

Fig. 10.6 Weight loss of filter paper caused by thermophilic fungi in flasks incubated at 45°C. All flasks contained 7-g filter paper with mineral nutrients providing a C : N ratio of 200 : 1. They were inoculated with *Chaetomium thermophile* for 4 weeks, then further treatments were applied: none, no further treatment; + Scyt, inoculated with *Scytalidium*; + T.l., inoculated with *Thermomyces lanuginosus*; + Thermo, inoculated with *Thermoascus aurantiacus*; + A. fum., inoculated with *Aspergillus fumigatus*; +N, extra nitrogen added (half the original amount) but no other fungus added. (Based on Deacon, 1985.)

containing filter paper cellulose (7 g) and nitrate nitrogen (giving a C:N, 200:1) were inoculated with thermophilic fungi and incubated at 45°C. In the initial 4 weeks, *Chaetomium thermophile* degraded 1 g of the original filter paper but its growth then stopped and there was no further degradation after 7 weeks. (Actually, the 1-g weight loss underestimates the substrate breakdown because some of the substrate carbon remains in the mycelial biomass, but this does not affect the interpretation.) The failure of *Chaetomium* to degrade the cellulose after 4 weeks was caused by nitrogen-depletion, because the addition of extra nitrogen at this stage led to further breakdown (Fig. 10.6).

In identical experiments, *Chaetomium* was allowed to grow for 4 weeks (by which time it was nitrogen-depleted) then other thermophilic fungi were added, but **with no extra nitrogen**. As shown in Fig. 10.6, *Scytalidium thermophilum* could not degrade the cellulose in these conditions, after *Chaetomium* had grown, although it can degrade cellulose in pure culture. In contrast, *Thermoascus aurantiacus* or *Aspergillus fumigatus* could grow over the nitrogen-depleted *Chaetomium* colony and cause further breakdown of the cellulose. So, we can conclude that *Scytalidium* cannot replace *Chaetomium* once this has exhausted the nitrogen supply, whereas *Thermoascus* and *A. fumigatus* can replace *Chaetomium* and in some way gain access to nitrogen, perhaps by inducing the *Chaetomium* hyphae to lyse. We noted earlier that *A. fumigatus* is a relatively late colonizer of composts, after *Chaetomium* and other cellulolytic fungi have become established. Part of its success in doing so might be related to its ability to recycle the nitrogen previously used by *Chaetomium*.

The basidiomycota (*Agaricus bisporus*, *Coprinus cinereus*, etc.) typically occur late in the succession of fungal activities in natural materials, and many of them have been shown to use proteins as nitrogen sources. Indeed, *A. bisporus* can grow in culture media when nitrogen is supplied only in the form of living or heat-killed bacteria. We noted earlier that *Phanaerochaete* is induced to synthesize lignin-degrading enzymes in response to nitrogen-limitation; and we noted that basidiomycota such as *Coprinus* disrupt the hyphae of other fungi on contact, which perhaps gives a source of organic nitrogen. So, it seems that nitrogen avail- ability is a key factor in fungal successions, and that the later colonizers have special abilities to obtain the nitrogen that earlier colonizers have utilized. These points are relevant to biotechnology. There is increasing interest in the recycling of spent mushroom composts, to save costs and overcome the environmental problems of disposal. If the compost is to be reused for *A. bisporus* then it must be supplemented with the materials depleted in the previous cropping cycle. An interesting alternative is to reuse the compost for growth of other edible species. In experimental studies, the 'oyster fungus', *Pleurotus ostreatus*, has been grown on cotton waste used previously for cropping of *Volvariella volvacea* (the paddy-straw mushroom, commonly grown in the Orient). The cropping efficiency of *Pleurotus* (calculated as kilograms of fresh weight of fruitbodies per 100 kg dry weight of the substratum) was 63% on the reused material, matching the 60% cropping efficiency of *Pleurotus* on fresh cotton waste. *Pleurotus* has a potentially high market value because of its subtle taste and, incidentally, its reputation as an aphrodisiac. Perhaps in this case the study of successions will be driven by more than just nitrogen!

Fungal communities in herbivore dung

The dung of herbivores is, at least in some ways, ideal for studies of fungal decomposer activities. Rabbit pellets have been used most frequently because they are relatively uniform natural substrata, and by feeding caged rabbits on herbage containing the spores of selected coprophilous fungi it is possible to manipulate the fungal community without resorting to disruptive treatments. There is also an important environmental significance in these studies because vast amounts of plant material are cycled annually through herbivores.

Herbivore dung supports a characteristic community of fungi. Their reproductive structures appear in a sequence roughly corresponding to the illustrations shown in Fig. 9.5 (see Chapter 9). Sporangiophores of zygomycota (*Pilaira*, *Pilobolus*, *Phycomyces*, etc.) appear in the first few days, followed by fruiting structures of ascomycota (perithecia of *Sordaria*, apothecia of *Ascobolus*, etc.) and conidiophores of deuteromycota. Then, toadstools of basidiomycota like *Coprinus* and *Bolbitius* appear. This sequence had been assumed to reflect a succession of fungal growth phases — the zygomycota entering first as pioneers that exploit the simple, soluble nutrients, then the ascomycota and deuteromycota invading to exploit cellulose, etc., and finally the basidiomycota invading and growing on the lignin-enriched material. However, Harper and Webster (1964) showed that most or all of these fungi pass through the gut as spores and are deposited with the dung. Their spores germinate at roughly equivalent times (4–12 h) after treatment with proteolytic enzymes at 37°C to simulate passage through the rabbit gut, and their hyphal growth rates are similar in culture. So, it seems that all these fungi might colonize the dung initially, and they differ only in the time required to produce their sporing stages.

Consistent with this, Harper and Webster inoculated individual fungi into sterilized rabbit pellets and found that the fungi produced their reproductive structures after different times, corresponding closely to the times of appearance in the mixed community on natural dung (Table 10.2). However, the **duration** of fruiting by the early fruiting fungi was truncated in natural dung compared with in the pure inoculations, and in further experiments *Coprinus* was found to be the main cause of this because it antagonized the other fungi. The cautionary point raised by these studies is that sequences of fruiting do not necessarily correspond to sequences of somatic growth. Instead, the fruiting sequences reflect different fungal 'strategies' — the pioneer zygomycota have a relatively short phase of growth on the more readily utilizable substrates and then sporulate, whereas the ascomycota, deuteromycota and especially basidiomycota have longer phases of growth during which they degrade the major structural polymers, before they use their mycelial resources for fruiting.

Fungal decomposers in the root zone

Roots are difficult to study in soil so there have been few detailed investigations of their fungal decomposer communities. Yet, this is an important subject because roots provide the major continuous input of organic nutrients for soil fungi, including root-infecting pathogens and their potential biocontrol agents. Here, we focus on the roots of cereals and (non-cereal) grasses, which have been studied most intensively and provide clear evidence of decomposer activities in microsites.

Waid (1957) made a detailed study of the fungi on the roots of ryegrass (*Lolium perenne*) by sampling roots of different ages (evidenced by their transition from white to brown as the root cortex senesced) and by examining individual roots at different distances behind their tips. By a combination of microscopic observation of roots and dissection of the root tissues, which were plated onto agar, he obtained a composite picture of the changing fungal community on both a temporal and a spatial scale (Fig. 10.7). The main features are listed below, but using more recent evidence to explain some of the points that were not known at that time.

1 The growing root tips were virtually free from fungal hyphae, but an increasingly complex fungal community developed with distance (age) behind the tips.

2 Even in young root zones, and persisting into the old root zones, a fungus with hyaline (colourless) hyphae was seen in the innermost cortical

Table 10.2 Times of first appearance of reproductive structures of coprophilous fungi on rabbit pellets. (Data from Harper & Webster, 1964.)

	On sterilized dung reinoculated with single species	On natural dung (mixed culture of fungi)
Pilaira anomala (zygomycota)	2	2
Pilobolus crystallinus (zygomycota)	4	4
Ascobolus spp. (ascomycota)	7–13	9–12
Sordaria fimicola (ascomycota)	9	9
Coprinus heptemerus (basidiomycota)	7–8	9–13
C. patouillardii (basidiomycota)	14	35

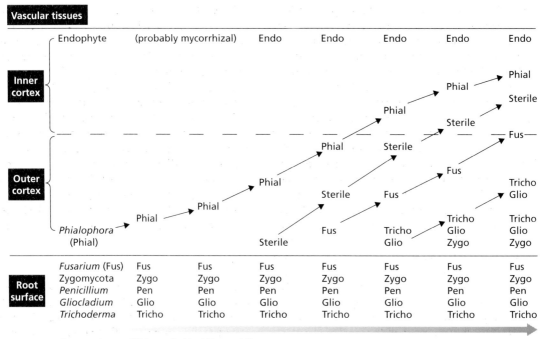

Fig. 10.7 Diagrammatic representation of changes in the fungal population of ryegrass (*Lolium perenne*) roots with increasing age and distance behind the root tip. Based on Waid (1957) and more recent evidence. An endophyte (probably an arbuscular mycorrhizal fungus) colonizes early and is found in the inner cortex of roots of all ages. A dark mycelial fungus (probably *Phialophora graminicola*) is found progressively deeper in the cortex with age and distance from the root tip. There are sequential waves of colonization of the cortex by a sterile hyaline fungus (unidentified, shown as 'sterile') and *Fusarium culmorum* (Fus). The outer cortex is invaded at a late stage by the typical root surface fungi, including *Gliocladium roseum* (Glio) and *Trichoderma* (Tricho). The root surface bears substantial populations of *Penicillium* spp. and zygomycota (Zygo) such as *Mucor* and *Mortierella* spp.

cells. This almost certainly was an arbuscular mycorrhizal fungus growing in the living host cells (see Chapter 12) because these fungi characteristically proliferate in the inner cortex of cereal and grass roots.

3 Behind the tips, *Mucor* and *Penicillium* spp. were found on the root surface, and they remained associated with the surface of roots of all ages. These fungi are assumed to utilize the soluble nutrients that leak from root cells, or possibly the root surface mucilage (a pectin-like material — see Chapter 12). However, their hyphae cannot be seen or identified easily in the root zone, so it is possible that they have an initial phase of growth or periodic phases of growth and then persist as spores on or near the root surface.

4 A fungus with darkly pigmented hyphae was found on the root surface in young root zones, and it was seen to occur progressively deeper in the cortex as the roots aged. This fungus almost certainly was *Phialophora graminicola* which is abundant on grass roots. It is a weak parasite that exploits a narrow window of opportunity, invading the root cortical cells as they start to senesce, but ahead of purely saprotrophic species (see Chapter 11, Fig. 11.12).

5 As the darkly pigmented fungus progressed inwards, it was followed by a sequence of fungi. Among these were an unidentified fungus that grew as sterile hyaline hyphae, and *Fusarium culmorum* which is a characteristic colonizer of dead and dying tissues but also can be pathogenic to cereals and grasses in conditions of water-stress.

6 In the old root zones, *Trichoderma* spp. and *Gliocladium roseum* colonized the outer cortex. These are characteristic soil fungi and are known to antagonize and overgrow other fungi on agar plates (see Chapter 11). So, they probably grew into the cortex as secondary (opportunistic) invaders.

We know now that this progression of fungi in grass roots is not the result of pathogenic attack. Instead, the cortex of cereal and grass roots senesces naturally behind the growing root tip, even in aseptic conditions. This can be shown by using nuclear or cytoplasmic staining to assess cell viability (Plate 10.1, facing p. 210). The senescence begins in the root epidermis when the root hairs start to die, and it progresses inwards through the cortex, layer by layer, with increasing distance behind the root tip. But, the root itself remains functional, extending at the tip while the older root regions serve a transport function (through the xylem and phloem) with a largely redundant, senescing root cortex (Henry & Deacon, 1981).

The interesting feature of the work on ryegrass roots is that a sequence of fungal activities occurs in microsites—the root diameter is much less than 1 mm, and within this there are at least five recognizable phases of activity by different types of fungus. We return to this topic in Chapter 11, where we will see that *Phialophora* is a natural biocontrol agent of the take-all fungus in grasslands.

Fungal communities in decaying wood

Wood is a difficult material to degrade, so it is largely unavailable to most fungi. It consists mainly of cellulose (40–50% dry weight), hemicelluloses (25–40%) and lignin (20–35%), with very low levels of readily available sugars. It has a very low nitrogen content (commonly a C:N ratio of about 500:1) and low phosphorus content. It also contains potentially fungitoxic compounds, particularly in the heartwood. In broad-leaved trees these are usually tannins, whereas in conifers they are phenolics such as terpenes, stilbenes, flavonoids and tropolones. The most toxic of the tropolones are thujaplicins which act as unfolders of oxidative phosphorylation; they are particularly abundant in cedarwood, making this a naturally decay-resistant wood for high-quality garden furnishings, etc.

Newly felled trees contain sugars in the columns of ray parenchyma which run longitudinally in the trunk. These columns are colonized rapidly by pioneer fungi which can stain the wood and reduce its commercial value (e.g. *Trichoderma* spp.), although the blue-stain fungus *Chlorosplenium aeruginascens* (ascomycota) which colonizes the fallen branches of oak has been used to make highly priced decorative pieces termed 'Tunbridge ware'. Once this brief phase of activity is over, the wood is accessible only to the typical wood-decay fungi (ascomycota and especially basidiomycota). These are grouped in three categories according to their mode of attack and conditions in which they grow: the soft-rot, brown-rot and white-rot fungi.

Soft-rot fungi

Soft-rot fungi are the least specialized for wood decay because they include several ascomycota and deuteromycota that commonly degrade cellulosic materials in soil (e.g. *Chaetomium*, *Ceratocystis* spp.) or estuarine waters (*Lulworthia*, *Halosphaeria*, *Pleospora* spp.). Nevertheless, they are important agents in the decay of wet wood, including fence posts, window frames, the timbers of cooling towers and all wood in marine and estuarine environments.

Soft-rot fungi have a characteristic decay pattern, shown in Fig. 10.8. Their hyphae grow in the lumen of individual woody cells and produce fine branches that penetrate the wall to produce decay cavities in the major cellulosic wall layer, termed the S2 layer. They degrade cellulose and hemicelluloses, but have little or no effect on lignin, which remains more or less intact. The decay is caused by typical cellulase enzymes (see Chapter 5) and the limited diffusion of these leads to the formation of rhomboidal cavities around the individual hyphae. All the soft-rot fungi need high nitrogen levels for wood decay: typically about 1% nitrogen content in the wood. This is why they are the characteristic decay fungi of fence posts because they recruit nitrogen from the surrounding soil.

Brown-rot fungi

Brown-rot fungi are predominantly basidiomycota, such as *Piptoporus betulinus* which produces

(a)

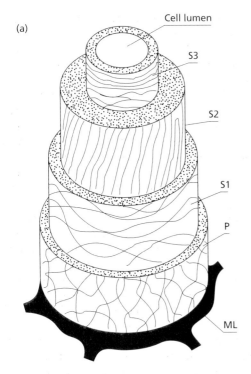

(b)

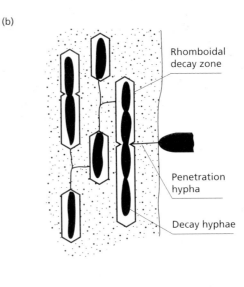

Fig. 10.8 (a) Diagrammatic representation of the cell wall layers in woody tissue, showing the arrangement of the cellulose microfibrils. ML, middle lamella between adjacent cells; P, thin primary wall with loosely and irregularly arranged microfibrils; S1–S3, secondary wall layers.

(b) Characteristic decay pattern of a soft-rot fungus in the S2 layer. The fungus penetrates by narrow hyphae, then forms broader hyphae in planes of weakness in the wall and these hyphae produce rhomboidal cavities where the cellulose has been enzymatically degraded.

bracket-shaped fruitbodies on the trunks of dead birch trees, and *Serpula lacrymans*, the 'dry-rot' fungus of buildings in Europe (see Chapters 4 and 6). *S. lacrymans* posed an enigma for a long time because it was unknown from any natural environment, only from buildings. But, now it has been found to occur rarely in the Himalayan forests of northern India. Dry rot has been recorded in Europe since about 1765, before there was any export of timber from India. So this 'rare' fungus seems to have arrived in Europe as air-borne basidiospores and then flourished in buildings where the climatic conditions are similar to those in its natural habitat (Singh *et al.*, 1993).

The term brown rot refers to the characteristic colour of the decayed wood, because these fungi degrade most of the cellulose and hemicelluloses but leave the lignin more or less intact as a brown,

chemically modified framework. The hyphae grow in the cell lumen, as in the case of soft-rot fungi, but the decay pattern is quite different because the S2 wall layer is almost completely degraded, even far away from the hyphae. Also, the decay is typically irregular, with some groups of wood cells heavily degraded and others only slightly so. This causes the wood to crack along lines of weakness as it dries, giving a characteristic brick-like pattern.

The remarkable ability of brown-rot fungi to cause a generalized decay of cellulose would be difficult to explain in terms of diffusion of cellulase enzymes. In fact, the cellulases produced by brown-rot fungi have little effect on cellulose *in vitro*, unlike the cellulases of soft-rot fungi. Instead, the brown-rot fungi degrade cellulose by an oxidative process, involving the production of hydrogen peroxide (H_2O_2) during the breakdown of hemicelluloses. Being a small molecule, H_2O_2 can diffuse through the woody cell walls to cause a generalized decay. In support of this, the characteristic decay pattern of brown-rot fungi can be mimicked experimentally by treating wood with H_2O_2 alone, and at least one of these fungi, *Poria placenta*, has been shown to degrade cellulose only if hemicelluloses also are present, as substrates for generating H_2O_2. This mode of attack is an efficient way of using the scarce nitrogen resources in wood, because it does not require the release of large amounts of extracellular enzymes.

White-rot fungi

The white-rot fungi include both ascomycota (e.g. *Xylaria hypoxylon*, the familiar candle-snuff fungus on dead logs) and basidiomycota (e.g. *Coriolus versicolor*, commonly seen as clusters of small, leathery, fan-shaped brackets on stumps). They degrade cellulose, hemicelluloses and lignin more or less simultaneously, so that the wood remains white as it decays. The patterns of colonization vary, but if the hyphae are present in the cell lumen then they cause a progressive thinning of the woody cell walls, starting in the innermost S3 layer and working outwards.

These fungi seem to use conventional cellulase enzymes for wood decay, but they are extremely efficient in their use of nitrogen. For example, the hyphae of *C. versicolor* have been found to have a nitrogen content of 4% when grown on laboratory media of C:N ratio, 32:1; but they had only 0.2% nitrogen content when grown on a medium of C:N, 1600:1. In nitrogen-poor conditions this fungus may preferentially allocate nitrogen to the production of extracellular enzymes and essential cell components, and also efficiently recycle the nitrogen in its mycelia (Levi & Cowling, 1969). They might benefit also from the associated growth of nitrogen-fixing bacteria in wood.

The most notable feature of white-rot fungi, however, is their ability to degrade lignin completely. This is a complex, three-dimensional polymer composed of phenylpropane units (six-carbon rings with three-carbon side chains) bonded to one another in various ways (Fig. 10.9). The breakdown of such a polymer would require a multitude of enzymes if these were of a conventional type. But, this seems implausible in any case

because an extremely diverse range of intermediates have been detected during the course of lignin degradation by white-rot fungi, inconsistent with the normal modes of action of enzymes. The resolution of this problem has come in recent years, with the discovery that lignin is oxidized by white-rot fungi. The process is termed 'enzymatic combustion' because, once initiated, it is largely uncontrolled (Kirk & Farrell, 1987). Although some of the details are still unclear, several enzymes have been found to be involved, some of them serving to generate oxidants and some catalysing the cleavage of the aromatic rings.

The major enzyme that initiates ring cleavage is **laccase**, which catalyses the addition of a second hydroxyl group to phenolic compounds. The ring can then be opened between two adjacent carbon atoms that bear the hydroxyl groups, as shown in Fig. 10.9. This process occurs while the ring is still attached to the lignin molecule. It is termed *ortho* fission, in contrast to *meta* fission which bacteria employ to cleave the phenolic rings of pesticide molecules (where the ring is opened at a different position in relation to the hydroxyl groups).

The other enzymes are involved mainly in generating or transferring oxidants. They include **glucose oxidase** which generates H_2O_2 from glucose, **manganese peroxidase** which oxidizes manganese (II) to manganese (III) which can then oxidize organic molecules, and **ligninase** (lignin peroxidase) which catalyses the transfer of singlet oxygen from H_2O_2 to aromatic rings and is one of the main initiators of attack on the lignin framework. These initial oxidations involving single electron transfer generate highly unstable conditions, setting off a chain of chemical oxidations.

Clearly, the degradation of lignin is highly dependent on a supply of oxygen, so it does not occur in waterlogged conditions. It also poses potential hazards to the fungus because some of the oxidative intermediates can be fungitoxic. The white-rot fungi generate such compounds from phenylpropane units *in vitro*, but detoxify them by polymerization into melanin-like pigments. This does not seem to be common in wood, except where two colonies of wood-decay fungi meet. Then they produce dense, heavily melanized 'zone lines' in the region of contact (Fig. 10.10). When

(a) Structure of lignin

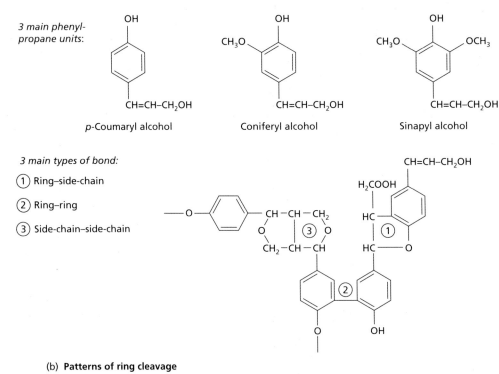

3 main phenyl-propane units:

p-Coumaryl alcohol   Coniferyl alcohol   Sinapyl alcohol

3 main types of bond:

① Ring–side-chain

② Ring–ring

③ Side-chain–side-chain

(b) Patterns of ring cleavage

Protocatechuic acid

Ortho fission   **Or**   *Meta* fission

Acetate
+
Succinate

2 Pyruvate
+
Formate

Fig. 10.9 (a) Structure of lignin, consisting of three types of phenylpropane unit, linked into a complex three-dimensional polymer by three main types of bond. (b) The method of enzymatic opening of aromatic rings during lignin breakdown (while the rings are attached to the polymer) or during breakdown of pesticides and other xenobiotics, shown for a simple ring of protocatechuic acid. Initially, the ring is substituted with two hydroxyl groups on adjacent carbon atoms, then it is opened between these carbons (*ortho* fission by fungi) or adjacent to one of them (*meta* fission which is a plasmid-encoded function of bacteria).

fungi are isolated from either side of a zone line they are found to belong to different species or, more commonly, different strains of a single species that are somatically incompatible with one another.

Biotechnology of wood-decay fungi

Lignocellulose is abundant as a by-product of the wood-processing industries and also in crop residues, so there is the potential to use it as a cheap commercial substrate. For example, if the

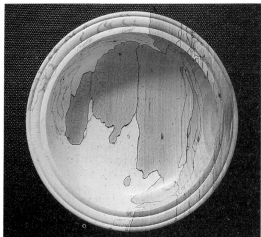

Fig. 10.10 (a) Part of a decaying tree stump showing dark zone lines at the junctions of fungal colonies in the wood. (b) Decorative bowl made from beech wood with dark zone lines.

cellulose could be degraded to sugars, these could be used to produce fuel alcohol by microbial fermentation, as an alternative to fossil fuels. This prospect has stimulated research on delignification, especially by the white-rot fungus *Phanaerochaete chrysosporium* which grows rapidly in

submerged liquid culture and, unusually for basidiomycota, produces abundant conidia. In near-optimum culture conditions it can degrade as much as 200 mg lignin per 1 g of mycelial biomass per day. However, it also degrades and utilizes the cellulose component of lignocellulose, defeating the object of the exercise. Genetic engineering offers a potential solution to this, and was made possible by the discovery that *Phanaerochaete* produces ligninase in the early stationary phase of batch culture (see Chapter 3). The enzyme production was strongly promoted by nitrogen-limitation and further stimulated by addition of lignin. By comparing the messenger ribonucleic acid bands produced in these conditions and corresponding non-inducing conditions, the gene for ligninase was identified (see Chapter 8) then sequenced and cloned into *Escherichia coli*. The ligninase produced by the recombinant bacterium acts on a range of lignin 'model compounds' *in vitro* so there is a prospect of using high-yielding recombinant micro-organisms for the delignification process.

References

Andrews, J.H. (1992) Fungal life-history strategies. In: *The Fungal Community: its Organization and Role in the Ecosystem* (eds G.C. Carroll & D.T. Wicklow), pp. 119–45. Marcel Dekker, New York.

Carroll, G.C. & Wicklow, D.T. (1992) *The Fungal Community: its Organization and Role in the Ecosystem.* Marcel Dekker, New York.

Chang, Y. & Hudson, H.J. (1967) The fungi of wheat straw compost. (two papers, I and II). *Transactions of the British Mycological Society*, **50**, 649–66, 667–77.

Cooney, D.G. & Emerson, R. (1964) *Thermophilic Fungi*. Freeman, San Francisco.

Deacon, J.W. (1979) Cellulose decomposition by *Pythium* and its relevance to substrate-groups of fungi. *Transactions of the British Mycological Society*, **72**, 469–77.

Deacon, J.W. (1985) Decomposition of filter paper cellulose by thermophilic fungi acting singly, in combination, and in sequence. *Transactions of the British Mycological Society*, **85**, 663–9.

Garrett, S.D. (1951) Ecological groups of soil fungi: a survey of substrate relationships. *New Phytologist*, **50**, 149–66.

Harper, J.L. & Webster, J. (1964) An experimental analysis of the coprophilous fungus succession. *Transactions of the British Mycological Society*, **47**, 511–30.

Henry, C.M. & Deacon, J.W. (1981) Natural (nonpathogenic) death of the cortex of wheat and barley seminal roots, as evidenced by nuclear staining with acridine orange. *Plant and Soil*, **60**, 255–74.

Hudson, H.J. (1968) The ecology of fungi on plant remains above the soil. *New Phytologist*, **67**, 837–74.

Kirk, T.K. & Farrell, R.L. (1987) Enzymatic 'combustion': the microbial degradation of lignin. *Annual Review of Microbiology*, **41**, 465–505.

Levi, M.P. & Cowling, E.B. (1969) Role of nitrogen in wood deterioration. VII. Physiological adaptation of wood-destroying and other fungi to substrates deficient in nitrogen. *Phytopathology*, **59**, 460–8.

Singh, J., Bech-Andersen, J., Elborne, S.A., Singh, S., Walker, B. & Goldie, F. (1993) The search for wild dry rot fungus (*Serpula lacrymans*) in the Himalayas. *The Mycologist*, **7**, 124–30.

Tribe, H.T. (1960) Decomposition of buried cellulose film, with special reference to the ecology of certain soil fungi. In: *The Ecology of Soil Fungi* (eds D. Parkinson & J.S. Waid), pp. 246–56. Liverpool University Press, Liverpool.

Waid, J.S. (1957) Distribution of fungi within the decomposing tissues of ryegrass roots. *Transactions of the British Mycological Society*, **40**, 391–406.

Chapter 11
Fungal interactions: mechanisms, relevance and practical exploitation

Fungi interact with each other and with other micro-organisms in many ways. We consider some of the major types of interaction in this chapter, in terms of the mechanisms involved, their consequences for fungal activities and the potential for exploiting them in practice. This chapter extends the discussion of fungal decomposer activities in Chapter 10, but also includes examples of plant disease control (see Chapter 12).

The terminology of species interactions is fraught with difficulties, but mycologists generally recognize three types of interaction.

1 The ability of one species to exclude another by **competition** (sometimes called **exploitation competition**), i.e. by being faster or more efficient in exploiting a resource (space, substrate, etc.).

2 The ability of one species to exclude or replace another by **antagonism** (sometimes called **interference competition** or **combat**), i.e. by directly affecting another organism through antibiotic production, etc.

3 The ability of two species to interact to the benefit of one (**commensalism**) or both (**mutualism**) and with no negative impact on the other species.

It must be emphasized that these types of interaction grade into one another: fungi can interact differently in different conditions.

The role of antibiotics

An antibiotic can be defined somewhat arbitrarily, as a diffusible secondary metabolite of one (micro-)organism that inhibits another (micro-)organism at a concentration of $100\,\mu g\,ml^{-1}$ or less. This definition serves to exclude general metabolic by-products such as carbon dioxide and organic acids. It restricts the term to specific highly active compounds with specific cellular targets.

The best-known antibiotics from fungi are the penicillins (see Chapter 6), cephalosporins and griseofulvin (see Chapter 14) which are used clinically. However, fungi produce many other antibiotics that act against other fungi, bacteria or both, but which have not been exploited commercially because of their instability, general toxicity or other undesirable properties. Table 11.1 shows the number of antibiotics discovered from fungi by 1974, and many more have been found since then. Such compilations must be interpreted with caution because they are biased towards commercial screening for potential pharmaceuticals, and towards the fungi most often screened for this purpose. Nevertheless, they accord with general experience that the production of diffusible antibiotics seems to be quite rare among aseptate fungi, more common among the septate fungi and especially common in some genera of the deuteromycota such as *Penicillium, Aspergillus, Fusarium* and *Trichoderma*.

The significance of antibiosis in nature is difficult to assess, because antibiotics are likely to be produced in small amounts and in microsites. In this respect, a substantial advance was made by Thomashow *et al.* (1990), who used high-performance liquid chromatography to detect an antifungal antibiotic (phenazine-1-carboxylic acid) produced by a strain of *Pseudomonas* in the rhizosphere of wheat seedlings in soil—one of the first direct demonstrations of antibiotic production in a complex natural environment. The levels were low ($27–43\,ng\,g^{-1}$ of root with adhering soil) even on young seedlings grown from seeds coated with more than 10^8 *Pseudomonas* cells (see Chapter 14).

Table 11.1 Taxonomic distribution of antibiotic-producing fungi. (Based on Berdy, 1974.)

Fungal group	Number of antibiotics described
Chytridiomycota, oomycota, zygomycota	14
Ascomycota	61
Basidiomycota	140
Deuteromycota	553
Penicillium	123
Aspergillus	115
Fusarium	46
Trichoderma	13

From laboratory studies we can predict the conditions in which antibiotics would be produced (see Chapter 6): during nutrient-limitation equivalent to early stationary phase in batch culture systems or during continuous growth in substrate-limited conditions in chemostats (so that their synthesis is not nutrient-repressed). Two general predictions stem from this.

1 Antibiotic production would not be important as a means of **gaining** access to a resource, because a fungus must already be growing, with a surfeit of the necessary precursor metabolites, in order to produce antibiotics (see Chapter 6).

2 Antibiotic production is most likely to be a means of **defending** an existing resource, in conditions where growth is substrate-limited.

For example, antibiotics might help *Penicillium* or *Aspergillus* spp. to defend their starch-rich or oil-rich food sources in stored commodities (see Chapter 7). The antibiotic patulin (see Chapter 6, Fig. 6.9), which also is a mycotoxin, might enable *Penicillium expansum* to defend its sugar-rich territory in rotting apples (see Chapter 12). Antibiotics also might be produced by cellulolytic fungi, but around their individual hyphae to defend the zones of enzymic erosion against opportunistic microbes (see Chapter 5).

Antibiotics in biocontrol: the role of *Trichoderma*

Apart from their exploitation in medicine, most current interest in antibiotics has centred on their potential roles in biocontrol of plant pathogens. *Trichoderma* spp. (deuteromycota, Fig. 11.1) have

heavily dominated this field (Chet, 1987). They were among the first fungi to be shown to produce antibiotics (Weindling, 1934), and now they are available in several commercial biocontrol formulations—to control *Botrytis cinerea* on grapes, silver leaf disease of plum trees, seedling diseases of glasshouse ornamental crops, etc. They work with varying degrees of success. For example, the product Trichodex (*Trichoderma harzianum*) is not as effective as fungicides for controlling *Botrytis* 'grey mould' on grapes in field conditions, but can be used effectively as an alternating treatment with fungicides (O'Neill *et al.*, 1996).

Trichoderma spp. produce both volatile and non-volatile antibiotics, active against fungi or bacteria or both (Fig. 11.2). The non-volatile (water-soluble) compounds include **trichodermin, suzukacillin** and **alamethicine**. The major volatile antibiotic, which gives a 'coconut' smell to many *Trichoderma* strains, is **6-pentyl-α-pyrone** (6-PAP). However, *T. virens* (known as *Gliocladium virens* until recently) produces a different spectrum of antibiotics — all strains of this species seem to produce **viridin** and its reduction product **viridiol**, but some strains also produce **gliovirin** and **heptelidic acid**, whereas others produce **gliotoxin** instead of these.

Antibiotics are widely considered to be important in biocontrol by *Trichoderma*. For example, 6-PAP (Fig. 11.2) is produced by some (but not all) strains of *T. viride, T. harzianum* and *T. hamatum*, and the 6-PAP producers are the more antagonistic to plant pathogens *in vitro* and in small-scale seedling bioassays. In the case of *T. virens* (formerly *G. virens*) the gliovirin-producing strains inhibit the growth of *Pythium ultimum* but not *Rhizoctonia solani in vitro*, and when applied experimentally to cotton seeds only the gliovirin-producing strains controlled seedling disease caused by *P. ultimum*. Conversely, gliotoxin (Fig. 11.2) is more active against *Rhizoctonia* than against *Pythium in vitro*, and gliotoxin-producing strains are best for control of *Rhizoctonia* on seedlings. Antibiotic-deficient mutants of both types of strain were less effective than the wild-type in disease control (Howell *et al.*, 1993).

Although these studies are valid for biocontrol, they are too artificial to be of value for understanding the behaviour of *Trichoderma* in natural situations. In all biocontrol studies the *Trichoderma* strains were inoculated at high levels and in nutri-

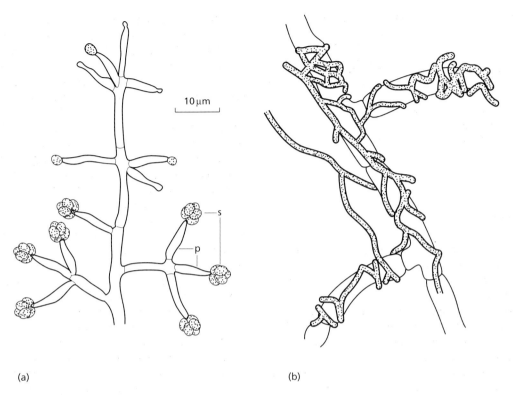

(a)

(b)

Fig. 11.1 The genus *Trichoderma*, showing (a) typical branching pattern of the conidiophore, bearing clusters of phialides (p) with spores (s) at their heads; (b) coiling of hyphae round the wider hyphae of *Rhizoctonia solani*.

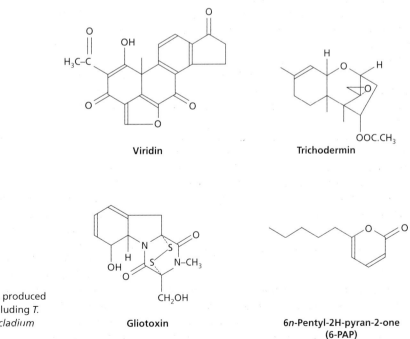

Viridin

Trichodermin

Gliotoxin

6n-Pentyl-2H-pyran-2-one
(6-PAP)

Fig. 11.2 Some antibiotics produced by *Trichoderma* species (including *T. virens*, formerly called *Gliocladium virens*).

ent-rich conditions, so they could exploit the food source before other organisms did so. Indeed, *Trichoderma* is only effective in soil or horticultural potting mixes if it is incorporated in a wheat-bran carrier; it does not give control in the absence of a 'dedicated' food base, because then it does not produce sufficient, if any, antibiotics. This supports the view that the natural role of antibiotics is to defend a food base (defence of territory) and not to capture a food source from spores in soil — the normal form in which *Trichoderma* persists. Therefore, *Trichoderma* spp. must have other attributes for resource capture. These fungi grow rapidly in culture—up to 25-mm hyphal extension per day at room temperature. They secrete chitinase and β-1,3-glucanase when grown in the presence of other fungal hyphae or on hyphal walls in culture. They also coil round the hyphae of other fungi on agar plates (Fig. 11.1) and eventually (24–48 h) penetrate the hyphae beneath the coils. *Trichoderma* spp. are common secondary invaders of organic matter in nature. We saw this in relation to ryegrass roots in Chapter 10. They overgrow the colonies of many other fungi on agar plates, and they are among the few fungi that can grow from soil particles sprinkled onto pre-colonized agar plates (Table 11.2). Thus, the *Trichoderma* spp. are classic examples of secondary opportunistic invaders (see Table 10.1), with a range of attributes suited to this role — the production of wall-lytic enzymes and antibiotics, coupled with hyphal coiling and hyphal penetration which are features associated with mycoparasites (see later).

Hyphal interference

Several basidiomycota antagonize other fungi, including other basidiomycota, on contact. This was discovered during *in vitro* studies on the interactions of fungi from herbivore dung (see Chapter 10). It helps to explain how the basidiomycota can dominate ultimately the fungal community, truncating the fruiting of other fungi, and also why the basidiomycota tend to be mutually exclusive — both *Coprinus heptemerus* and *Bolbitius vitellinus* occur on dung, but never in the same dung pellets because these two species are mutually antagonistic. Other basidiomycota were then found to

Table 11.2 Number of British soils yielding different mycoparasites (or organisms with mycoparasite-like features) when soil particles were placed on agar plates completely colonized by other fungi*. (Data from Mulligan & Deacon, 1992.)

Mycoparasite	Precolonizing fungus	Soils yielding the mycoparasite (maximum 28)
Pythium oligandrum	*Fusarium culmorum*	18
	Trichoderma aureoviride	3
	Rhizoctonia solani	0
	Botrytis cinerea	12
Gliocladium roseum	*F. culmorum*	20
	T. aureoviride	27
	R. solani	26
	B. cinerea	25
Trichoderma harzianum	*F. culmorum*	2
	T. aureoviride	0
	R. solani	22
	B. cinerea	13
Papulaspora sp.	*F. culmorum*	0
	T. aureoviride	11
	R. solani	3
	B. cinerea	19

* The mycoparasites either grew completely across the established fungal colonies or not at all. The results show that all the mycoparasites are common in soil (detected in at least 18 of the 28 soils tested at random); the results also suggest that the mycoparasites may invade organic materials selectively depending on the fungi that are already established there.

exhibit the same contact antagonism, which was termed **hyphal interference** (Ikediugwu & Webster, 1970).

The mechanism of hyphal interference is still unknown. It occurs rapidly, a few minutes after hyphal contact, and often the effect is localized to a single hyphal compartment that comes in contact with an 'interfering' hypha. The first visible sign is localized vacuolation and loss of turgor in the affected hyphal compartment, and if a dye such as neutral red is added at this stage it is taken into the affected compartment but does not enter the healthy hyphae (Fig. 11.3), indicating that hyphal interference causes loss of membrane integrity. In electron micrographs the affected cytoplasm is seen to be degenerate, the mitochondria are swollen, and a wide gap—the extraplasmalemmal zone—is seen between the retracted plasma membrane and the hyphal wall in contact with an interfering hypha; sometimes the damage is contained within a hyphal compartment by zones of dense, coagulated cytoplasm on either side of the contact point (Ikediugwu, 1976). However, these features reveal little about the antagonistic mechanism because similar ultrastructural changes can be caused by a range of toxicants. To date, the only clue to the mechanism of interference is that a poorly diffusible factor is involved, because hyphal interference sometimes occurs between hyphae separated by a Cellophane membrane up to 50 μm wide.

Hyphal interference by basidiomycota does not seem to be a parasitic mechanism, because the aggressor hypha does not enter the damaged hypha. Instead, this seems to be a highly efficient mechanism for inactivating other hyphae that are potential competitors for the same substrates; like using a machete to clear a path through a jungle rather than producing diffusible toxins to kill all the trees.

Control of pine root rot by hyphal interference (see Plate 11.1, opposite p. 210)

The fungus *Heterobasidion annosum* (basidiomycota) is the most important pathogen of coniferous trees in the northern hemisphere. It grows slowly but progressively along the woody roots, rotting them and spreading from tree to tree by root contact. At a later stage it can spread into the base of the trunk, causing a butt rot which destroys some of the most valuable timber. When established in a site, the fungus is almost impossible to eradicate except by mechanical extraction of the infected stumps and major roots, which provide the inoculum for infection of any newly planted trees. So, control measures have focused on preventing it from becoming established, especially in newly afforested sites.

Heterobasidion produces air-borne basidiospores from bracket-shaped fruitbodies at the bases of infected trees. These spores pose little threat in

Fig. 11.3 Hyphal interference by basidiomycota. Hyphae of *Heterobasidion annosum* have taken up the dye neutral red due to membrane damage where they have been crossed by hyphae of *Phlebiopsis gigantea*. Note the localization of the damage to the contacted hyphal compartments of *Heterobasidion*.

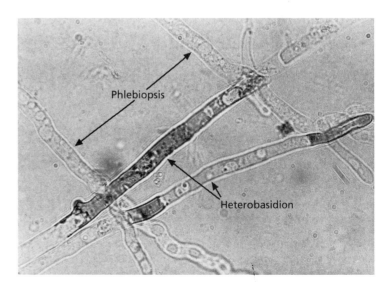

undisturbed forests because they have insufficient food reserves to initiate infection of woody roots when washed into the root zone. However, the situation is different in commercial forestry, where trees are felled for harvest or thinned to create the desired plant density as the plantation develops. The tissues of the exposed stump surfaces can remain alive for several months, but with declining resistance to infection. They provide a highly selective environment for pathogens like *Heterobasidion*, which can colonize from basidiospores and then grow down into the dying roots and infect the adjacent healthy trees (Fig. 11.4 and Plate 11.1). The simplest way to avoid this is to kill the stump surface tissues by applying phytotoxic chemicals, so that the dead tissues are colonized rapidly by saprotrophs, to the exclusion of *Heterobasidion*. This is the common practice in many forests (and is necessary for broad-leaved trees, to prevent them regenerating from the stumps). The chemicals used include urea and borates. However, they are

environmentally undesirable, especially if the forests are in catchment areas for domestic water supplies.

Rishbeth (1963) developed an alternative method in which spores of *Phlebiopsis gigantea* (basidiomycota) are applied to freshly exposed stump surfaces. *P. gigantea* is a weak parasite of pines, so it is favoured more than saprotrophs on the fresh stump surfaces. It excludes *Heterobasidion*, and it can grow rapidly down the attached roots with their declining resistance, but it poses no threat to the adjacent healthy trees. Moreover, by growing down into the roots *P. gigantea* can help to control any existing pockets of infection by *Heterobasidion* because it prevents the pathogen from growing up to the stump surface and sporulating there (Fig. 11.4). So, it has a partial eradicant effect as well as a protective effect. This is explained by the finding that *P. gigantea* shows strong interference against *Heterobasidion* on agar plates (see Fig. 11.3), although this was not known

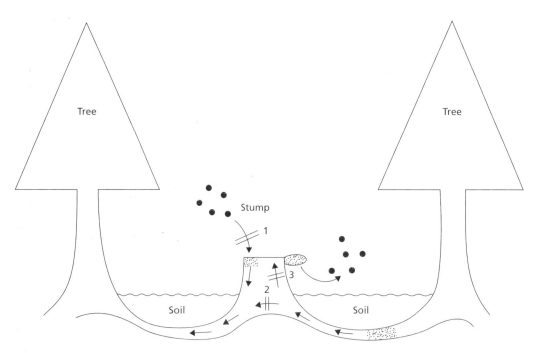

Fig. 11.4 Mode of action of *Phlebiopsis gigantea* in controlling *Heterobasidion annosum* (stippled) in recently felled pine stumps. 1, *P. gigantea* is established on the stump surface and prevents colonization from airborne basidiospores of *H. annosum*; 2, when *P. gigantea* has colonized the dying stump and root tissues it prevents the spread of *H. annosum* from existing foci of infection in the root zone so it cannot spread to healthy tree roots; 3, *P. gigantea* also prevents *H. annosum* from growing up to the stump surface to sporulate. (From Deacon, 1983.)

(a)

(b)

Plate 9.1 (a) Ectomycorrhizas on birch roots, seen as inflated, branched short roots bearing a light-coloured fungal sheath (see Chapter 10). (b) Rings of toadstools of ectomycorrhizal fungi around a young birch tree in an experimental site; toadstools of *Hebeloma* spp. and *Lactarius pubescens* are seen in the outer zones but a few toadstools of *Leccinum* are seen closer to the tree.

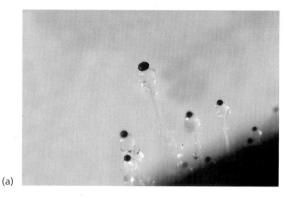

(a)

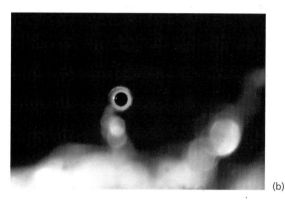

(b)

Plate 9.2 Sporing structures of *Pilobolus* on horse dung. (a) Sporangiophores, each with a terminal vesicle and a black sporangium; a ring of yellow carotenoid pigment is seen at the base of the vesicle of the longest (mature) sporangiophore. (b) A sporangiophore that has orientated towards a light source and is seen end-on, showing that the vesicle beneath the sporangium acts as a lens and has 'magnified' the yellow pigment, used for phototropism.

Plate 10.1 Wheat seminal root in a region about 3-weeks-old, stained with acridine orange and viewed under a fluorescence microscope to detect nuclei. Nuclei are present only in the innermost (sixth) cortical cell layer, next to the endodermis; the outer five cortical cell layers have died by natural senescence.

Nuclei absent

(a)

(b)

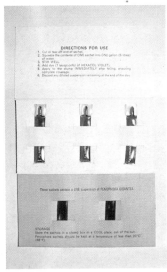

(c)

(d)

Plate 11.1 (a) Patch of trees killed by *Heterobasidion annosum* in a plantation of pines in Suffolk, UK. (b) Bracket-shaped fruitbodies (basidiocarps) at the base of a tree killed by *H. annosum*. (c) Sachets containing spores of the biocontrol agent *Phlebiopsis* (formerly termed *Peniophora*) *gigantea* in a sucrose-dye solution, for application to freshly cut pine stumps. (d) A stump after treatment with the spore suspension of *P. gigantea*; the base of the felled tree can also be seen. (Photographs kindly provided by the late Dr John Rishbeth.)

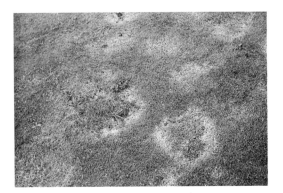

Plate 11.2 Take-all patch disease of turf caused by *Gaeumannomyces graminis* var. *avenae* (GGA). In the absence of effective control by competing fungi, GGA attacks turf composed of the susceptible bent grasses (*Agrostis* species), causing circular patches of dead grass up to 1 m diameter. The centres of the patches are recolonized by the coarser, more resistant grasses and by dicotyledonous weeds. This adversely affects the playing surface on golf-course greens, etc.

Plate 12.1 Panama wilt of bananas caused by *Fusarium oxysporum* f. sp. *cubense*. (a) Young plant of Cavendish cultivar (about 1 m tall) in naturally infested field soil. The leaf on the left shows progressive yellowing towards the midrib, followed by necrosis at the leaf margin. Many of the leaves have already died and collapsed. (b) Older plants with characteristic hanging sheaths of dead leaves.

(a)

(b)

(a)

(b)

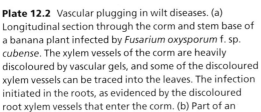

Plate 12.2 Vascular plugging in wilt diseases. (a) Longitudinal section through the corm and stem base of a banana plant infected by *Fusarium oxysporum* f. sp. *cubense*. The xylem vessels of the corm are heavily discoloured by vascular gels, and some of the discoloured xylem vessels can be traced into the leaves. The infection initiated in the roots, as evidenced by the discoloured root xylem vessels that enter the corm. (b) Part of an infected banana corm cut transversely to show an infected root trace (arrow) and a ring of discoloured xylem vessels in the corm. At the top of the photograph is a developing 'sucker' (S) which will give rise to a new plant and would be used as planting material to propagate the crop. The infection is seen to be spreading into this sucker (arrowheads).

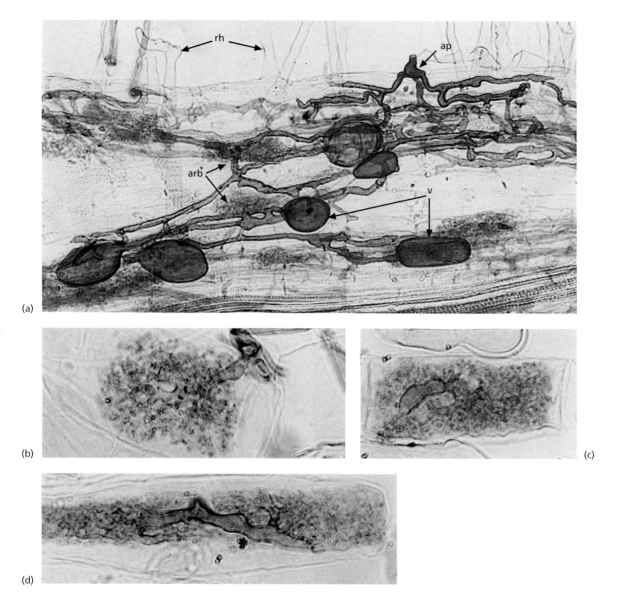

(a)

(b)

(c)

(d)

Plate 12.3 Arbuscular mycorrhizal fungus in naturally infected clover roots. The roots were heated in strong alkali to remove the host cytoplasmic components and to soften the tissues, then stained with trypan blue to reveal the fungal structures. (a) Partly crushed root showing an appressorium-like entry point (ap) and hyphae ramifying between the root cortical cells; several vesicles (v) and arbuscules (arb) are visible. The plant root hairs (rh) are not stained. (b–d) Three arbuscules in crushed root cortical cells at high magnification; only the main arbuscular branches are discernible because the finer branches are below the limit of resolution.

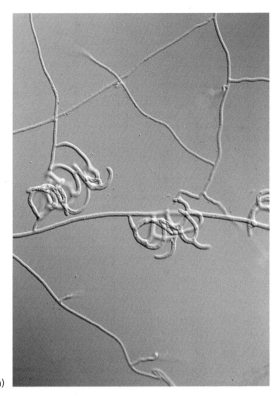

(a)

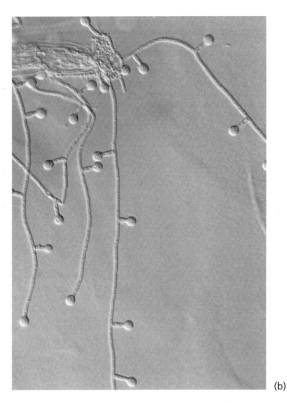

(b)

Plate 13.1 Nematode-trapping fungi. (a) Adhesive network of *Arthrobotrys oligospora*. (b) Adhesive knobs of *Monacrosporium ellipsosporum* on hyphae growing from a parasitized nematode (top left). (Photographs courtesy of B.A. Jaffee; from Jaffee, 1992.)

Plate 13.2 Zoospores of the nematode-parasitic fungus *Catenaria anguillulae* accumulating and encysting at the excretory pore of the nematode *Panagrellus redivivus*. The single posterior flagellum of one zoospore can be seen. (From Deacon & Saxena, 1997.)

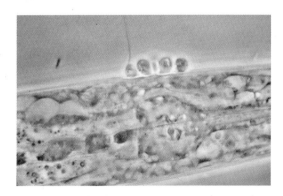

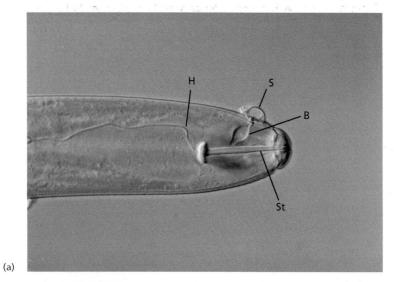

(a)

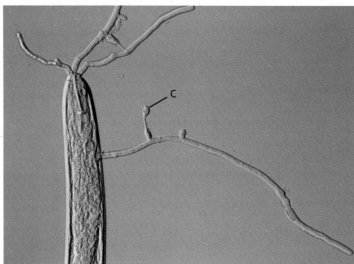

(b)

Plate 13.3 The endoparasitic fungus *Hirsutella rhossiliensis* on a nematode host. (a) Infection from a spore (S) that adhered near the nematode mouth. The fungus has penetrated the nematode cuticle and formed an infection bulb (B) within the host; a narrow infection hypha (H) can be seen growing from the base of the nematode stylet (St). (b) Fungal hyphae growing out from a fully colonized nematode and producing adhesive conidia (C). (Photographs courtesy of B.A. Jaffee; from Jaffee, 1992.)

at the time when *P. gigantea* was first developed as a biocontrol agent.

The history of this development is also notable, because it was done almost single-handedly by Rishbeth. Fortunately, *P. gigantea* is one of the relatively few basidiomycota that sporulates readily in culture; it produces brick-shaped conidia by hyphal fragmentation behind the colony margin. In early attempts to develop a biocontrol formulation, Rishbeth tried to produce tablets of dried conidia, but most of the spores lost their viability. Yet, if they were kept in liquid suspension they germinated and their storage life also was short. To overcome this, the spores were suspended in sucrose solution at a water potential that prevents their germination. In this way they can be marketed as sachets of dense spore suspensions that can be stored for at least 6 months in a refrigerator. The suspension is then diluted and applied to stump surfaces during routine felling and thinning operations; it incorporates a dye, bromocresol purple, so that foresters can check that the surfaces have been inoculated. This biocontrol method has proved highly effective and is used in pine forests over much of Europe and North America. However, it is only effective for pines, because *P. gigantea* is specialized as a weak parasite of pines; it is not effective on other conifers such as spruce, for which stump protection chemicals have to be used.

Mycoparasitism

Mycoparasites are defined simply as fungi that parasitize other fungi. Thus, according to the definition of parasites in general, they obtain some or all of their nutrient requirements from the living, functioning cells or tissues of another organism (in this case a fungal host) with which they live in intimate association. They can do this either by killing the host cells then feeding off them (**necrotrophic** parasites) or by absorbing nutrients from living host cells (**biotrophic** parasites).

We will see in Chapter 12 that these definitions and distinctions, although clear in principle, can be difficult to apply in practice, and especially for mycoparasites because fungal hyphae have a limited life-span in any case—they age and senesce in the older regions, or in response to nutrient stress. So, there is a grey area between 'true' mycoparasitism and other forms of antagonism such as contact-inhibition or antibiosis, and also between mycoparasitism and the ability of several fungi (the so-called fungicolous fungi) to grow on toadstools and other fruiting bodies of limited life-span.

Biotrophic mycoparasites

There are several types of biotrophic mycoparasite, with different feeding mechanisms (Jeffries & Young, 1994), but the most common and distinctive group are the **haustorial biotrophs**. These fungi penetrate living host hyphae to produce a specialized nutrient-absorbing haustorium inside the host wall but separated from the cell contents by a host cell membrane (Fig. 11.5). The host remains alive while the parasite feeds from it and produces a limited mycelium and sporulates on the host surface. Often this type of parasitism causes little damage, as long as the host has an adequate food supply.

This form of parasitism is shown by several zygomycota (*Piptocephalis*, *Dispira* and *Dimargalis* spp.) that have elongated, few-spored sporangia (merosporangia; see Chapter 1, Fig. 1.7). With only few exceptions, they parasitize other zygomycota such as *Mucor* and *Pilaira* on dung or in soil. Most of these mycoparasites can be grown in laboratory media containing extracts of host or non-host hyphae. The need for hyphal extracts can be replaced by supplying relatively high concentrations of vitamins (especially thiamine) and amino acids, and by providing glycerol instead of glucose as the carbon source. These nutrients could be expected to occur in the host hyphae. In any case, the haustorial biotrophs seem to depend entirely on their fungal hosts in nature. Their spores are triggered to germinate near host hyphae, and the germ-tubes show pronounced tropism towards the host. Then, the germ-tube tip produces an appressorium on the host surface and a penetration peg enters the host to form a haustorium. The sequence is similar to that for rust fungi discussed in Chapter 4.

For *Piptocephalis virginiana* there is evidence of specific recognition in the infection process (Manocha & Chen, 1990). This fungus infects only some members of the order Mucorales, so that the

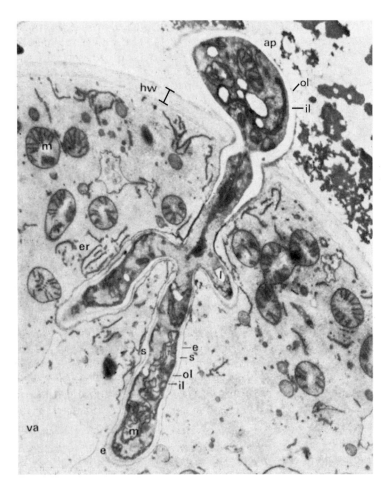

Fig. 11.5 Electron micrograph of an appressorium (ap) and a branched haustorium of the mycoparasite *Piptocephalis unispora* (zygomycota) in a fungal host *Cokeromyces recurvatus* (zygomycota). Hw, host wall; ol and il, outer layer and inner layer of the *Piptocephalis* wall; s, sheath, a fluid-filled layer surrounding the haustorium and bounded by a host membrane termed the extra-haustorial membrane (e); va, host vacuole; m, mitochondria; er, endoplasmic reticulum of the host fungus; l, lipid in the haustorium. (Courtesy of P. Jeffries; from Jeffries & Young, 1976.)

hosts can be compared with closely related non-hosts. The walls of the hosts were found to have two surface-located glycoproteins which could be removed by treating the hyphae with sodium hydroxide or proteinase, leading to impaired attachment and appressorium formation by the mycoparasite. Hyphae treated in this way were found to bind lectins that recognize fucose, *N*-acetylgalactosamine and galactose. In this respect the treated hyphae were identical to non-host hyphae, but the untreated host hyphae did not bind these lectins. So, it is suggested that the glyco-proteins that occur normally on host hyphae cover the fucose, *N*-acetylgalactosamine and galactose residues (which somehow interfere with infection) and enable the parasite to attach and infect. In support of this, the normal hosts are found to be resistant to parasitism if they are grown in liquid

culture media, but susceptible when grown on agar plates. The hyphae from liquid culture were found to have the three sugar residues on their surface, presumably because the glycoproteins that cover them were not produced in these conditions. We noted in Chapter 2 that hyphal walls can be covered with extrahyphal materials and that the production of these is strongly influenced by environmental conditions.

If the attachment of a mycoparasite can be hindered by specific sugar residues, then can it also be promoted by others? The answer seems to be 'yes' in the case of *P. virginiana*, because attachment was impaired if the parasite hyphae were pre-incubated in solutions of glucose or *N*-acetylglu-cosamine. This indicates that the mycoparasite has surface components that recognize these sugars on the host. But, both the host hyphae and non-host hyphae were found to have these sugars (again assessed by the use of lectins that recognize specific sugar residues). So, we can summarize the host–parasite recognition system in a simple (but speculative) model: the hyphae of both hosts and non-hosts seem to have exposed glucose and N-acetylglucosamine residues that mediate the attachment of the mycoparasite, but non-hosts have other exposed sugar residues that interfere with this process, whereas these other sugar residues are obscured by two surface glycoproteins on host hyphae.

Control of potato black scurf by Verticillium biguttatum

V. biguttatum (deuteromycota) is a haustorial biotroph with potential for use in biocontrol of *Rhizoctonia solani* (basidiomycota). It is a naturally occurring parasite of *R. solani*, especially of the strains that cause black scurf of potatoes — the familiar black or brown crusts on the tuber surface. These scurfy patches are sclerotial masses of *Rhizoctonia* and they significantly reduce the marketability of the crop even though the damage is only cosmetic. Like all the haustorial biotrophs, *V. biguttatum* has relatively little effect on growth of its fungal host. It penetrates from germinating spores, produces haustoria (Fig. 11.6) and then forms a limited mycelium outside the host, where

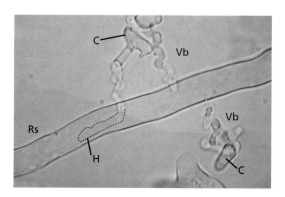

Fig. 11.6 Hypha of *Rhizoctonia solani* (Rs) infected by the biotrophic mycoparasite *Verticillium biguttatum* (Vb). Conidia (C) of Vb have germinated and the germ tubes grew in a spiral manner towards the *Rhizoctonia* hypha, then penetrated the living host hypha to produce club-shaped haustoria (H), the outline of one is shown. (From van den Boogert & Deacon, 1994.)

it sporulates. The infected host hyphae grow more or less unimpeded, but even a localized infection of a host colony can markedly suppress the production of sclerotia over the colony (Fig. 11.7). Apparently, *V. biguttatum* creates a nutrient-sink towards itself within the host mycelial network, counteracting the normal nutrient sink towards developing sclerotia, discussed in Chapter 4.

V. biguttatum can be grown easily on normal laboratory media, and it produces abundant spores for use in biocontrol. These spores have been shown to reduce black scurf in field experiments, but there is one significant limitation: *V. biguttatum* needs a relatively high temperature (at least 12°C) for growth, whereas *Rhizoctonia* can start to grow at about 4°C so it colonizes potato plants earlier in the growing season. This problem would not be important if potato tubers could be inoculated later in the growing season because sclerotia are produced only when the potato skins begin to mature. Agronomists have developed potential ways of doing this in field conditions (van den Boogert *et al.*, 1994). They are beyond the scope of this text but they illustrate how the exploitation of fungi for biocontrol of diseases often depends on changes in cropping practice or crop technology.

Necrotrophic mycoparasites

Necrotrophic mycoparasites are quite different from the biotrophs discussed above. They have characteristically wide host ranges, and in many cases they produce inhibitory toxins or other metabolites as part of the parasitic process. But, it can be difficult to define the boundaries of necrotrophic mycoparasitism, because many antagonistic fungi (e.g. *Trichoderma*) might feed off other fungi that they have killed by antibiosis etc. Below, we consider three examples to illustrate these points.

Antagonism by Gliocladium roseum

G. roseum (deuteromycota) is a common soil fungus that often colonizes organic matter in the later stages of decomposition, as we saw for grass roots in Chapter 10. It is also one of the few fungi that overgrow other fungal colonies on agar if soil particles are sprinkled on pre-colonized agar plates (see Table 11.2). *G. roseum* can kill some

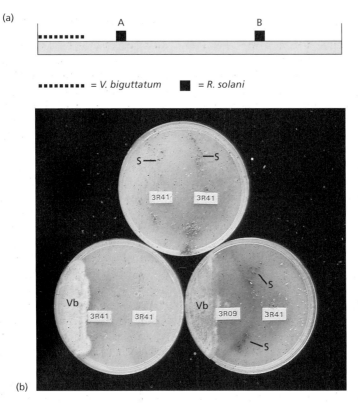

Fig. 11.7 Effect of the biotrophic mycoparasite *Verticillium biguttatum* (Vb) on production of sclerotia by its fungal host, *Rhizoctonia solani*. (a) Experimental design: plates of mineral nutrient agar containing powdered cellulose as the carbon source were inoculated at two positions (A and B) with inoculum blocks of *R. solani*, then spores of Vb were added to the margin of some agar plates. (b) Top: control plate double-inoculated with *R. solani* strain 3R41 but without *V. biguttatum*. *R. solani* grew from both inoculum blocks and the colonies anastomosed where they met, then they produced many sclerotia (S) behind the zone of fusion. Bottom left: similar plate double-inoculated with *R. solani* strain 3R41 but *V. biguttatum* was added at the margin of one colony. Both *R. solani* colonies grew from the inoculum blocks and anastomosed. Parasitism of one of the colonies suppressed the production of sclerotia over the whole agar plate, presumably by creating a nutrient sink in the host hyphal network. Bottom right: plate inoculated with strains 3R09 and 3R41 of *R. solani*, and spores of *V. biguttatum* were placed on strain 3R09. The two *R. solani* strains anastomosed where they met but this led to cytoplasmic death because of vegetative incompatibility (seen as a dark zone between the colonies). *V. biguttatum* suppressed sclerotium production on the inoculated colony (3R09), but not on strain 3R41. The same effect was seen when *V. biguttatum* was added to strain 3R41 and not 3R09, showing that the suppressive effect cannot cross a cytoplasmic incompatibility barrier. (From van den Boogert & Deacon, 1994.)

fungi by producing diffusible inhibitors in nutrient-rich conditions, but on water agar it antagonizes the hyphae after contact, causing localized vacuolation and loss of turgor about 30–90 min after contact. This is slower than hyphal interference by basidiomycota, which can take only a few minutes. *G. roseum* also branches and coils round the damaged compartments, unlike

the basidiomycota, so it seems to use them as a nutrient source. Thus, in nature *G. roseum* is probably a secondary (opportunistic) invader of decomposing organic matter, gaining some of its nutrients by antagonizing living hyphae, some by exploiting dead hyphae and some from the underlying substratum. Like *Trichoderma*, it is a difficult fungus to categorize.

Antagonism by Talaromyces flavus

T. flavus (ascomycota) has a *Penicillium*-like asexual stage. It first attracted attention as a parasite of *R. solani* because it coils round the hyphae on agar plates, penetrating them and causing localized disruption. However, recent interest has focused on its potential to control the vascular wilt pathogen *Verticillium dahliae* (see Chapter 12). It invades the melanized microsclerotia of *Verticillium* on diseased roots, and sporulates on the surface of these structures. *T. flavus* produces up to four antibiotics in culture, one of them being an antifungal compound, **talaron**. However, it exerts its main effect—at least on the hyphae of *V. dahliae*—by secreting the enzyme **glucose oxidase**, which generates hydrogen peroxide from glucose (Kim *et al.*, 1988). Consistent with this, *Talaromyces* was found to antagonize *V. dahliae* only when glucose was present in culture media, and not when other common sugars were supplied instead of glucose. A commercial source of glucose oxidase had the same effect on *V. dahliae* hyphae, but only in the presence of glucose; these effects could be reproduced by low concentrations of hydrogen peroxide alone. It seems that *Talaromyces* has several potential mechanisms for antagonizing other fungi; they might act separately or in combination.

Mycoparasitic Pythium *spp.*

Most *Pythium* spp. are plant pathogens (see Chapter 12), but six species are non-pathogenic to plants and instead they parasitize other fungi. The most common of these is *P. oligandrum* which has distinctive spiny-walled oogonia (Fig. 11.8). Like *Trichoderma* and *Gliocladium*, mentioned earlier, it can grow from soil crumbs onto agar previously colonized by other fungi (see Table 11.2; Fig. 11.9). Using this as a method of detection, *P. oligandrum* has been found in nearly 50% of the soils sampled in Britain, especially from agricultural sites. It is equally common in the USA, continental Europe and New Zealand. An interesting feature seen in Table 11.2 is that different 'host' fungi on the pre-colonized agar plates tend to select for different mycoparasites from soil. So, perhaps the necrotrophic mycoparasites have a degree of 'host-specificity' in nature, insofar as all of them are secondary opportunistic invaders of decomposing organic matter, but they might preferentially colonize materials in which different fungi have grown.

P. oligandrum is a remarkably aggressive mycoparasite in culture, behaving differently from the mycoparasites discussed above. It does not produce antibiotics or other diffusible toxins but it

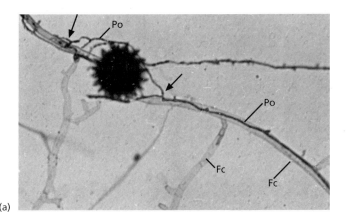

(a)

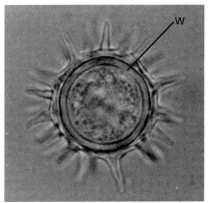

(b)

Fig. 11.8 (a) Parasitism of a hypha of *Fusarium culmorum* (Fc) by narrow hyphae of *Pythium oligandrum* (Po) on water agar. Entry points are marked by arrows. *P. oligandrum* has produced a spiny oogonium from the host hyphal nutrients. The hyphae were stained with cotton blue, showing that only *P. oligandrum* has hyphal contents; the host hyphae have no cytoplasm. (b) A spiny oogonium of *P. oligandrum* containing a mature oospore; the oospore wall (w) is clearly seen. There are no antheridia because *P. oligandrum* typically produces oospores by parthenogenesis.

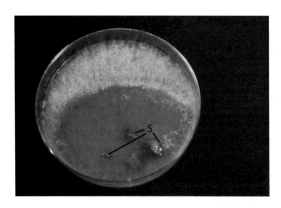

Fig. 11.9 Agar plate previously colonized by a fungus (*Phialophora* species) and then 'seeded' with crumbs of fresh soil (S). *Pythium oligandrum* has grown from the soil crumbs, destroying the hyphae of *Phialophora* and causing its white aerial mycelium to collapse.

can destroy the hyphae of several other fungi within 3–5 min of contact. It does this in two ways (Fig. 11.10).

1 It can cause the host hyphal tips to lyse within 60–90 s after they have made contact with a *P. oligandrum* hypha: the host protoplasm is expelled forcibly from a pore in the wall at the contact site, then *P. oligandrum* produces branches that grow in the spilled contents.

2 If a tip of *P. oligandrum* contacts a subapical region of a host hypha the mycoparasite tip grows on, but a branch develops at the contact site within 3–5 min and penetrates the host hypha. Then, the mycoparasite grows within the host, breaking through the septa and destroying each host compartment in turn, like a train in a metro system.

Some fungi are not invaded by *P. oligandrum* until 1 or 2 h after contact, and during this time the mycoparasite coils round the host hypha from branches that originate at the initial contact point. It then penetrates from beneath the coils. This is commonly seen in interactions with plant-pathogenic *Pythium* spp. and the older hyphae of *R. solani*. Some other fungi (e.g. many basidiomycota or antibiotic-producing species) seem to be immune or avoid being parasitized because they antagonize the mycoparasite.

Little is known about the fundamental mechanisms of these interactions, but they must be related closely to the mechanisms of apical growth and hyphal branching discussed in Chapter 3. For example, it is possible that *P. oligandrum* induces changes in the actin cytoskeleton at the tips of host hyphae, causing protoplasm to be released through a breach in the plastic hyphal wall. The branching of *P. oligandrum* at the point of contact with a host (after the *P. oligandrum* tip has grown on) might occur in response to a specific recognition signal (it does not do this after contact with root hairs) or in response to a temporary slowing of the original tip—an apical disturbance like that induced by osmotic shock.

P. oligandrum and the other mycoparasitic *Pythium* spp. have the potential to be used for biocontrol of plant pathogens, especially as seed treatments for control of seedling pathogens. *P. oligandrum* grows rapidly on cheap commercial substrates such as molasses, and it produces large numbers of oospores in shaken liquid culture. These oospores have been applied to seeds of cress and sugar beet, using a commercial seed-coating process, and they gave good protection against seedling disease caused by *P. ultimum* in experimental conditions (McQuilken *et al.*, 1990). There are also two commercial powder formulations of *P. oligandrum*, but they are not used widely. The major problem is that the oospores germinate slowly and poorly because of their constitutive dormancy (see Chapter 9). Only a maximum 10–40% of the oospores freshly harvested from culture broth will germinate readily, and this value reduces to 5% or less after the oospores have been dried for storage.

P. oligandrum also has a role as a natural biocontrol agent, as Martin and Hancock (1986) showed in the irrigated cotton crops of California. Some fields were found to be naturally suppressive to seedling disease caused by *P. ultimum* whereas other fields were heavily diseased (disease-conducive soils). When soils from the two types of field were analysed chemically, the suppressive soils had a generally higher chloride content in the topsoil, caused by evapotranspiration of water when irrigation was stopped near harvest time, which brought dissolved salts to the soil surface. These suppressive soils also had consistently higher resident populations of *P. oligandrum*. Experiments confirmed the role of *P. oligandrum* in disease suppression (Fig. 11.11). When fresh, green cotton leaves were added to conducive soils the leaves were colonized rapidly by *P. ultimum*,

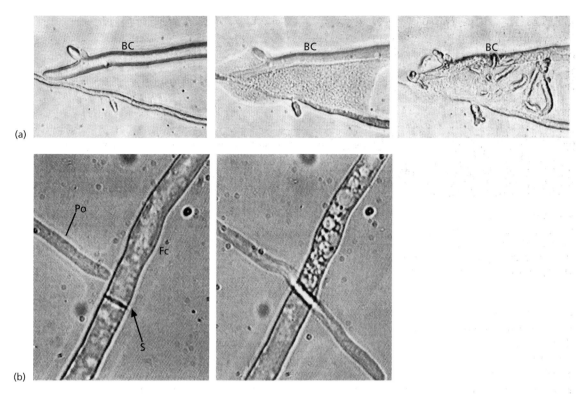

(a)

(b)

Fig. 11.10 Effect of *Pythium oligandrum* on susceptible host fungi. From Laing & Deacon (1991). (a) A hyphal tip of *Botrytis cinerea* (Bc) has contacted a hypha of *P. oligandrum* (first frame) and lysed about 4 min later (centre frame), releasing its hyphal contents. By 36 min after contact (third frame) the *P. oligandrum* hypha has formed numerous branches in the spilled host contents and also has penetrated the host hypha. (b) A hyphal tip of *P. oligandrum* (Po) has contacted a sub-apical region of a hypha of *Fusarium culmorum* (Fc) just behind a host septum (s). Five minutes later, the hyphal tip of *P. oligandrum* has grown on, but it branched and penetrated the Fc hypha at the initial point of contact, causing cytoplasmic coagulation in the penetrated host compartment. *P. oligandrum* subsequently grew through the damaged compartment and colonized the Fc hypha through successive septa.

which formed large numbers of oospores. Leaves added to suppressive soils were also colonized initially by *P. ultimum* but *P. oligandrum* then invaded the leaves 1 or 2 days later and became the dominant fungus, suppressing the growth and production of oospores by the pathogen. So, it seems that fresh crop residues added to soil at the end of the cropping season would be colonized by *P. ultimum*, increasing the population of oospores for infection of a subsequent crop. But, in soils of higher chloride content the mycoparasite is favoured over *P. ultimum*, making these soils disease-suppressive so fewer oospores of the pathogen are produced for infection of the next crop. This was confirmed by experiments in which either chloride or *P. oligandrum* oospores were added to disease-conducive soils at the levels known to occur in suppressive soils. Then, the conducive soils became disease-suppressive. This is a classic example of naturally occurring disease control, caused by a mycoparasite, but the mycoparasite is favoured by some environmental factor — in this case, a relatively high soil chloride content.

Competition

The term competition, as defined at the start of this chapter, describes all the types of interaction in which one organism gains ascendancy over another because it arrives sooner, grows faster or uses the substrate more efficiently, etc. It is probably the most common type of interaction in natural environments, but this is difficult to prove. Here we consider three cases where the evidence is strong.

Control of take-all by *Phialophora graminicola*

As noted in earlier chapters, the take-all fungus *Gaeumannomyces graminis* attacks the roots of cereals and grasses. The variety *tritici* causes major yield losses of wheat grown continuously, without crop rotation, because the fungus survives in soil in the dead host tissues. The larger pieces of crop residue provide a food base for infection of the next wheat crop. However, during a 1- or 2-year break from cereals these residues fragment and decay, so that the fungus has a negligible food base. In these conditions it behaves like *P. graminicola*, described in Chapter 10: it must wait for the root cortical cells to senesce and colonize them to increase its food base progressively for infection of the living root tissues. *G. graminis* also produces air-borne ascospores. In most conditions they are

of little importance because of their low food reserves.

G. graminis var. avenae attacks oats (see Chapter 8) and the fine turf grasses (*Agrostis* spp.) on golf-course greens and other playing surfaces, causing take-all patch disease (Plate 11.2, facing p. 210). It behaves similarly to *G. graminis var. tritici* except for its different host range. However, the interesting point is that take-all patch of turf grasses is a rare disease, occurring only in specific conditions. It is seen when established turf is limed heavily to correct for over-acidity. This happens after several years of applying ammonium fertilizers to the grass, because ammonium is absorbed by roots and exchanged for hydrogen ions (H+), causing progressive lowering of the soil pH. The disease also occurs in newly planted turf, especially if the soil has been fumigated to destroy any pathogens. In both of these circumstances the pathogen infects from air-borne ascospores that are washed into the root zone, and the disease patches spread rapidly for 2 or 3 years and then stop spreading. These effects are explained by the fact that *P. graminicola* (Fig. 11.12) and similar weak parasites are natu-

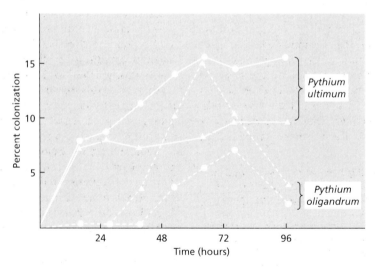

Fig. 11.11 Colonization of pieces of cotton leaf residue by *Pythium ultimum* (solid lines) and *Pythium oligandrum* (broken lines) from natural inoculum in soil. The soil was a disease-conducive soil and was either untreated (circles) or supplemented with NaCl (triangles) at a level equivalent to the salt level found in naturally suppressive soils in Californian cotton fields. In all cases the pathogen *P. ultimum* colonises faster than *P. oligandrum*, but the rate of colonization by *P. ultimum*

declines once *P. oligandrum* starts to colonize. *P. oligandrum* is favoured by the higher salt content, giving additional control of *P. ultimum*. (Note that the later decline of *P. oligandrum* is an artefact—this fungus rapidly converts its mycelial nutrients into dormant oospores that do not give rise to colonies when the residues are plated to assess the population levels.) (Reproduced from Martin & Hancock, 1986.)

rally abundant on grass roots. They are specialized to exploit the naturally senescing root cortical tissues, so they compete with *G. graminis* for the food base that it requires when infecting from ascospores with minimal nutrient reserves. These fungi are destroyed by soil fumigation, and their populations also decline progressively as the turf pH is lowered (Table 11.3). So, the liming of turf creates conditions ideal for infection by *G. graminis* in the absence of its natural competitors. Many experiments have confirmed this, showing that *P. graminicola* and other weak parasites like *Idriella bolleyi* (see Chapter 3, Fig. 3.8) can prevent take-all infection of both cereals and grasses. They do so by competition for the naturally senescing root cortical tissues—a form of biocontrol termed **competitive niche exclusion**. The remarkable feature in this case is that the fine turf grasses (*Agrostis* spp.) are highly susceptible to take-all disease, and yet the disease is rare, because it is kept in check naturally by the resident biocontrol agents of the root zone.

Control of *Pithomyces chartarum*, the cause of facial eczema

P. chartarum (deuteromycota) is a saprotroph which grows on the accumulated dead leaf sheaths at the bases of pasture grasses. It is common in parts of New Zealand, Australia and South Africa, where it causes facial eczema of sheep. In this condition, the animals develop blistering sores when their faces are exposed to sunlight. However, this is a secondary symptom. The primary cause is a mycotoxin, **sporidesmin**, which is present only in the spores of *Pithomyces* and which is ingested when sheep graze on pastures where the fungus is sporulating. It causes necrosis of the liver and scarring and partial blockage of the bile duct, so that the partial breakdown products of chlorophyll, ingested in the feed, accumulate in the blood. They are photoactive compounds so they cause photosensitization of the skin where it is not covered by hair.

P. chartarum requires quite specific conditions for sporulation—a combination of relatively high temperature and high humidity over a period of days. So, in countries where facial eczema is common the farmers are warned by radio broadcast that the climatic conditions are suitable and the animals can be removed temporarily from the pastures. Only some fields on a farm are 'facial-eczema' fields, suggesting that *Pithomyces* might be controlled naturally by some agents. Although nothing is known about this, there are some clues from the use of fungicides, because *Pithomyces* can be controlled by the benzimidazole compounds (see Chapter 14). As shown in Fig. 11.13, these fungicides not only suppressed *Pithomyces*, assessed by spore production from the pastures, but also stimulated the levels of some other fungi such as *Alternaria tenuis*. This is consistent with laboratory studies where the growth of *Pithomyces* was strongly inhibited by even low fungicide levels, whereas *Alternaria* was unaffected by even large fungicide doses. *Alternaria* is one of the common fungi on leaf surfaces (see Chapter 7) and also on above-ground plant remains. The fact that its population increases markedly after fungicide applications in the field indicates that it has expanded its territory, at the expense of the fungi such as *Pithomyces* which were inhibited by the fungicide. In other words, *Alternaria* and *Pithomyces* are part of a fungal community with overlapping niches, so that any factor which suppresses one of them will enable the others to take its place. The benzimidazole fungicides are no longer effective in controlling *Pithomyces* in parts of New Zealand because fungicide-resistant mutants have developed in response to repeated fungicide applications. However, it seems likely that *Alternaria* (and other fungi) will have delayed the development of resistance since the early 1960s when these fungicides were first applied. They

Table 11.3 Relationship between turf pH and incidence of *Phialophora graminicola*, assessed by planting wheat assay seedings in cores of turf from golf-course greens. (Data from Deacon, 1973.)

pH	Per cent of wheat assay roots with *P. graminicola*	
	Yorkshire sites	**East Anglian sites**
<4.0	<1	0
4.0–4.5	3	22
4.5–5.0	21	50
5.0–5.5	25	60
5.5–6.0	48	88
>6.0	100	99

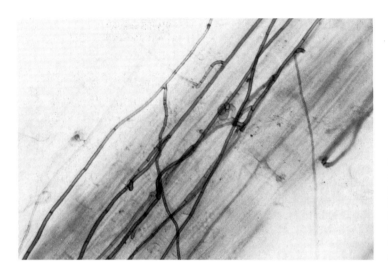

Fig. 11.12 Wheat seminal root colonized by *Phialophora graminicola*, showing dark runner hyphae but no vascular discoloration (compare with Fig. 8.10, where the take-all fungus caused intense vascular damage in similar conditions).

would have done so by filling the niche of *Pithomyces* so that this fungus could not re-establish easily in fungicide-treated pastures.

Control of ice-nucleation active bacteria on leaves

Both examples above relate to competitive niche exclusion. To reinforce them we can deal briefly with a third example, where competition has been shown to occur between strains of a single species. The example involves a bacterium, *Pseudomonas syringae*, some strains of which have natural surface proteins with ice-nucleation activity. The presence of these bacteria on plant leaves promotes the formation of ice crystals when the air temperature falls slightly below 0°C; in fact, these bacteria are used in artificial snow-making machines! If these bacteria are absent then the temperature can cool to −4°C or less before icing occurs, and this difference can be important for frost-sensitive plants. It has been shown that mutant strains (ice−) lacking the surface protein can be pre-inoculated onto leaves to exclude the wild-type (ice+) strains and give significant protection against frost injury. Of most interest, however, an ice− strain is significantly more effective against its ice+ parent than against a non-parental ice+ strain. In other words, even within a single species there seem to be a diversity of wild-type strains that occupy slightly different microhabitats (or have different patterns of resource utilization on a leaf). Competitive niche exclusion operates most effectively when an ice− mutant is similar to a wild-type strain in all other respects (Lindow, 1992).

Commensalism/mutualism

If we exclude lichens as organisms in their own right, then there are few well-established examples of fungi that live in association with other fungi or with other micro-organisms to their mutual benefit (**mutualism**) or to the benefit of one and not to the detriment of the other (**commensalism**). These associations may well exist, but they are not well documented. Here, we consider only one example.

The non-cellulolytic fungus *Thermomyces lanuginosus* is found during much of the high-temperature phase of composts while cellulolytic fungi such as *Chaetomiun thermophile* and *Humicola insolens* are active (see Chapter 10). It is assumed that *Thermomyces* grows in these conditions by using sugars made available by the cellulolytic fungi. However, this is difficult to demonstrate in composts, so most of the evidence has come from *in vitro* studies. The methods used were similar to those described in Chapter 10: flasks of sterile filter paper with nitrate and other mineral nutrients were inoculated with *C. thermophile* (Ct) alone, *T. lanuginosus* alone (Tl) or *Chaetomium* in combination with *T. lanuginosus* (Ct + Tl), and the loss of dry weight of the flask contents was assessed at different times (Fig. 11.14).

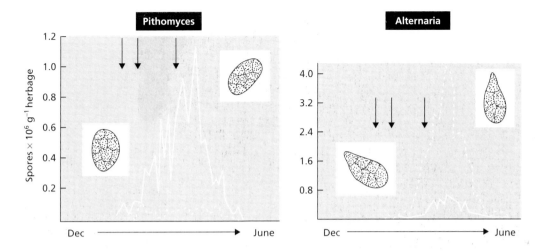

Fig. 11.13 Numbers of spores of *Pithomyces chartarum* and *Alternaria tenuis* in New Zealand pasture during a growing season. Solid line, untreated pasture; broken line, pasture treated with the fungicide benomyl at three times marked by arrows. The distinctive spore (conidia) shapes of these fungi are shown. (Based on McKenzie, 1971.)

and nitrogen source from the association. Yet, there was no evidence of parasitism, and this was confirmed by study of hyphal interactions on agar.

3 *Thermomyces* in some way enhanced the break-

Thermomyces could not grow alone in these conditions, because it cannot degrade filter paper cellulose and it cannot use nitrate as a nitrogen source. *Thermomyces* also grew very poorly with *Chaetomium* at 37°C (well below its temperature optimum – see Chapter 7, Fig. 7.2) and did not affect *Chaetomium*: about 1 g of cellulose was degraded by Ct alone and by Ct + Tl after 4 weeks. However, *Thermomyces* grew very well in combination with *Chaetomium* at 45°C, and the weight loss was always significantly greater for Ct + Tl than for Ct alone. Figure 11.14 shows that this was true whether nitrogen was used at a 'standard' level (C:N ratio of 175:1) or double level (C:N of 88:1). In each case, the early rate of decomposition was the same for Ct and Ct + Tl, but the rate of decomposition was maintained for longer by Ct + Tl than by Ct alone. The interpretation of even simple experiments like these is very difficult, but a number of points can be made.

1 The interaction of *Thermomyces* and *Chaetomium* is temperature dependent.

2 At the higher temperature *Thermomyces* grew well in association with *Chaetomium*, and it must have obtained both a carbon source (sugars)

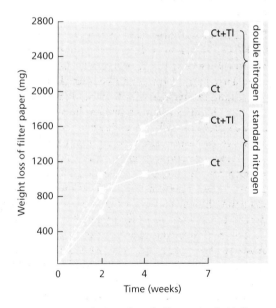

Fig. 11.14 Weight loss of sterile filter paper (initially 7-g dry weight) in flasks inoculated with *Chaetomium thermophile* alone (Ct) or *C. thermophile* and the non-cellulolytic fungus *Thermomyces lanuginosus* (Ct + Tl) and incubated at 45°C. Mineral nutrients were supplied to the flasks with a standard amount of nitrogen (C:N ratio 175:1) or double nitrogen (C:N = 88:1). (From Deacon, 1985.)

down of cellulose by *Chaetomium*. This might be expected from knowledge of the regulation of cellulase enzymes (see Chapter 5), because any sugars that accumulate would slow the rate of enzyme action and also repress the synthesis of further enzymes. So, by using some of these sugars *Thermomyces* might overcome this negative feedback.

4 *Thermomyces* must also have influenced the efficiency of nitrogen usage so that nitrogen was available to maintain the rate of cellulose breakdown for longer than when *Chaetomium* was growing alone. We saw in Chapter 10 that *Chaetomium* soon becomes nitrogen-limited, presumably because it cannot recycle nitrogen efficiently.

It would be interesting to know if *Chaetomium* benefits in some way from the interaction, perhaps by increasing its own biomass or by staving off its replacement by other fungi in composts. However, there is no evidence on this point, so the example must be described as commensalism. It shows, at least, that fungi do not always have negative impacts on one another.

References

Berdy, J. (1974) Recent developments of antibiotic research and classification of antibiotics according to chemical structure. *Advances in Applied Microbiology*, **18**, 309–406.

Chet, I. (1987) *Trichoderma*—application, mode of action, and potential as a biocontrol agent of soil-borne plant pathogenic fungi. In: *Innovative Approaches to Plant Disease Control* (ed. I. Chet), pp. 137–60. Wiley, New York.

Deacon, J.W. (1973) Factors affecting occurrence of the Ophiobolus patch disease of turf and its control by *Phialophora radicicola*. *Plant Pathology*, **22**, 149–55.[Note: *Ophiobolus* is now called *Gaeumannomyces*; P. *radicicola* is now called *P. graminicola*]

Deacon, J.W. (1983) *Microbial Control of Plant Pests and Diseases*. Van Nostrand Reinhold, Wokingham.

Deacon, J.W. (1985) Decomposition of filter paper cellulose by thermophilic fungi acting singly, in combination, and in sequence. *Transactions of the British Mycological Society*, **85**, 663–9.

Howell, C.R., Stipanovic, R.D. & Lumsden, R.D. (1993) Antibiotic production by strains of *Gliocladium virens* and its relation to the biocontrol of cotton seedling diseases. *Biocontrol Science and Technology*, **3**, 435–41.

Ikediugwu, F.E.O. (1976) The interface in hyphal interference by *Peniophora gigantea* against *Heterobasidion*

annosum. *Transactions of the British Mycological Society*, **66**, 291–6, 281–90. [Note that *Peniophora* is now called *Phlebiopsis*.]

Ikediugwu, F.E.O. & Webster, J. (1970) Antagonism between *Coprinus heptemerus* and other coprophilous fungi. *Transactions of the British Mycological Society*, **54**, 181–204.

Jeffries, P. & Young, T.W.K. (1976) Ultrastructure of infection of *Cokeromyces recurvatus* by *Piptocephalis unispora* (Mucorales). *Archives of Microbiology*, **109**, 277–88.

Jeffries, P. & Young, T.W.K. (1994) *Interfungal Parasitic Relationships*. CAB International, Wallingford.

Kim, K.K., Fravel, D.R. & Papavizas, G.C. (1988) Identification of a metabolite produced by *Talaromyces flavus* as glucose oxidase and its role in the biocontrol of *Verticillium dahliae*. *Phytopathology*, **78**, 488–92.

Laing, S.A.K. & Deacon, J.W. (1991) Video microscopical comparison of mycoparasitism by *Pythium oligandrum*, *P. nunn* and an unnamed *Pythium* species. *Mycological Research*, **95**, 469–79.

Lindow, S. (1992) Ice⁻ strains of *Pseudomonas syringae* introduced to control ice nucleation active strains on potato. In: *Biological Control of Plant Diseases; Progress and Challenges for the Future* (eds E.C. Tjamos, G.C. Papavizas & R.J. Cook), pp. 169–74. Plenum Press, New York.

McKenzie, E.H.C. (1971) Seasonal changes in fungal spore numbers in ryegrass-white clover pasture, and the effects of benomyl on pasture fungi. *New Zealand Journal of Agricultural Research*, **14**, 379–92.

McQuilken, M.P., Whipps, J.M. & Cooke, R.C. (1990) Control of damping-off in cress and sugar-beet by commercial seed-coating with *Pythium oligandrum*. *Plant Pathology*, **39**, 452–62.

Manocha, M.S. & Chen, Y. (1990) Specificity of attachment of fungal parasites to their hosts. *Canadian Journal of Microbiology*, **36**, 69–76.

Martin, F.M. & Hancock, J.G. (1986) Association of chemical and biological factors in soils suppressive to *Pythium ultimum*. *Phytopathology*, **76**, 1221–31.

Mulligan, D.F.C. & Deacon, J.W. (1992) Detection of presumptive mycoparasites in soil placed on host-colonized agar plates. *Mycological Research*, **96**, 605–8.

O'Neill, T.M., Elad, Y., Shtienberg, D. & Cohen, A. (1996) Control of grapevine grey mould with *Trichoderma harzianum* T39. *Biocontrol Science and Technology*, **6**, 139–46.

Rishbeth, J. (1963) Stump protection against *Fomes annosus*. III. Inoculation with *Peniophora gigantea*. *Annals of Applied Biology*, **52**, 63–77. [Note: *Fomes annosus* is now called *Heterobasidion annosum*; *Peniophora* is now called *Phlebiopsis*.]

Thomashow, L.S., Weller, D.M., Bonsall, R.F. & Pierson III, L.S. (1990) Production of the antibiotic phenazine-1-carboxylic acid by fluorescent *Pseudomonas* species in the rhizosphere of wheat. *Applied and Environmental Microbiology*, **56**, 908–12.

van den Boogert, P.H.J.F & Deacon, J.W. (1994) Biotrophic mycoparasitism by *Verticillium biguttatum* on *Rhizoctonia solani*. *European Journal of Plant Pathology*, **100**, 137–56.

van den Boogert, P.H.J.F., Kastelein, P. & Luttikholt, A.J.G. (1994) Green-crop harvesting, a mechanical haulm destruction method with potential for disease control of tuber pathogens of potato. In: *Seed Treatment: Progress and Prospects* (ed. T. Martin), pp. 237–46. British Crop Protection Council, Farnham.

Weindling, R. (1934) Studies on a lethal principle effective in the parasitic action of *Trichoderma lignorum* on *Rhizoctonia solani* and other soil fungi. *Phytopathology*, **24**, 1153–79.

Chapter 12
Fungi as plant parasites

Fungi are pre-eminent as plant pathogens. About 70% of the major crop diseases are caused by fungi, with an economic cost of billions of dollars a year. Much of this chapter will be devoted to these pathogens and their adaptations. We also deal with non-pathogenic parasites, including the important mycorrhizal fungi which help to sustain much of the vegetation on earth.

It is worth defining some terms again. A **parasite** is an organism that gains all or part of its nutritional requirements from the functioning tissues of another organism (host) with which it lives in intimate association. Some people prefer to use the term **symbiont** instead of parasite, justifying this on historical grounds, but symbiosis now has a different meaning in everyday use, and no sensible plant pathologist would dream of telling a farmer that his crop has been destroyed by a symbiont! Semantics aside, the important point is that parasitism is a feeding relationship and it does not imply harm or benefit. For example, mycorrhizal fungi often benefit the host by aiding mineral nutrient uptake, but at a cost to the plant in terms of a drain on photosynthate, and if mineral nutrients are plentiful then the cost can outweigh the advantage. The inconspicuous fungal endophytes also might be beneficial in protecting plants from pathogenic attack or insect feeding (see Chapter 6), but not invariably so. The term 'parasitism' adequately describes all these relationships, and the term **pathogen** can then be used for parasites that cause overt disease.

Major groups of plant-pathogenic fungi

There are various ways of grouping the fungal pathogens of plants, but the following scheme will best serve our needs. We deal with these groups in the following sections and use them where appropriate to illustrate principles of plant pathology.

1 Pathogens of immature or compromised tissues. These fungi are characterized by their broad host range and rapid, aggressive invasion, often involving extensive destruction of tissues by enzymes or toxins.

2 Pathogens of mature, non-compromised tissues. These fungi typically show a high degree of host-specialization.

(a) **Necrotrophic pathogens**: characterized by tissue invasion, induction of host cell death and the ability to tolerate or overcome host-resistance mechanisms.

(b) **Biotrophic pathogens**: characterized by limited tissue invasion, and maintenance of host cell viability; they do not induce the host-resistance mechanisms.

3 Endophytes that incidentally cause host death (specifically, the vascular wilt pathogens).

Pathogens of immature or compromised tissues

A wide range of fungi fall into this category. In the past they have been termed 'unspecialized' pathogens, but it is more accurate to say that they are specialized to exploit immature, wounded or senescent plant tissues, as this is their major means of growth. Their host ranges are characteristically wide, because they exploit tissues that do not exhibit the major defence mechanisms that vary between plant types. Nevertheless, the different pathogens within this group tend to be associated with specific types of disease. The following examples illustrate these points.

Seedling pathogens

A quite narrow range of pathogens cause seedling diseases, sometimes termed **damping-off** diseases. They include *Pythium* spp. which attack the emerging root tips, *Rhizoctonia solani* which characteristically attacks the young shoot base, *Sclerotium rolfsii* which also attacks basal stem tissues, especially in the warmer climates, and *Fusarium* spp. which are seed-rot and seedling pathogens, notably of cereals. All these fungi grow rapidly and can overwhelm the host in a short time. However, infection often depends on predisposing environmental factors, such as wet soil conditions for the *Pythium* spp. In waterlogged soils the seeds and emerging tissues release large amounts of nutrients which trigger germination of the soil-borne spores and also support pre-penetration growth. In fact, a large degree of pre-penetration growth is often essential for seedling pathogens, and inherent differences in nutrient exudation from seeds of different cultivars are strongly correlated with disease susceptibility. As one example, Hayman (1970) compared two breeding lines of cotton, one of which (termed W) was more susceptible to *R. solani* than was the other, termed G. Fifty seeds of each type were allowed to germinate in aseptic conditions for 48 h, then the exudates (mainly amino acids and sugars) were collected and inoculated with *R. solani*; the exudates from W seeds supported 66 mg (dry weight) of mycelial growth, whereas those from G seeds supported only 19 mg. Furthermore, when G seeds were soaked in exudate from W seeds and then sown in soil, they suffered twice as much damping-off as when G seeds were soaked in G exudate. Planting of W seeds next to G seeds also increased the amount of disease of the G seeds. The practical implications of this are that crop cultivars must be selected for inherently low seed exudation rates, and seeds must be sown in conditions that do not enhance nutrient exudation.

Many types of seedling disease can be controlled by broad-spectrum seed-applied fungicides (see Chapter 14). This is economically viable because these compounds are cheap, but there is increasing effort to find effective biocontrol agents as alternatives. Among the most promising organisms for this are bacteria (*Bacillus*, *Pseudomonas* spp.) that produce antibiotics or that rapidly utilize the nutrient exudates. *Pythium oligandrum* and *Trichoderma* spp. also are being explored as seed and seedling protectants (see Chapter 11). A problem, however, is the speed at which seedling pathogens can initiate infection. In particular, the soil-borne sporangia of the common pathogen *Pythium ultimum* can germinate within 1–2 h in response to volatile metabolites from germinating seeds (probably ethanol but possibly also acetaldehyde), and the seed tissues can be heavily colonized within 6–12 h of planting in *Pythium*-infested soil. Biocontrol agents would need to work extremely fast against such pathogens (Nelson, 1987, 1992).

Many of the principles above apply also to *S. rolfsii*, although this fungus typically attacks older seedlings (say 3–4 weeks old) in warm regions where the rainfall is seasonal. The sclerotia of this pathogen (see Chapter 4) survive in soil over the dry period and are triggered to germinate after the first rains. Then, the fungus makes substantial pre-penetration growth before it attacks the newly sown crop. The key to understanding this pathogen lies in the sclerotia, as shown by the experiment in Fig. 12.1 and Table 12.1. If sclerotia are harvested fresh from laboratory cultures they show **hyphal germination**, whereby only a few, sparse hyphae emerge. The fungus can then infect living plant tissues at a short distance from the sclerotia (0.5 cm) but not from further away. However, sclerotia that have been air-dried will show **eruptive germination** when remoistened: all the sclerotial reserves are used to produce a mass of hyphae, which are organized into mycelial cords and can infect a plant from up to 3.5 cm distance. The fungus is most effective, however, in the presence of freshly decomposing organic matter (e.g. crop residues from a previous season which begin to decompose after the first rains). Volatile compounds, especially methanol, are released from freshly decomposing plant tissues due to the actions of plant pectic enzymes which persist in the dried tissues. These volatile compounds trigger sclerotia to germinate eruptively and also serve as nutrients for the fungus, orientating its growth. In these conditions the sclerotia can infect plant seedlings even 6 cm away. These experimental results seem to parallel closely the behaviour of the fungus in cropping systems, where *S. rolfsii* characteristically colonizes crop residues

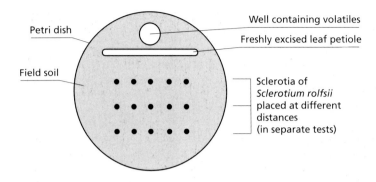

Fig. 12.1 Experimental system to study the ability of *Sclerotium rolfsii* to infect living plant tissue (detached leaf petioles) when sclerotia were placed at a distance from the host tissue. In some treatments the well contained volatile compounds released from dried lucerne hay that had been remoistened. See Table 12.1.

Table 12.1 Germination of sclerotia of *Sclerotium rolfsii* and infection of leaf petioles when sclerotia were placed at different distances from petioles in the experiment shown in Fig. 12.1. (Data from Punja & Grogan, 1981.)

	Distance of placement of sclerotia from leaf petiole (cm)						
	0	1	2	3	4	5	6
NON-DRIED SCLEROTIA							
Volatiles absent							
Per cent germination	19	16	11	11	19	14	15
Per cent infection	12	0	0	0	0	0	0
Volatiles present							
Per cent germination	90	73	67	51	17	14	9
Per cent infection	90	71	54	39	0	0	0
DRIED SCLEROTIA							
Volatiles absent							
Per cent germination	78	75	71	63	74	68	73
Per cent infection	78	75	60	56	0	0	0
Volatiles present							
Per cent germination	100	100	96	95	100	98	87
Per cent infection	100	100	91	87	100	82	16

from surviving sclerotia and uses the residues as a food base to support infection of the basal stem tissues of young plants. It has an extremely aggressive mode of attack on the plants, producing pectic enzymes to degrade the tissues, aided by the production of copious amounts of oxalic acid (see later).

As a final example of pathogens that attack juvenile plants we can consider the *Rhizoctonia* strains that cause stunting diseases of cereals; they include 'anastomosis group 8' of *R. solani*, and an unidentified *Rhizoctonia* that causes crater disease in South Africa (Fig. 12.2). Patches of stunted cereal plants first appear about 3–4 weeks after sowing, and from this time onwards the stunted plants make little further growth whereas the surrounding crop grows normally. The roots of the stunted plants typically have brown, spearpointed, rotted ends, where the fungus has destroyed the root tips, or in crater disease the roots bear large sclerotium-like beads of the fungus at the points where they are rotted (see Fig. 12.2). In all cases these diseases are found in particular conditions: where the soil is not ploughed and instead only the top few centimetres of soil are disturbed, and where the soils themselves have inherently poor structure (poor soil aggregate stability) so they tend to impede root growth. The roots are attacked at the point where they meet the undisturbed soil, probably because root impedance increases the rate of nutrient exudation. In addition, *Rhizoctonia* can form an undisturbed

(a)

(b)

Fig. 12.2 Crater disease of wheat in South Africa, caused by *Rhizoctonia* infection of the roots. (a) Disease patches seen from the air. (b) Root systems of young (3 week) wheat seedlings: a healthy root system on the right, and a diseased root system on the left, where the discoloured roots bear sclerotium-like 'nodules' of *Rhizoctonia* which have girdled and killed the roots. (From Deacon & Scott, 1985.)

mycelial network in the non-ploughed soil, and might translocate nutrients through this network to the points of attack on the young roots. The plants never recover from this early attack, even though cereals can produce new roots continuously during much of the growing season. The reason is that the first-formed roots of cereals, termed seminal roots, penetrate deeply into soil and can draw water from the lower soil layers when the upper layers begin to dry later in the season. These important roots are destroyed early by *Rhizoctonia*.

All the examples in this section demonstrate the important role of nutrients in facilitating infection by seedling pathogens, because nutrients support pre-penetration growth. In many cases the environmental conditions that favour infection are related to this effect.

Decline and replant diseases

A number of perennial fruit crops, including strawberry, apple and avocado, show a progressive decline in yield as the crop ages, until production is no longer economic. If these plants are removed and replaced by young plants then the new plants often grow poorly or fail to establish at all (replant problems). Thus, **decline and replant diseases** are closely related, indicating a common cause.

Several pathogens have been associated with these diseases, but there is a common pattern in all cases: the root tips and young 'feeder roots' have only a short life-span before they are destroyed. The difficulty in investigating these diseases is that 'root turnover' is a normal feature, even in healthy plants: the individual root tips can die and be replaced by new tips periodically. However, in plants exhibiting decline or replant diseases this balance is altered so that the rate of root tip decay eventually exceeds the rate of new tip production. One of the common causes seems to be attack by *Pythium* spp. in orchard crops such as apples, although these *Pythium* spp. (e.g. *P. sylvaticum*) do not commonly produce zoospores. On avocado, however, the principal cause is *Phytophthora cinnamomi*, which also causes a serious (and related) dieback disease of eucalypts in Australia. The involvement of these fungi was identified mainly by using the acylalanine fungicides (e.g. metalaxyl; see Chapter 14) which act specifically against oomycota. When applied experimentally (but often at uneconomic levels) these fungicides can cause spectacular improvements in plant growth. Similar experimental studies suggest that *Pythium* causes up to 10% yield reduction in wheat crops in the Pacific north-west of the USA, even though the crops are apparently healthy. *Pythium arrhenomanes* (and to a lesser degree the closely related species *P. graminicola*) similarly contributes to progressive decline of sugarcane crops.

These examples illustrate two important and related points.

1 The young tissues remain susceptible to infection by these pathogens throughout the life of a plant.

2 Many of these pathogens can have long-term associations with their hosts, and develop at least a degree of host-adaptation. For example, *P. graminicola* and *P. arrhenomanes* are characteristically, but not exclusively, associated with graminaceous hosts. In this case the host-adaptation has been related to the efficiency of zoospore encystment on 'host' compared with 'non-host' (non-graminaceous) plants (Table 12.2). In contrast, the host-adaptation of *Phytophthora sojae*, a pathogen of soybeans, is associated with zoospore taxis to the flavonoids of this crop (see Chapter 9).

In any case, the long-held view that these pathogens are unspecialized must now be dis-carded; they are specialized for attacking immature plant tissues.

Pathogens of fruits and other fleshy tissues

Ripening fruits and other fleshy plant tissues can be rotted rapidly by pathogens in appropriate conditions. Among the best-studied pathogens of this type are *Botrytis cinerea* which causes grey mould of soft fruits such as strawberries, raspberries and grapes, *Penicillium italicum* and *P. digitatum* which rot citrus fruits, *P. expansum* which rots apples, and *Monilinia fructigena* (sexual state, *Sclerotinia fructi-gena*) which rots apples, pears, peaches and some other fruits (Fig. 12.3). *Botrytis* typically initiates its growth on the senescing flower remains (it is also a problem in the commercial cut-flower industry) and grows from them to invade the living tissues of the ripening fruit. The other fruit-rotting fungi are essentially wound pathogens: *Penicillium* enters through wounds caused during harvesting and handling, whereas *Monilinia* usually initiates attack in the field, through wounds caused by wasps or birds. In any case, these pathogens are unable to rot the fruit until it passes through the phase termed climacteric, when the acid content

Table 12.2 Attraction and encystment of zoospores on roots of wild grass plants and wild dicotyledonous plants. (From Mitchell & Deacon, 1986.)

	Pythium graminicola			*Pythium aphanidermatum*		
	Grass (A)	Dicot (B)	Index (A/B)	Grass (A)	Dicot (B)	Index (A/B)
Experiment 1						
Zoospores attracted	399	289	**1.38**	315	315	**1.00**
Zoospores encysted	279	28	**9.96***	248	223	**1.11**
Mean of nine experiments with various grass and dicot species						
Zoospores attracted			**1.25**			**1.01**
Zoospores encysted			**3.62***			**1.13**

*Significantly different from the corresponding value for *P. aphanidermatum*.
Comparisons were made between *Pythium* spp. that characteristically infect the Gramineae (usually *P. graminicola*) and *Pythium* spp. of broad host range (usually *P. aphanidermatum*). In each experiment the roots of wild grasses or wild dicotyledonous plants were placed in zoospore suspension and assessed for: (i) number of zoospores that accumulated near the root tips; and (ii) zoospores that encysted near the root tips. An index was obtained as shown. The experiments showed that there is a small difference in chemotaxis of the two types of fungus to the different types of root, but there is a much bigger difference in encystment. The zoospores of graminicolous (narrow host range) *Pythium* spp. encyst on their host roots but only poorly on non-host roots, whereas the *Pythium* spp. of broad host range encyst equally well on grass or dicotyledonous roots.

(a)

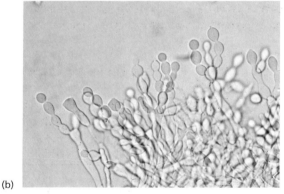

(b)

Fig. 12.3 (a) Apple naturally infected by *Monilinia fructigena*, lying on an orchard floor; it is covered in white sporing pustules of the fungus. (b) Part of a sporing pustule seen under a microscope, showing the branched chains of bud-like conidia arising from the tips of conidiophores.

declines and sugar levels start to rise. A similarly rapid rot is caused by *Sclerotinia sclerotiorum* on the maturing seed heads of sunflower, and by it or *S. minor* on the older leaves of lettuce in contact with soil (the disease called 'lettuce drop').

A common feature of all these diseases is that the tissues are rotted by **pectic enzymes** which degrade the **middle lamella**, or cementing layer, between plant cell walls (Fig. 12.4). This is true also for the seedling pathogens discussed earlier. The pectin of the middle lamella consists of chains of α-linked galacturonic acid residues, some of which are methylated, as well as mixed polymers of galacturonic acid, mannose and lesser amounts of other sugars. The enzymes involved are of three major types.

1 Pectin methyl esterase, often produced by the plant itself and which demethylates pectin. Methanol released in this way can act as a germination trigger and potential carbon source for *Sclerotium rolfsii* (see earlier).

2 Pectic lyase (PL), a chain-splitting enzyme with both endo- and exo-acting forms. It cleaves the bonds between the sugar residues in a characteristic way, eliminating water during the process.

3 Polygalacturonase (PG), another chain-splitting enzyme with both endo- and exo-forms. Unlike PL, this enzyme is a hydrolase — it uses water to add hydrogen ions (H^+) to one sugar residue and hydroxide ions (OH^-) to the other residue during cleavage of the bond.

Fungi are induced to synthesize these enzymes in the presence of pectic substrates, and the enzymes probably act in concert. This was shown when the genes encoding the equivalent enzymes of *Erwinia carotovora* (the bacterium that causes potato black-leg) were engineered separately into *Escherichia coli*, and the recipient cells required a minimum of one exo-PL, two endo-PLs and one endo-PG in order to rot potato tissues. The pectic enzymes of fungi exist as a range of isomers which separate according to size and net charge during gel electrophoresis. The resulting pectic zymograms can be used to distinguish population subgroups of fungi (see Chapter 8), but the isomers probably do not differ significantly in function.

The dependence on pectic enzymes as a principal mechanism of pathogenesis helps to explain why some pathogens are restricted to the rotting of fruits and why others infect only juvenile plant tissues (e.g. *Pythium* spp. and *R. solani*). As plant tissues age, the pectic compounds increasingly become cross-linked with calcium, which renders them resistant to degradation, perhaps by restricting access by the endo-acting enzymes. For

Junctions between plant cells

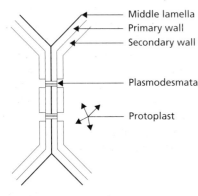

Middle lamella (composed mainly of pectic compounds)
Primary wall (composed mainly of hemicellulose (mixed polymers))
Secondary wall (composed mainly of cellulose)

Plasmodesmata (protoplasmic connections through the walls)

Protoplast

Pectic compounds

Composed mainly of **pectic acids** and **pectinic acids**.

Pectic acids: straight chains of α-1,4-galacturonic acid.   **Pectinic acids**: like pectic acids but some residues are methylated.

Pectic enzymes

1 Pectinmethylesterase hydrolyses methyl esters of pectinic acid to produce pectic acid.
2 Polygalacturonases (*exo-* and *endo-* types) hydrolyse pectic acid chains to galacturonic acid.
3 Pectin lyase (*exo-* and *endo-* types). Cleave pectic acid chains to monomers by eliminating water in the process.

Fig. 12.4 Occurrence and enzymatic breakdown of pectic compounds in plant cell walls.

example, the hypocotyls of bean seedlings (*Phaseolus vulgaris*) show a marked increase in resistance to infection by *R. solani* at about 2 weeks after seed germination, coinciding with enhanced calcification of the middle lamella. The pathogens that rot older tissues (*S. rolfsii*, and *S. sclerotiorum* on seed heads) seem to overcome this by producing large amounts of oxalic acid, which they also do in laboratory culture. Oxalic acid can combine with calcium, removing it from association with the pectic compounds and rendering them susceptible to enzymic attack.

Initially it was thought that the principal role of pectic enzymes in pathogenesis was to separate cells, causing disruption of the fine cytoplasmic bridges (plasmodesmata) that link cells through their walls. More recent studies suggest other roles. Pectic enzymes can be directly toxic to plant protoplasts *in vitro*, and some of the partial breakdown products of pectin — the oligomers composed of four or five galacturonic acid residues that are generated by endo-PG — are especially toxic, causing rapid cell death. Other partial breakdown products of pectic compounds can induce plant resistance responses. Plant cells also contain a PG-inhibiting protein, as part of a possible general resistance mechanism; although the evidence is unclear at present, the distribution of this protein might contribute to the differential susceptibility of tissues.

Differences in the activities of pectic enzymes are clearly seen in comparisons of different fruit-rot fungi such as *P. expansum* and *M. fructigena* on apples. *Penicillium* causes a light-coloured, soft, watery rot, which can be removed completely as a lozenge, leaving firm, healthy tissue beneath it (but, it is wise not to eat this, because *P. expansum* produces a mycotoxin!). *Monilinia* causes a mid-brown, firm and dry rot which cannot be scooped away from the healthy tissue. A simple (although perhaps too simple) explanation of this difference is that the main chain-splitting enzymes of *Monilinia* are inhibited in the apple tissue, whereas they are active in the case of *Penicillium*, leading to complete breakdown of the pectic compounds and thus separation of cells at the margin of the rot where they are not held together by fungal hyphae. The difference in browning reaction — a typical wound response of plant cells—is related to oxidation of phenolic compounds, which are known to be enzyme inhibitors. Apparently, *Penicillium* prevents their production or accumulation, enabling its enzymes to degrade completely the pectic materials. The roles of pectic enzymes in plant pathogenesis were reviewed by Collmer & Keen (1986).

Diseases of senescence: the stalk-rot pathogens

Some fungi that damage seedlings can also grow on senescing plant tissues, exploiting the declining host resistance. We saw this in the invasion of cotton leaves by *Pythium ultimum* in soil (see Chapter 11). Other pathogens characteristically invade senescing tissues near harvest time of a crop, and important examples of this are the stalk-rot pathogens of maize and sorghum. Some of these fungi (e.g. *Macrophomina phaseolina*) will attack many crops whereas others are quite host-specialized (*Diplodia maydis*, *Gibberella zeae*, *Colletotrichum graminicola*). But, their common behaviour with respect to senescence allows them all to be considered as a group.

These stalk-rot pathogens are extremely common where maize and sorghum are grown; but, they cause serious disease in only some sites or seasons, and in others they cause hardly any

damage. The reason is that they respond to the plant's physiology, and particularly when stress conditions predispose the plants to infection. These stresses are of various types, including temporary drought, poor light, mineral nutrient deficiency, insect (stem-borer) damage and even high grain-set (heavy demand for filling of the grain). Moreover, the stress must occur at a specific time, the most crucial time being when the plants have flowered and are in the early stages of grain-filling. In these conditions the entire crop can be infected, resulting in premature death and collapse (Fig. 12.5). A unifying hypothesis to explain all these effects was developed early in this century and refined by Dodd (1980) as a 'photosynthate stress–translocation balance' hypothesis. It is based on the well-established fact that the filling of the grain in maize and sorghum places a heavy demand on plant sugars. Only about 80% of this can be supplied by photosynthesis in the leaves, the other 20% being supplied from sugar storage reserves, principally in the base of the stem. So, as the plant approaches maturity its reserve sugars are depleted in a controlled manner. However, any stress conditions that temporarily reduce the rate of photosynthesis cause more sugars to be removed from the stem tissues. This leads rapidly to premature senescence of the stalk tissues, so that weak parasites that were already present but held in check are able to invade and destroy the stalk base. Ironically, these diseases can be controlled very easily by removing the developing grain, but this is hardly a practical solution! Instead, growers need to predict the likelihood of stress conditions on their farms and adjust the fertilizer and plant spacing accordingly, to minimize the chances of late-season stresses such as drought. So far as the fungi are concerned, they are specialized, natural invaders of senescing tissues which sooner or later will exploit the plant; they become pathogens only when the plant senesces early in response to stress.

Host-specialized necrotrophic pathogens

Although the boundaries are not precise, the pathogens considered here differ from those above in that they typically invade mature, healthy tissues of plants in non-stressed conditions. As a

(a)

4cm

(b)

Fig. 12.5 Maize root and stalk rot complex. (a) Field of maize plants that died prematurely from the root rot-stalk rot complex in a year of unusually poor rainfall in South Africa. (b) A plant stem base from the field, showing extensive rotting of the roots and basal stalk tissues. The main pathogen involved was *Phialophora zeicola*, which initiates attack in the roots then spreads to the stalk, but other pathogens can initiate attack in the basal stalk tissues. (From Deacon & Scott, 1983.)

consequence, they are host-specialized because they must circumvent, tolerate or overcome the specific resistance factors of a particular host. These resistance factors can involve physical barriers or fungitoxic chemicals, and they are either pre-formed or formed in response to infection. There are also more general factors that influence the success of infection, such as environmental influences and inoculum level of the pathogen. We begin with an example of these factors because they tell us much about how fungi are adapted for a parasitic lifestyle.

The role of inoculum: *Botrytis fabae* and *B. cinerea*

B. cinerea and *B. fabae* are closely related fungi, but *B. cinerea* invades the compromised tissues of many plants, whereas *B. fabae* is a host-specialized pathogen of broad bean (*Vicia faba*) and a few close relatives of this. The ovoid conidia of *B. fabae* are 20–25 µm long, compared with only 10–15 µm for

B. cinerea — a roughly sixfold difference in spore volume. This difference can affect the efficiency of impaction onto leaves, especially onto the wide leaves of broad bean plants (see Chapter 9). It also represents a substantial difference in endogenous food reserves to support infection, discussed below.

The spores of both fungi will germinate when placed in water drops on the surface of broad bean leaves, but *B. cinerea* usually fails to penetrate the leaf cuticle whereas *B. fabae* penetrates and causes an immediate host reaction—the death of the penetrated cell and its immediate neighbours, giving a brown, necrotic spot (Fig. 12.6). This reaction is termed the **hypersensitive response**, and the fungitoxic metabolites that accumulate in these dead cells are often sufficient to prevent further growth of any casual invader (discussed later). The difference in initial penetration by the *Botrytis* spp. is largely explained by their nutrient reserves. *B. cinerea* will penetrate and cause a necrotic spot if sugars are added to the inoculum drop, or some-

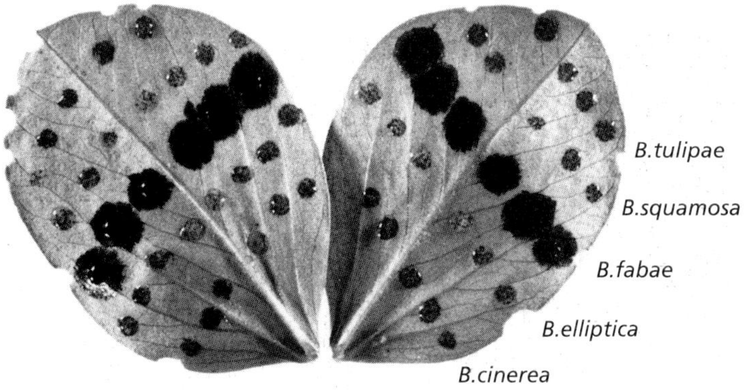

Fig. 12.6 Leaves of broad bean (*Vicia faba*) 5 days after inoculation with drops of water containing spores of different *Botrytis* species. Only *B. fabae* has caused severe spreading lesions, but all species have caused localized necrosis of the leaf tissues. (Photograph courtesy of J. Mansfield.)

B.tulipae

B.squamosa

B.fabae

B.elliptica

B.cinerea

times even if the normal leaf surface micro-organisms are inactivated by antibiotics so that *B. cinerea* can use the leaf-surface exudates. Conversely, aged spores of *B. fabae*, taken from old parts of agar colonies, can fail to penetrate because their endogenous reserves have been depleted, but they penetrate if supplied with exogenous sugars. Thus, *B. cinerea* has evolved a strategy as a general invader of senescent tissues: it produces many more spores from a given amount of resource but they depend on nutrients released from compromised host tissues. *B. fabae* produces fewer spores from the same amount of resource, but each of them has sufficient reserves to initiate infection of the healthy host.

Such differences in the infectivity of pathogens can be quantified, to indicate the likelihood of infection of plants in different conditions. Figure 12.7 shows this for *B. cinerea* and *B. fabae* when water drops containing different numbers of spores were placed on broad bean leaves and the presence of absence of infection (necrotic spots) beneath the drops was recorded. The appropriate plot of the results is as logarithm of dosage (spore number) against percentage response expressed as a probit value (probit of 5 = 50%). Then, the skewed sigmoid curve that would result from plotting actual spore numbers against percentage infection becomes a straight line, and the estimated dose that gives 50% probability of infection (ED_{50} value) corresponds to a probit of 5. The same type of plot is used widely in plant and animal experiments to obtain LD_{50} values for **lethal** toxicity of chemicals, etc. *B. fabae* was found to have an ED_{50} of 4 in water drops, corresponding to a 16% chance of any single spore being able to initiate infection.

B. cinerea had an ED_{50} of about 500, corresponding to a 0.14% chance of a single spore initiating infection. This simple method of inoculating with successive dilutions of spores can give information of direct field relevance. For example, several different *Colletotrichum* spp. are found on rubber leaves in Malaysia, and by inoculating the spores at different dilutions in droplets Wastie and Sankar (1970) were able to predict: (i) the species most likely to infect from low spore numbers in field conditions; and (ii) the age of leaves that were most infectible. It was found that the young, newly emerging leaves soon became resistant as the thickness of the cuticle developed—an example of a pre-formed resistance factor.

The determinants of host range

Host-adapted necrotrophic pathogens can infect a number of taxonomically related plants, at least in artificial inoculations, but they seldom infect unrelated plants. For example, Fig. 12.6 shows that, of several different *Botrytis* spp. inoculated onto bean leaves, only *B. fabae* caused spreading lesions on this host; the others were either weak parasites (e.g. *B. cinerea*) or they were pathogens of different types of plant (e.g. *B. tulipae*). So, we can regard these pathogens as family adapted, genus-adapted, etc., even if they are economically damaging to only a single crop species. As another example, the pathogenic forms of *Gaeumannomyces graminis* infect roots of wheat, barley, oats, rye, maize, rice and several turf and weed grasses in field conditions (although to different degrees) but do not infect plants outside of the Gramineae. *G. graminis* is thus a family adapted parasite. These

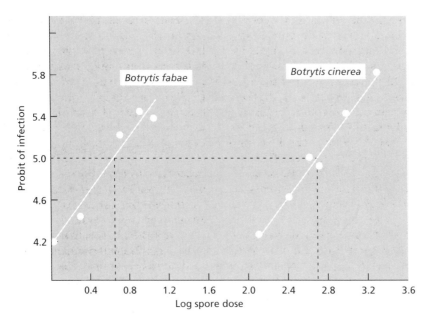

Fig. 12.7 Log dose/probit response plots of infection of broad bean leaves by *Botrytis fabae* and *Botrytis cinerea*. Based on Wastie (1962). Standard sized water drops containing different numbers of spores were placed on bean leaves and infection was scored as a necrotic spot (as in Fig. 12.6). The original curves of spore number in a drop against percentage of drops giving infection were sigmoidal, but become straight lines when the logarithm of spore dose is plotted against the probit value of percentage infection (the probit value for 50% is 5.0). The ED_{50} (estimated dose of spores giving 50% probability of infection) is obtained from the curves as shown.

examples raise two related questions: (i) how is a fungus adapted to parasitize its host group; and (ii) why can it not infect other plants? We begin to address these questions here, but they spill over into later sections.

The family Gramineae is a useful starting point because its members are parasitized by a wide range of fungi that do not grow on other plants (Table 12.3). Conversely, the Gramineae are not parasitized by some fungi of extremely wide host range, including all the *Phytophthora* spp. such as *P. cinnamomi* which has more than 1000 hosts, and the common vascular wilt pathogens *Fusarium oxysporum*, *Verticillium albo-atrum* and *V. dahliae* — the Gramineae has its own vascular wilt fungus, *Hymenula cerealis*.

The general answer to why pathogens are adapted to a genus or family may be that the plants within that taxon have similar types of defence chemicals. The Gramineae often respond to infection by producing lignin-like phenolic compounds

Table 12.3 Examples of family specific soil-borne parasites of Gramineae.

Gaeumannomyces graminis (roots of wheat, oats, turf grasses)
Phialophora graminicola (roots of grasses), *P. zeicola* (roots, stalks of maize)
Magnaporthe rhizophila (roots of millet)
Leptosphaeria namari, *L. korrae* (rhizomatous/stoloniferous grasses)
Fusarium culmorum (wheat, barley stem bases)
Pseudocercosporella herpotrichoides (wheat, rye stem bases)
Rhizoctonia cerealis (wheat stem bases)
R. solani anastomosis group 8 (wheat, barley roots)
Pythium graminicola and *P. arrhenomanes* (roots of sugarcane, maize, small-grain cereals)
Hymenula cerealis (vascular wilt pathogen of wheat)
Idriella bolleyi (weak parasite of wheat, barley, grasses)
Heterodera avenae (cyst nematode, especially of oats)
Striga hermonthica (root-parasitic vascular plant, especially on maize)

rather than other types of antifungal compounds, discussed later. In addition, the Gramineae have different wall-matrix components compared with many other plants: instead of pectin in the middle lamella (see Fig. 12.4), they have mixed polymers of β-1,4-xylan and α-1,3-arabinose with lesser amounts of non-cellulosic β-1,4- and β-1,3-linked glucans. Recognizing this, Cooper *et al.* (1988) investigated the enzymes produced by three stem-base pathogens of cereals (*Rhizoctonia cerealis* which causes sharp eyespot, *Fusarium culmorum* which causes *Fusarium* foot rot and *Pseudocercosporella herpotrichoides* which causes 'true' eyespot) when grown on cereal cell walls in culture. All three fungi produced arabinase, xylanase and β-1,3-glucanase, but only small amounts of pectic enzymes. In contrast, when pathogens of non-grass plants (*F. oxysporum, V. albo-atrum*) are grown on walls of their hosts they produce large amounts of pectic enzymes. So, the cereal pathogens seem to be adapted to degrade the typical wall components of their hosts. Indeed, the cereal pathogens produced arabinase and glucanase constitutively, even in the absence of host wall material, and the production of these enzymes was not repressed by sugars. These are unusual features because polymer-degrading enzymes in general, including pectic enzymes, are both inducible and repressible. We saw in Table

12.2 that zoospores of the grass-adapted *Pythium* spp. (*P. graminicola* and *P. arrhenomanes*) encyst in larger numbers on grass than on non-grass roots, in conditions where *Pythium* spp. of broader host range show no differential response. This also could be related to differences in the plant polysaccharides, because the root surface 'slime' behind the root cap is derived by swelling and extrusion of the matrix materials of the root cell walls. In other words, it would be derived from pectic materials in most plants but from the mixed polymers of xylose, arabinose and glucose in the Gramineae. Zoospores of the grass-adapted *Pythium* spp. seem to recognize this difference.

In a few cases, host-adapted parasitism has been related to stimulation of germination. A classic example involves *Sclerotium cepivorum*, which causes white rot of onion (*Allium cepa*), garlic and a few other *Allium* spp. As we saw in Chapter 9, the sclerotia are triggered to germinate by sulphur-containing metabolites of the hosts. This eruptive germination produces a mass of hyphae and thus a large inoculum for infection. Esler and Coley-Smith (1984) investigated whether this is the sole determinant of host range. When sclerotia alone were added to soil, only a narrow range of plants became infected (Table 12.4), and these were the plants that produce the germination triggers. When sclerotia were added to soil with garlic

Table 12.4 Infection of different plants by *Sclerotium cepivorum* when different forms of inoculum were used. Intensity of infection is shown as 0 (no infection), +, ++, or +++ (heavy infection). (Adapted from Esler & Coley-Smith, 1984.)

Plant type	Inoculum and test conditions			
	Sclerotia in soil	**Sclerotia plus garlic extract in soil**	**Pre-germinated sclerotia in sterile conditions**	**Mycelial blocks in sterile conditions**
Allium spp. that trigger germination of sclerotia (onion, garlic) (Liliaceae)	+++	+++	+++	+++
Ornamental *Allium* spp. that do not trigger germination of sclerotia (Liliaceae)	0	+++	+++	+++
Other members of the Liliaceae	0	+++	+++	+++
Non-related plants				
Clover (Leguminosae)	0	0	0	++
Flax (Linaceae)	0	0	0	++
Tomato (Solanaceae)	0	0	+	+++
Cabbage (Cruciferae)	0	0	+	+++

extract (to trigger germination) a wider range of *Allium* spp. and related ornamental plants were infected. So, these plants are potential hosts but they escape infection because they do not release the germination triggers. However, an even wider range of plants, including cabbage and tomato, could be infected when inoculated with *S. cepivorum* on nutrient-rich agar discs. These are non-hosts from quite different families (Cruciferae, Solanaceae), but their young tissues were overwhelmed by an abnormally high inoculum level of the pathogen. This is typical of many necrotrophic fungi: even if they are host-adapted they have a general pathogenic ability so that they will overwhelm the natural resistance factors if they infect from a massive, abnormal inoculum level.

Pre-formed host-resistance factors

Most plants have structural barriers that help to prevent infection by casual invaders. They include such factors as a thick leaf cuticle or corky layers of stems which only host-adapted pathogens can breach. The enzyme **cutinase** could be important for overcoming some of these barriers, especially for necrotrophic fungi that invade directly through the leaf surface rather than through stomata. There is a mass of circumstantial evidence for this point, although it was challenged recently when targeted gene disruption was used to block cutinase production. The gene-disrupted strains of three leaf pathogens had no detectable cutinase activity *in vitro*, but this had no effect on the pathogenicity of two of the fungi and it only reduced, but did not eliminate, the pathogenicity of a third fungus (VanEtten *et al.*, 1994). This finding has led to the recognition that some of the pathogenicity genes (including, perhaps, a gene for cutinase production) might only be expressed during infection and not in laboratory culture. There might also be multiple forms of these genes (gene redundancy), some of which are expressed *in vitro* but others only in the host.

Pre-formed fungitoxic compounds have been identified in several plants, and may have evolved as a general resistance mechanism. For this reason they have been termed **phytoanticipins**. We saw examples of this in the antifungal saponins of oats and tomato (see Chapter 8). In both cases the host-adapted pathogen was able to detoxify the saponin by producing a specific enzyme, but other pathogens cannot detoxify it and so cannot infect that host. As another example, onions with red or yellow skins are resistant to *Colletotrichum circinans*, whereas white-skinned onions are susceptible and develop 'onion smudge', named after the darkly pigmented spores and hyphae that develop on the infected bulbs. The pigments are not significant in themselves, but they are associated with high levels of the phenolic compounds, catechol or protocatechuic acid, which prevent the spores from germinating. A similar pre-formed resistance compound might occur in green banana skins, preventing infection by *Colletotrichum musae* until the fruit ripens. This is why ripe bananas have small brown spots on their surface — the lesions are caused by *C. musae* which was present on the green fruits but did not infect until the fruit was ripe.

Defence mechanisms that develop in response to infection

Wall papillae

The most common type of structural response to infection is the development of a **papilla**—a localized thickening of the host wall where a hypha penetrates a cell. In extreme cases the papilla continues to develop as the hypha grows (Fig. 12.8) so that it encases the penetration hypha. In electron micrographs, papillae are seen as massive localized accumulations of wall material in which membranes and other degenerate cytoplasmic components have been trapped. Cytochemical stains usually reveal the presence of callose, but in Gramineae the papillae often show staining reactions for lignin and suberin, so they have been termed lignitubers.

The significance of papillae in plant resistance is unclear, but they might at least delay the invading hypha while other defences are activated. The lignin-like precursors in papillae of the Gramineae may be particularly important, because lignin can form complexes with the wall polymers, preventing their enzymic degradation. Ride and Pearce (1979) demonstrated this for cereal leaves inoculated with *B. cinerea* and several other fungi that induced the production of lignified papillae during attempted penetration. The normal leaf cell

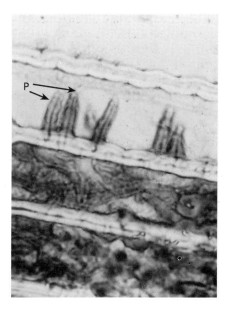

Fig. 12.8 Papillae (P) produced by a living wheat root cell during invasion by *Gaeumannomyces graminis*. The host cells lower in the photograph have been colonized by the pathogen.

walls were digested easily by commercial wall-degrading enzymes, but the papillae and the surrounding 'haloes' of lignified wall material were not digested. Thus, the lignified papillae could have three resistance-related functions:

1 to increase the wall thickness that a fungus must penetrate while depending on its endogenous energy reserves;

2 to render the wall resistant to digestion by the parasite enzymes;

3 to confer a locally toxic environment caused by the phenolic precursors of lignin and suberin.

A host-adapted pathogen infecting from a sufficient food base could be expected to overcome this defence — the take-all fungus is a classic example. But, non-adapted pathogens may be inhibited, and weak parasites such as *Phialophora graminicola* (see Chapter 10) are known to be stopped by lignified papillae during their attempted invasion of living cells, so they must wait for the cells to senesce.

Oxidative burst

One of the earliest events in the attempted penetration of leaves and other above-ground parts of plants is an oxidative burst, remarkably similar to the oxidative burst of phagocytes during the immune response in humans. It has been studied mainly in relation to the hypersensitive response (e.g. see Fig. 12.6), but it seems to be a general reaction of plant cells to trauma caused by attempted parasitic invasion or chemical factors. Lamb *et al.* (1994) have shown that it involves an extremely rapid (2–3 min) production of hydrogen peroxide at or near the plant cell surface, by the reaction of oxygen with nicotinamide adenine dinucleotide phosphate (NADPH) to produce superoxide:

$$O_2 + NADPH \rightarrow O_2^- + NADP^+ + H^+$$

Then, O_2^- is transmuted to hydrogen peroxide by a plasma membrane oxidase. The hydrogen peroxide then causes cross-linking of the plant cell wall proteins, which could strengthen the wall against attack by wall-degrading enzymes. A similar mechanism would be expected to occur in papillae, helping to slow or stop pathogen invasion.

Phytoalexins

Many plants respond to attempted infection by producing fungitoxic compounds, collectively termed **phytoalexins**. Their chemical structures are diverse (Fig. 12.9), but they tend to be similar within a plant family. Phytoalexins might thus be important in conferring resistance to pathogenic attack in general (most plants are resistant to most potential pathogens), and they might also be the key defence molecules that host-adapted parasites can overcome.

Phytoalexins are secondary metabolites produced by three major pathways in plants: (i) the acetate–malonate route; (ii) the acetate–mevalonate route; and (iii) the shikimic acid route (see Chapter 6 for discussion of these in fungi). In the hypersensitive response, an invaded cell and its immediate neighbours die, and phytoalexins accumulate in these dead cells, producing a local fungitoxic environment which may be sufficient to prevent any further growth. A similar response involving phytoalexin accumulation can often be induced by artificial wounding or application of a toxic chemical to the plant surface. It is always preceded by an oxidative burst, and hydrogen

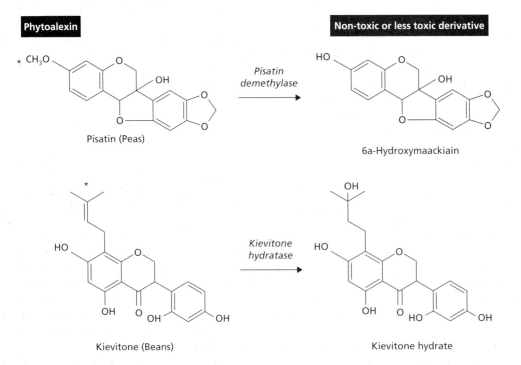

CH₃CH₂CH=CHC=C.CO.C CCH=CHCOOH

Wyerone acid (Broad beans)

Ipomeamarone (Sweet potatoes)

Detoxification by pathogens (at position *)

| **Phytoalexin** | | **Non-toxic or less toxic derivative** |

Pisatin (Peas) → *Pisatin demethylase* → 6a-Hydroxymaackiain

Kievitone (Beans) → *Kievitone hydratase* → Kievitone hydrate

Fig. 12.9 Some phytoalexins and examples of their detoxification by enzymes of host-specialized pathogens.

peroxide is thought to be one of the initial signals that activates the defence genes involved in phytoalexin production.

The sequence of events leading to phytoalexin accumulation has been studied by exploiting the discovery that low-molecular-weight compounds consisting of a few β-linked sugar residues can act as powerful **elicitors** of the hypersensitive response. These oligosaccharides are released from the extracellular polysaccharide of fungal hyphae, probably by the actions of β-glucanase enzymes that reside in plant cell walls. When applied to plant cells or tissues *in vitro*, the elicitors

cause an early oxidative burst and the induction of plant enzymes of the phenylpropanoid pathway, including **phenylalanine ammonia lyase** which is involved in phytoalexin production, and synthesis by the plant of **chitinase** and **β-1,3-glucanase** which can degrade fungal walls.

Clearly, this cascade of cellular responses to invasion will create a hostile environment for a potential pathogen. So, how does a host-adapted necrotrophic pathogen overcome or avoid them? Studies on the *Botrytis*–broad bean interaction have shown that both *B. fabae* and *B. cinerea* can cause a hypersensitive response in appropriate conditions (see earlier) and the phytoalexin **wyerone acid** accumulates in the dead cells. But, *B. fabae* is somewhat more tolerant of wyerone acid than is *B. cinerea in vitro*, and *B. fabae* also causes less wyerone acid to accumulate in the lesion, perhaps by detoxifying it. *B. cinerea* never spreads further from the necrotic spot, whereas *B. fabae* is checked temporarily but then regrows and spreads through the surrounding tissues, killing them progressively and presumably detoxifying the phytoalexin as it grows. The resulting large brown lesions give rise to the name for this disease — chocolate spot. The mechanisms of detoxification of phytoalexins have been reported for some other fungi. The pathogen of pea plants, *Nectria haematococca* detoxifies the pea phytoalexin **pisatin** by demethylation (Fig. 12.9), while other strains of this pathogen infect bean plants and detoxify one of the bean phytoalexins **kievitone** by hydration (Fig. 12.9). In both cases the ability of the pathogen to infect the plant is correlated with the ability to detoxify the phytoalexin. Moreover, when the pisatin demethylase gene from *N. haematococca* was transformed into *Cochliobolus heterostrophus*, a maize pathogen, this became a pathogen of pea leaves. So, there is very strong evidence that host-adapted pathogens have evolved mechanisms for overcoming the inhibitory effects of their hosts' phytoalexins, and that even the acquisition of a single phytoalexin-detoxifying gene can change the host range of a fungus.

There is, however, an interesting twist to the story. In the *Nectria*–bean pathosystem, all strains of *N. haematococca* were found to produce the hydrating enzyme, kievitone hydratase, but some strains do not secrete it from the hyphae and they are non-pathogenic; only the strains that secrete the enzyme are pathogenic. In the pea system the relationship between pathogenicity and production of the detoxifying enzyme was supported by many lines of evidence, but targeted disruption of the gene for pisatin demethylase led to strains that were still pathogenic. The reason for this is still unclear. VanEtten *et al.* (1995) reviewed all the recent work on detoxification of phytoanticipins and phytoalexins. They suggested that pathogens may have evolved several mechanisms for overcoming the host defence reactions, so that some of these mechanisms, although still expressed, may be redundant. Perhaps of more fundamental interest, VanEtten *et al.* discovered that the gene that codes for pisatin demethylase in *N. haematococca* is on a supernumerary, **dispensable minichromosome.** This chromosome can be lost entirely without affecting the ability of the fungus to grow in culture, although it is necessary for pathogenicity (VanEtten *et al.*, 1994). It is interesting to draw a parallel (even if it is a loose one) between this and the roles of plasmids in plant-associated bacteria. The ability of *Rhizobium* to form nitrogen-fixing root nodules is conferred by a *Sym* (symbiotic) plasmid, but *Rhizobium* can grow well on agar or even in the rhizosphere without the plasmid. Similarly, the ability of *Agrobacterium tumefaciens* to cause crown gall disease is conferred by the T_i (tumour-inducing) plasmid, but *Agrobacterium* grows as a normal rhizosphere inhabitant without the plasmid.

Vascular wilt pathogens and other endophytes

Vascular wilt pathogens cause some of the most devastating plant diseases. For example, Dutch elm disease (*Ophiostoma ulmi* and *O. novo-ulmi*) has destroyed much of the European elm population in the past 30 years, as it did earlier in the USA. Similarly, Panama disease of banana (*Fusarium oxysporum*) almost destroyed the banana export industry of Central America and the Caribbean before a resistant banana cultivar was discovered quite fortuitously. Hop wilt, caused by *Verticillium albo-atrum*, has caused similar major problems in Britain. Yet, these fungi can be present as avirulent forms or even sometimes as virulent strains without causing serious disease. For most of the time, they seem to live as harmless endophytes in plant tissues. We shall look at the evidence for this, because it contrasts with the common perception of these fungi as aggressive pathogens. But, to begin, we outline their classic aggressive behaviour.

The principal vascular wilt fungi are *F. oxysporum*, *V. albo-atrum* and *V. dahliae*. The pathogenic strains of *F. oxysporum* look identical to one another but cause disease of different hosts, so they are classified as special forms, latinized to *formae speciales* (singular *forma specialis*) and abbreviated to ff. ssp. or f. sp. (singular). For example, the pathogen of bananas is termed *F. oxysporum* f. sp. *cubense*, that of tomato is *F. oxysporum* f. sp. *lycopersici*, and so on, for the 80 or so host-specific pathogenic forms of *F. oxysporum* (Armstrong & Armstrong, 1981). There are also saprotrophic strains of *F. oxysporum*, commonly found in the rhizosphere. In contrast, *V. albo-atrum* and *V. dahliae* are less host-specific because a strain isolated from one host often will infect several other hosts by artificial inoculation, although there can be different degrees of host-related virulence.

All these vascular wilt fungi initiate disease when they enter the xylem vessels, either through wounds or by growth through the cortex of a young root region before the endodermis has developed; in most and perhaps all cases the fully developed endodermis prevents them from reaching the xylem. They produce conidia which are carried up the xylem vessels in the transpiration stream until they become lodged on the perforated vessel end walls. Then the conidia germinate and the germ-tubes grow through the perforations and produce further conidia for carriage to the next vessel end wall. This systemic colonization of the xylem leads to various symptoms such as yellowing and necrosis of the leaves, epinasty (collapse of turgor at the base of the leaf stalk so that the leaves hang down), wilting and eventual death of the plant (Plate 12.1, facing p. 210). At a late stage in this process, as the plant is approaching death, the fungus grows out from the xylem and colonizes the parenchymatous tissues, where it forms its resting stages — chlamydospores of *F. oxysporum*, small sclerotia (microsclerotia) of *V. dahliae*, pigmented hyphae of *V. albo-atrum*. These resting structures are returned to the soil as the plant tissues decay. An essentially similar pattern is seen in Dutch elm disease, except that the fungus grows from the xylem into the young bark tissues and sporulates for dispersal by beetles (see Chapter 9).

Pathogenicity determinants

Toxins have long been implicated in the vascular wilt syndrome because they could explain some of the earliest symptoms such as leaf yellowing and loss of control of water balance in the leaves. The major toxin of *F. oxysporum* might be **fusaric acid**, which is produced in laboratory culture and has been detected in diseased plants. When applied to plant protoplasts *in vitro* it markedly alters their permeability to ions. *Ophiostoma novo-ulmi* also produces a toxin in culture — a hydrophobin termed **cerato-ulmin** which is produced only by pathogenic strains. When applied to the bases of

cut elm stems it produced leaf symptoms typical of Dutch elm disease; but more recent work using targeted gene disruption has failed to confirm its role — the mutant, non-producing strains were as pathogenic as the wild-type. We saw earlier that this approach does not always confirm the relationship between a pathogen function and disease, even when other work strongly suggests a relationship. It is possible that some pathogenesis-related genes (or gene copies) remain silent *in vitro* and are expressed only in the host. However, at present the role of toxins in the vascular wilt syndrome remains unproven.

A much clearer role can be assigned to **vascular plugging** because fungus-colonized xylem vessels typically contain gel-like material which becomes impregnated with, and stabilized by, brown phenolic compounds (Fig. 12.10; Plate 12.2, facing p. 210). The gels are composed of pectin, and for some time it was thought that the fungi cause this by releasing pectic enzymes to degrade the middle lamellae of the xylem parenchyma, consistent with their production of pectic enzymes in culture. Now, however, the gels are suggested to arise as a consequence of production of carbon dioxide by the fungus: the plant fixes this into organic acids which complex with cations in the middle lamella, causing the pectic compounds to swell and be extruded into the vessels through pits in the vessel walls. Gel plugging can serve to localize the fungus, preventing its further spread, but if enough vessels are plugged then the plant becomes water-stressed and dies. Some types of plant show a further reaction — they produce balloon-like swellings (**tyloses**) from parenchyma cells adjacent to the vessels, and these develop through the pits in the vessel walls. Their function seems to be to restrict the spread of the fungus until pectic gels have been produced. Lastly, phytoalexins are released into the xylem vessels by diffusion from adjacent living cells as part of a general trauma response.

Clearly, the development of vascular wilt diseases involves a complex interplay between the fungus and the plant. The fungal adaptations for this lifestyle include possible toxin production, an essentially single-cell growth habit involving yeast-like cells or rapid cycles of sporulation from germinating spores in the xylem, and the ability to cope with phytoalexins of a specific host. The plant

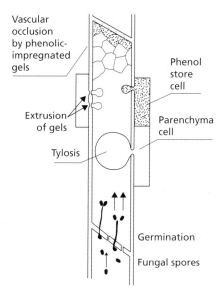

Vascular
occlusion
by phenolic-
impregnated
gels

Phenol
store
cell

Extrusion
of gels

Parenchyma
cell

Tylosis

Germination

Fungal spores

Fig. 12.10 Diagrammatic representation of pathogen spread and occlusion of the xylem vessels in vascular wilt diseases. Fungal spores are carried up the vessels in the water stream, become trapped on the perforated vessel end walls, germinate and pass through the perforations to produce further spores. The host can respond to invasion by producing tyloses from parenchymatous cells adjacent to the vessels. It also releases phenols into the vessels from phenol store cells, and the phenolics oxidize and polymerize in the vessels. Pectic gels are extruded into the vessels through bordered pits in the vessel walls (regions where the wall is thin because it consists only of primary wall, not overlaid by the secondary cellulosic wall). The gels coagulate and accumulate phenolics, forming occlusions ahead of the fungus. (Based on Beckman & Talboys, 1981.)

defence responses include the development of tyloses and gels, and the accumulation of phenolics and phytoalexins in the vessels. Beckman and Talboys (1981) proposed that the outcome of infection depends on the dynamics of these processes: if the fungus spreads quickly enough from its initial entry point it will outstrip the host defences, otherwise the infection will be contained locally. In some plant–fungus combinations, such as *F. oxysporum* f. sp. *pisi* on peas, the outcome of infection is governed by a 'gene-for-gene' relationship (see later). Specific cultivars of the crop can be bred for resistance to all current races of the pathogen, but the pathogen mutates to a new race that overcomes this resistance. Single genes in the pathogen

and plant seem to govern this, indicating the possible involvement of receptor-recognition systems that activate the plant defence response.

Environmental factors also play a major role in the development of vascular wilt diseases, presumably by influencing the host-pathogen dynamics. For example, the severe losses from Panama disease of bananas in the Americas early this century were exacerbated by the growing of bananas on poor sites: plantations established on so-called short-production sites were destroyed within a few years of planting, whereas plantations of the same susceptible banana cultivar, Gros Michel, remained economic for up to 20 years in sites with optimal soil and moisture conditions. In current terminology, the unfavourable sites would be termed disease-conducive and the others termed disease-suppressive. This has been related to the presence of competing or antagonistic micro-organisms that influence the pathogen population in soil (see Schneider, 1982). But, other factors also are involved. For example, the Panama disease problem in Central America was solved by replacing the older susceptible banana cultivar (Gros Michel) by the more wilt-resistant 'Cavendish' cultivars and, coincidentally, by replanting this cultivar only in the more favourable sites for banana growth. The new cultivars have now remained resistant for more than 40 years. Yet, in the subtropics (South Africa, Taiwan, Canary Islands, parts of Australia) the Cavendish cultivars are dying from Panama disease, and this has coincided with the spread of a new pathotype of *F. oxysporum* f. sp. *cubense*. The conditions in these countries are marginal for banana production because the winter temperatures drop to 12°C or lower. Bananas stop growing at this temperature (evidenced by their failure to produce new leaves), whereas *F. oxysporum* f. sp. *cubense* can grow at this temperature *in vitro*. So, the pathogen might initiate infection of a 'resistant' host while the resistance is not expressed. This could have longer-term implications for the host-pathogen dynamics, because the pathogen will already have a foothold by the time that the temperature rises.

Vascular wilt pathogens as endophytes

Endophytes can be defined as organisms that grow within living plants but not necessarily in the living cells, deriving nutrients without noticeable detriment to the plant. There are now many examples of this, discussed later, but the vascular wilt fungi also fit this definition. They grow almost exclusively in the non-living xylem vessels and only break out from these when the host is severely stressed near the end of its life. They are not tissue-invaders, and even their access to the xylem is achieved through wounds or through the immature tissues of the root tip. These features are of special note because they match a quite different phase of activity of these fungi — as harmless colonizers of the root cortex of perfectly healthy plants. Gerik and Huisman (1985) demonstrated this for pathogenic strains of *F. oxysporum* and *V. dahliae* which cause vascular wilt of cotton. Most often, these strains were found in the root cortex of healthy, field-grown cotton plants, where they could be detected by using polyclonal antibodies. *F. oxysporum* was found to be an early colonizer of the root surface and, to a lesser degree, of the outer cortex within 5 mm of the root tips; once it had established in this zone it maintained its population on or near the root surface further back from the tips. *V. dahliae* showed a different pattern of colonization: it was most often seen growing **between** the cells of the inner cortex of roots, just outside the endodermis, and it invaded some individual cortical cells or groups of cells to form a dense mycelial mass, but caused no obvious harm.

So, it seems that these vascular wilt fungi often grow as non-aggressive endophytes, adapted to utilize nutrients that leak from living host cells. Even when they grow in the xylem vessels they seem to utilize nutrients that leak from the surrounding living cells, and the disease that they cause may be almost incidental, resulting from the host's responses to contain them by vascular plugging. Perhaps the high degree of host-specificity of the *formae speciales* of *F. oxysporum* is related to tolerance of the different phytoalexins of their hosts; it is even possible that such tolerance is needed in the rhizosphere or when these fungi grow in the intercellular spaces (walls) of the root cortex.

Other endophytic parasites

Many other parasites—some of them pathogens—grow as endophytes. The best-known examples

are the smut fungi (basidiomycota) which grow systemically in the intercellular (wall) spaces of their hosts, usually causing no obvious symptoms until late in plant development when they produce masses of black spores from which the name 'smut' derives. In loose smut of barley and wheat (*Ustilago nuda*) the spores develop in the flowering heads in place of the grain and produce a black powdery mass at harvest time. In *Tilletia caries* (stinking smut of cereals, so-named because of the fish-like smell) the spores replace the grain but remain enclosed in the grain coating (pericarp) until the grain is threshed. In other smuts, such as *U. anemones* on anemone and buttercup, the spore masses develop beneath the surface of the stem or leaf before they mature and erupt through the plant surface.

A recent surge of interest in endophytes was spurred by the discovery that pasture grasses, especially ryegrass (*Lolium* spp.) and tall fescues (*Festuca* spp.), can be systemically infected by dark mycelial fungi whose hyphae grow sparsely in the wall spaces. Many of these fungi have not been induced to sporulate and may never do so in nature; but, some produce conidia in culture and are identifiable as *Acremonium* spp. As noted in Chapter 6, these fungi produce mycotoxins closely related to the ergot alkaloids, and they seem to act as insect antifeedants. The ergot fungus itself, *Claviceps purpurea* (ascomycota), is a pathogen that infects cereal flowers at the time when fertilization would lead to grain development. It then grows in place of the grain to produce large sclerotia, known as ergots, that contain high concentrations of alkaloids.

Yet other endophytes are receiving attention as possible biocontrol agents of plant pathogens. This stems from the finding that a challenge of plants with a high enough inoculum level of a weakly pathogenic fungus or a non-compatible pathogen can cause the plants to acquire resistance to subsequent challenge by a compatible pathogen. This acquired resistance is initially localized to the leaf that received the first challenge (**local acquired resistance**) but it can then spread to other parts of the plant (**systemic acquired resistance**, or SAR). So plants can, in effect, be immunized although the mechanism is quite different from the antibody-type response of vertebrates. The local acquired resistance involves similar factors to those in the hypersensitive response: an early oxidative burst, a marked increase in the activities of several pathogenesis-related proteins such as phenylalanine ammonia lyase, and the synthesis of fungal wall-degrading enzymes. In some way, as yet unknown, a signal is then sent to other parts of the plant and it activates about 40 'SAR' genes which then respond rapidly to subsequent attempted infections by even unrelated pathogens (Staskawicz *et al.*, 1995).

One example of SAR is already used commercially in Japan. The bases of stem cuttings of sweet potato (*Ipomoea batatas*) are challenged with a high inoculum level of a non-pathogenic species of *Fusarium* then planted into field sites, where they show increased resistance to infection by the vascular wilt pathogen, *F. oxysporum* f. sp. *batatas* (Komada, 1990). The method is practicable, because high agricultural subsidies in Japan make it economic to transplant sweet potato cuttings. Of interest is the fact that the initial challenge has to be strong enough to cause localized rotting of the stem base, and even then the acquired resistance lasts only for a few weeks; but, this is long enough to take the plants through their most susceptible stage for vascular wilt infection, and any infection after that time does not markedly affect the harvest yield.

Biotrophic pathogens

Biotrophic pathogens have an extended nutritional relationship with living host cells, so the success of this type of parasitism depends on two features of the pathogen:

1 the ability to avoid eliciting host cell death;
2 a mechanism for securing a nutrient supply from the living host tissues.

Clearly, on these criteria many of the endophytes would qualify as biotrophs, and indeed they are in some classifications of host–parasite interactions. A classic example is the fungus *Fulvia* (=*Cladosporium*) *fulvum* which causes blue mould of tomato plants. Unlike the ubiquitous *Cladosporium* spp. that grow as epiphytes on leaves (see Chapter 10), *F. fulvum* enters the host stomata from a germinating spore and forms an extensive intercellular network in the leaf, before emerging from the stomata as conidiophores which release further spores. It has no obvious means of feeding from

the living host cells except by scavenging leachates; but, it might enhance this rate of leakage and also use some of the cell wall polymers. In any case, it is highly host-adapted and has a **gene-for-gene relationship** with its host. A strain of *F. fulvum* can only feed from the host if it avoids inducing a hypersensitive response, and this is governed by 'avirulence' genes of the pathogen and corresponding resistance genes in the host.

Having dealt with this example, which might have implications for the behaviour of many endophytes, we now focus on the biotrophic pathogens that penetrate plant cells wholly or in part, and which cause crop losses by diverting host nutrients to the infection site.

The haustorial biotrophs

The most economically important biotrophs are the rust (basidiomycota), powdery mildew (ascomycota) and downy mildew (oomycota) fungi. In all cases, they establish localized infections by growing initially between the host cells and then invading through the walls of individual host cells to form haustoria. The initial stages of this were covered in Chapter 4, but now we focus on the haustorium itself. As shown in Fig. 12.11, the invading hypha penetrates the host cell wall and swells to form a haustorium, but this is always separated from the host cytoplasm by a host-derived membrane, termed the **extrahaustorial membrane.** The fungal wall also remains intact around the haustorium, and it is separated from the extrahaustorial membrane by a fluid matrix. There is a tight 'seal' in the neck region where the fungus enters the cell, and one consequence of this, experimentally, is that the whole haustorial complex (enveloping membrane, fluid matrix and the haustorium itself) can be isolated by digesting the host cell wall with enzymes. From studies on these isolated haustorial complexes of rust and powdery mildew fungi it has been shown that the extrahaustorial membrane lacks adenosine triphosphatase (ATPase) activity, unlike the rest of the host cell membrane and the fungal membrane. This almost certainly means that the fungus can drive nutrient uptake across its own (haustorium) membrane but the extrahaustorial membrane has no means of opposing this one-way flow of nutrients. In contrast, the membrane surrounding the

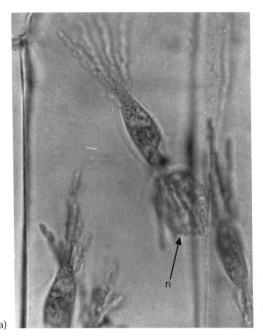

Fig. 12.11 (a) Haustoria of *Erysiphe graminis* in a cell of a cereal leaf. The haustoria of this fungus have characteristic finger-like projections; the host cell nucleus (n) is shown. (b) Diagrammatic representation of a haustorium in a host cell, showing how nutrients can be imported to the infected cell from other leaf cells and then absorbed by the fungus and translocated to the hyphae. The host cell membrane and fungal membrane have normal ATPase activity, so they can generate a transmembrane potential for nutrient uptake; the extrahaustorial membrane lacks ATPase activity so it cannot counteract the nutrient flow into the haustorium.

rest of the infected host cell has normal ATPase activity, so the infected cell can import nutrients from its neighbours, and the fungus can create a continuous drain on the plant's nutrients.

Much of the economic importance of rust and powdery mildew fungi stems from this ability to continuously draw nutrients from the host. These nutrients are used to support further growth but especially to support high rates of sporulation, so that the pathogen can spread to further sites. It is not uncommon for these fungi to have several infection cycles in a year, as the spores from an initial infection establish further infections, and so on. Sometimes, the host plant tolerates this with only moderately reduced growth, but it does not build up sufficient reserves to support good seed or fruit production. A single 'strategically' positioned lesion at the base of a cereal leaf, for example, can effectively prevent that leaf from being a net exporter of nutrients. This can affect not only shoot growth but also root growth,

because roots are relatively poor competitors for plant assimilates. Plants with a poorly developed root system are more prone to drought or mineral nutrient stress or to the effect of any pathogenic attack on the roots.

The downy mildew pathogens are haustorial biotrophs with similar behaviour to the rust and powdery mildew fungi, although they require more humid conditions for infection and sporulation. Examples include *Bremia lactucae* on lettuce, *Plasmopara viticola* on grapevine and *Pseudoperonospora humuli* on hops. However, there is a gradation of behaviour in the oomycota because some *Phytophthora* spp. which are closely related to the downy mildews begin their parasitic phase as haustorial biotrophs but then kill the host tissues and spread within them as necrotrophs (e.g. *P. infestans* on potato and *P. sojae* on soybean). Some other *Phytophthora* spp., such as *P. cinnamomi*, grow as necrotrophs with extremely broad host ranges, and all the *Phytophthora* spp. can be grown in axenic culture (i.e. separate from their hosts) whereas none of the downy mildew pathogens can be grown in this way.

Recognition systems

Biotrophs must avoid causing rapid death (the hypersensitive response) in order to feed on living cells. In all systems that have been examined there is a recognition system of the **gene-for-gene** type, whereby the host plant has **resistance genes** while the pathogen has corresponding **avirulence genes**. The host resistance genes code for a **receptor**, while the pathogen's avirulence genes code for an **elicitor** — a product that activates a resistance response. A race of the pathogen will be able to infect only if it does not produce the elicitor that corresponds to the host receptor. In fact there are batteries of these resistance and avirulence genes, but the outcome is always the same — the pathogen must avoid eliciting a response and thus must not produce any elicitor recognized by the host's receptors. It follows that host resistance is expressed as a dominant genetic trait, whereas virulence of the pathogen (failure to elicit a response) is a recessive trait.

There is now good evidence that the elicitors include fungal wall products, as we saw earlier. But, the situation is complicated by the finding that both *P. infestans* and *P. sojae* also produce suppressors of the hypersensitive response. In *P. infestans* they include small glucans and in *P. megasperma* they include a mannan glycoprotein. In some respects this would make sense for the *Phytophthora* spp. because they suppress cell death only for a time while they get established in the plant as biotrophs (suppressor activity?) and then they kill the cells. Corresponding information is much less easily obtained for the exclusively biotrophic fungi because they do not grow in culture as pathogenic forms (see Chapter 5). But, the endophyte *F. fulvum*, mentioned earlier, lends itself readily to experimental analysis. In this case, two elicitors have been identified by analysis of fluids in the intercellular spaces of inoculated leaves. The corresponding genes have been sequenced and the elicitor products identified as two small proteins of 28 and about 105 amino acids, respectively. These genes are strongly induced in the plant, as soon as the fungus grows through a stoma. An elicitor-binding site on the host (tomato) cell membrane also has been identified (Honee *et al.*, 1994). The corresponding resistance gene and also that of flax (for flax rust, *Melampsora lini*) have been cloned. They are very similar, with several highly conserved domains, and this is also true for receptors that recognize bacterial and viral pathogens. So, there is a family of plant cell-surface receptors that recognize specific pathogens.

Intracellular biotrophs and hormonal modifications

The last group of biotrophs to be considered are those that grow wholly within the cells of their hosts. These are the more primitive fungi and fungus-like organisms — the chytridiomycota (e.g. *Olpidium brassicae*), lagenids (e.g. *Lagena radicicola*) and plasmodiophorids (*Polymyxa* spp., *Spongospora subterranea* and *Plasmodiophora brassicae*). Most of them are common symptomless parasites which can be found in the roots of many plants. They have little economic significance except as vectors of plant viruses (see Chapter 9), although *S. subterranea* can cause the disfiguring powdery scab disease of potatoes and also 'crook root' of watercress. The reason why most of these fungi cause relatively little damage is that they do not cause a major diversion of plant nutrients towards the infection sites.

P. brassicae is an exception to this rule, but only in one respect — on some cruciferous crops it causes the damaging clubroot disease. The life cycle of this fungus was shown in Chapter 1, Fig. 1.4. The resting spores in soil germinate to release a zoospore which can infect the root hairs of many types of plant and produce a small, naked **primary plasmodium** in the root hair. This is perhaps much more common than is recognized because it goes unnoticed. Zoospores released from the primary plasmodia can infect the root cortex of crucifers (cabbage, cauliflower, etc.) and form larger secondary plasmodia. Then, the infected cells and surrounding cells undergo expansion (**hypertrophy**) and division (**hyperplasia**), leading to the development of large galls. The marked diversion of plant nutrients to these galls can seriously reduce the shoot growth and yield of crops with marketable tops such as cauliflower,

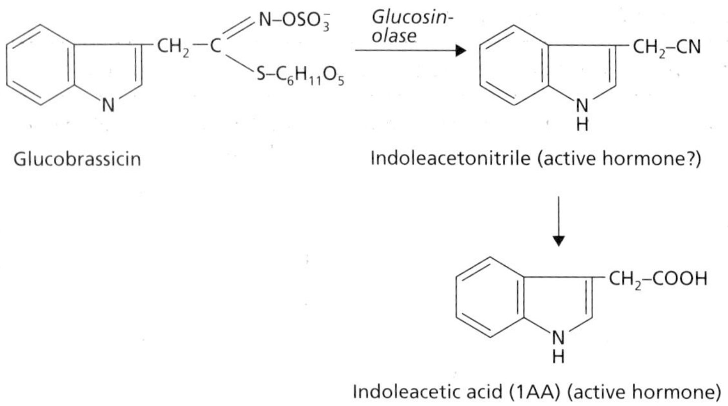

Fig. 12.12 Conversion of glucobrassicin to plant growth hormones as a result of infection of cruciferous plants by the clubroot pathogen, *Plasmodiophora brassicae*.

while the galls themselves make root crops such as swedes unmarketable. The development of galls involves hormonal imbalance, which is thought to be caused when a hormone precursor, **glucobrassicin**, in the host tissues is converted to the active hormone indolylacetic acid by the action of an enzyme glucosinolase, as shown in Fig. 12.12.

At one time it was common to speak of plants as being either susceptible or resistant to *P. brassicae* according to whether they developed galls. In truth, however, *P. brassicae* can produce secondary plasmodia in the roots of many cruciferous plants, including the common weed *Capsella bursa-pastoris* (shepherd's purse), without causing visible symptoms. Either these plants do not contain glucobrassicin or it is not converted to a hormone after infection. This raises an important cautionary note on which to end this discussion of pathogens. In wild plants *P. brassicae* is a parasite, not a pathogen. Indeed, most parasites live in balance with their hosts in natural plant communities. The spectacular diseases that we see in crop plants are a consequence of highly selective breeding, especially using single 'major' resistance genes which pathogens can overcome by single gene mutations. The recognition of this has led plant breeders to revert to 'horizontal' or 'durable' resistance which is multigenic and additive. Spectacular diseases also result from the ways in which crops are grown — as genetically uniform stands of plants through which pathogens can spread easily. This is less easy to correct in modern agriculture.

Mycorrhizas

'... under agricultural field conditions, plants do not, strictly speaking, have roots, they have mycorrhizas.'

This comment by the plant pathologist Stephen Wilhelm could apply equally well to natural plant communities. More than 70% of vascular plants (angiosperms and gymnosperms) are mycorrhizal, as are many pteridophytes (ferns and their allies) and some bryophytes (especially liverworts). The non-mycorrhizal plants tend to be those of open habitats, where competition for soil nutrients may not be a major problem, but in closed plant communities mycorrhizas are the norm.

The term mycorrhiza refers to an intimate association between a root or other underground organ and a fungus. Often, these associations are mutualistic because the fungus gains most or all of its organic carbon from the plant, while the plant obtains mineral nutrients from soil via the fungal hyphae. However, there are major variations on this theme, as we shall see.

There are several types of mycorrhiza, listed in Table 12.5 (Harley & Smith, 1983). Here we consider four of the most common types (Fig. 12.13) in terms of the host–parasite interaction and ecological and environmental significance.

Arbuscular mycorrhizas (AM)

AM are the most common type of mycorrhiza. They are found worldwide on most crop plants, wild herbaceous plants and trees, as well as on pteridophytes and some bryophytes. They have a very ancient origin, dating back to the early land plants (Simon *et al.*, 1993) and they could even have been a major factor in the colonization of

Table 12.5 The major types of mycorrhiza. (Modified from Harley, 1989.)

	Arbuscular	Ectotrophic*	Arbutoid	Monotropoid	Ericoid	Orchid
Main fungi involved	Zygomycota	Basidiomycota, ascomycota	Basidiomycota	Basidiomycota	Ascomycota, deuteromycota	Basidiomycota
Penetration of cells	+	0	+	+	+	+
Sheath	0	+	+	+	0	0
Hartig net	0	+	+	+	0	0
Haustoria	+	0	0	+	0	+ or 0
Coils in cells	+	0	+	0	+	+
Host plants	Bryophytes, pteridophytes, gymnosperms, angiosperms	Gymnosperms, angiosperms	Some Ericales	Monotropa	Ericales	Orchids

* Another mycorrhizal type (ectendomycorrhizas, not shown) has some features of ectomycorrhizas but the hyphae also can penetrate the cells.

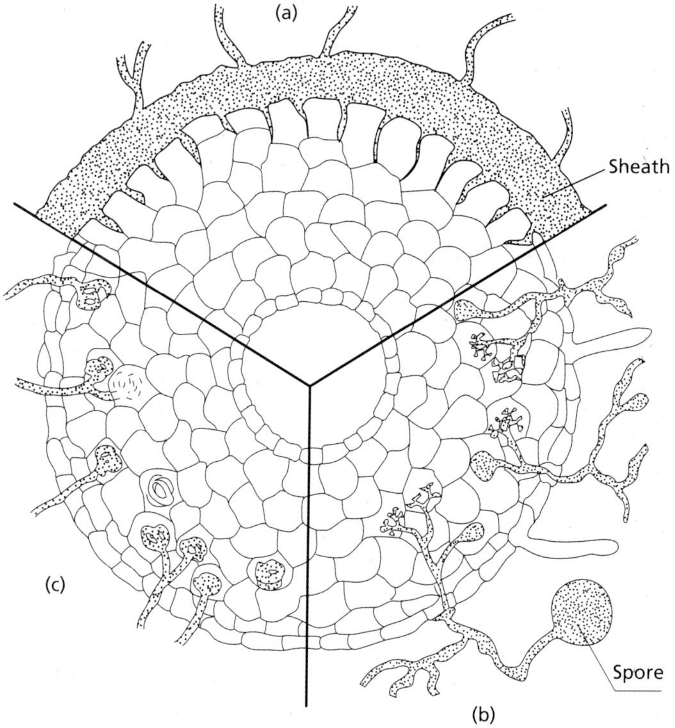

Fig. 12.13 Diagrammatic representation of three types of mycorrhiza in transverse section of a root. (a) Ectotrophic mycorrhiza of forest trees, showing a sheath of fungal tissue, and limited invasion of the root by intercellular hyphae (the Hartig net). See also Fig. 12.14. (b) Arbuscular mycorrhiza of many herbaceous plants and tropical trees, showing arbuscules and vesicles in the host cells (see Plate 12.3, facing p. 210). Large spores up to 400 μm diameter are formed on the hyphae that ramify in soil. (c) Endotrophic mycorrhiza of an orchid, showing active and partly digested hyphal coils within the host cells.

land. The fungi are members of the zygomycota, classified in six genera — *Glomus, Acaulospora, Gigaspora, Sclerocystis, Entrophospora* and *Scutellospora*. None of them can be grown in culture, away from their hosts. So, they are obligate biotrophic fungi, wholly dependent on plants for their survival.

Infection by these fungi does not cause any

visible changes in the roots, and they are not even seen unless the root tissues are cleared with strong alkali and then treated with trypan blue which stains the fungal walls. For this reason, these ubiquitous fungi were hardly recognized until the 1960s. They persist as large spores of up to 400 μm diameter in soil, and infect roots from germinating spores by forming an appressorium-like structure on the root surface (see Fig. 12.13; Plate 12.3, facing p. 210). From this, the hyphae grow between the root cortical cells and penetrate individual root cells to form **arbuscules**, which are extremely finely branched tree-like structures that occupy most of the cell volume. These are similar to the haustoria of biotrophic pathogens; they are surrounded entirely by the host cell membrane so that they remain separate from the host protoplasm. Some of these fungi also produce **vesicles** in the roots. These are swollen, lipid-rich bodies which probably have a storage function. Their presence led to the former common name, vesicular-arbuscular mycorrhiza, but the term arbuscular mycorrhiza is now preferred because not all of the fungi produce vesicles. The AM fungi grow from an infected root into soil, producing a ramifying system of broad, irregularly shaped, aseptate hyphae on which the spores are formed.

The arbuscules are almost certainly the major sites of nutrient exchange between the fungus and the host. Through them the fungus obtains carbohydrates from the plant, and the plant can obtain mineral nutrients that the hyphae have absorbed from soil. However, the mechanisms of nutrient exchange are largely unknown and, despite the superficial similarity, arbuscules differ from the haustoria of biotrophic pathogens in at least three ways.

1 There is no neck band or seal in the region where the arbuscule enters the host cell.

2 The host membrane which surrounds all the branches of the arbuscule has normal ATPase activity.

3 The individual arbuscules have only a limited life, of 14 days or so, before they autolyse or are digested by the host. Then they are seen as 'stumps' in the previously infected cells. However, the supply of mineral nutrients to the host is known to occur before the arbuscules are digested. It is thought that the two-way nutrient transfer is based largely on source-sink relationships across the fungus–plant interface. Electron micrographs show that polyphosphate granules are present in the mycelia but not in the arbuscules (although the fact that they are seen as **granules** is probably an artefact of chemical fixation). There is also a marked increase in alkaline phosphatase activity in the hyphae just before the plants show a major growth response to inoculation with these fungi. These points indicate that changes occur in phosphorus-related metabolism in the AM fungi at a time when phosphorus seems to be transferred to the plant.

Perhaps the most interesting feature of AM fungi is that they are biotrophic parasites with astonishingly wide host ranges — a strain that grows on tomato roots will also grow on sycamore trees or almost any other plant, although in nature there may be some host or habitat preferences. This contrasts markedly with the haustorial biotrophic pathogens (but not with the primitive intracellular biotrophs mentioned earlier) and it dispels the common perception that biotrophy is a highly host-specific and evolutionarily 'advanced' phenomenon. Quite the opposite is true — biotrophy is probably the aboriginal form of parasitism, but the rust and powdery mildew fungi might have evolved as biotrophs at a later stage in fungal evolution. We can thus view the AM fungi as having developed an early compatibility with plants, and having retained this to the present day as plants diversified and spread across the land masses. Nevertheless, several families of flowering plants are now typically (but not absolutely) non-mycorrhizal. They include the rushes (Juncaceae), sedges (Cyperaceae), crucifers (Cruciferae), Chenopodiaceae and Caryophyllaceae, which tend to be the primary colonizers of open land. Also, some plant families now form mycorrhizas with different types of fungus, discussed below, probably because these fungi better suit the plants' needs.

A recent discovery could provide new insights into this phenomenon (Gianinazzi-Pearson et al., 1994). Pea plants, which normally are strongly mycorrhizal, were chemically mutated for loss of ability to nodulate with *Rhizobium*, and some of these nod⁻ mutants were also unable to develop AM (myc⁻). Some nod⁻ mutants developed mycorrhizas as normal, but all myc⁻ mutants were nod⁻, suggesting that there is at least one 'symbiotic' gene which governs the ability to form intimate intracellular symbioses. Of interest, when inoculated with AM fungi, the myc⁻ mutants showed reactions resembling a hypersensitive response. Could a mutation of this type have occurred early in the evolutionary lineages that led to some of today's plant families, rendering most of the members of those families non-mycorrhizal?

Ectotrophic mycorrhizas

Ectotrophic mycorrhizas, or ectomycorrhizas, are found on coniferous and broad-leaved trees in temperate and boreal environments (pine, spruce, fir, oak, beech, birch, eucalypts, etc.). However, tropical trees and even some temperate trees (sycamore, ash, poplars) have AM, and some (e.g. willows) can have both types. The fungi involved are principally basidiomycota that form some of the common toadstools of forests (e.g. *Amanita*, *Boletus*, *Cortinarius*, *Hebeloma*, *Laccaria* spp.), but ectomycorrhizas also are formed by some ascomycota, including the truffle fungi.

The characteristic feature of ectomycorrhizas is the presence of a substantial sheath (**mantle**) of fungal tissue on the terminal, nutrient-absorbing rootlets (Fig. 12.14; see Plate 9.1), and the rootlets themselves often are short and stumpy, lacking root hairs. An extensive network of hyphae or mycelial cords (see Chapter 4) radiates from the surface of the sheath, while beneath the sheath the fungus invades between the root cortical cells to form a 'Hartig net'. Although the fungus is in close contact with the root cells in this region, there is no penetration of the host cells, hence the term ectotrophic (outside-feeding).

The lack of root hairs and the complete investment of the feeder roots by a fungal sheath mean that virtually all mineral nutrients that enter the plant must be channelled through the fungus. In turn, the fungus receives at least part of its organic carbon from the plant. Consistent with this, most of the ectomycorrhizal fungi can be grown in culture but they require simple organic carbon sources such as sugars; they have little or no ability to degrade cellulose and lignin, unlike the decomposer ascomycota and basidiomycota. Many of them can, however, degrade proteins and this may be significant for acquiring nitrogen in forest ecosystems. As in the case of AM fungi, there is a

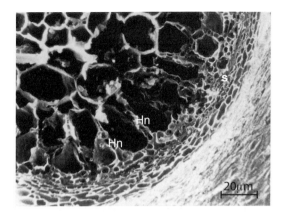

Fig. 12.14 Scanning electron micrograph of transverse section of an ectomycorrhiza, showing the fungal sheath (s) around the root and its extension between the outer root cortical cells as a Hartig net (Hn). (Courtesy of F.M. Fox.)

two-way nutrient exchange whereby the fungus obtains plant sugars and the plant can obtain mineral nutrients from the fungus. This must occur across the Hartig net but the details are poorly understood. A simple source-sink relationship could be aided by the conversion of plant sugars to the characteristic fungal carbohydrates in the sheath (see Chapter 6). The transfer of phosphates to the plant seems to occur only after an excess of phosphate has accumulated as polyphosphates in the sheath.

The ectomycorrhizal fungi do not show a high degree of host-specificity, at least in artificial inoculations in sterile laboratory culture. In field conditions, also, it is common to find mycorrhizas belonging to several different fungi on the root system of a single tree. Nevertheless, as we saw in Chapter 9 these different fungi might often occur in different parts of the root zone, and they can be grouped in very broad terms as either 'generalists' with wide host ranges especially on seedlings (e.g. *Laccaria* spp., *Hebeloma* spp., *Thelephora terrestris*, *Paxillus involutus*) or more host-restricted types which tend to predominate on mature trees (e.g. *Suillus luteus* on pines, *Suillus grevillei* on larch, and the truffles, *Tuber* spp., on oak or beech). These 'mature' types have been shown to differ from some of the 'juvenile' types by having strong proteolytic activities in culture. So, perhaps they are favoured when the availability of mineral nitrogen is limited and they can exploit the organic nitrogen in forest soils.

Ericoid mycorrhizas

The heathland plants in the family Ericaceae (*Erica, Calluna, Vaccinium*, etc.) form mycorrhizas with a distinct group of ascomycota and deuteromycota, one of which, *Hymenoscyphus ericae*, has been characterized and studied in detail. The septate hyphae of the fungus form coils within the root cells but outside of the host plasma membrane, and nutrient exchange is thought to occur primarily through these coils. The most notable feature of these mycorrhizas, however, is the strong evidence that the fungus provides its host with nitrogen in natural soils. This was shown initially by the uptake and transfer of ammonium labelled with ^{15}N when plants were grown in artificial laboratory conditions. But, when ^{15}N-ammonium was added to natural, acidic heathland soils the mycorrhizal plants actually took up less label than the non-mycorrhizal control plants, even though the mycorrhizal plants had accumulated more total nitrogen. Evidently, in these conditions the fungus was obtaining nitrogen and supplying it to the host from a different source, while the uptake of ammonium was simultaneously suppressed. This led to the discovery that the fungus secretes a proteinase with optimum activity at about pH 3, releasing amino acids from the soil organic matter. The rates of mineralization of nitrogen by saprotrophic decomposer fungi are extremely low in the cool, acidic sites where ericaceous plants typically occur. All members of the Ericaceae are strongly mycorrhizal, consistent with a major role of the mycorrhizal fungi in supplying mineral nutrients in these conditions.

Orchid mycorrhizas

Orchid mycorrhizas are entirely different from any of those discussed above, because the plant is actually parasitic on the fungus. The seeds of orchids are extremely small, with minimal nutrient reserves, and they are triggered to germinate when they are infected by an appropriate fungus. The fungi in these cases are species of *Rhizoctonia* (basidiomycota) or related fungi, and even the plant pathogen *R. solani* can form mycorrhizas

with orchids in experimental conditions. They grow on soil organic matter, degrading cellulose and other structural polymers, and they infect the orchid cells to produce hyphal coils, surrounded by the host cell membrane. But, the plant then draws sugars from the fungus, presumably as trehalose or other fungal carbohydrates (see Chapter 6). Consistent with this, orchid seeds can be triggered to germinate and the seedlings can be raised in commercial conditions by supplying them with trehalose. In natural conditions, many of the cells of the protocorm are filled with fungal coils, but these coils have only a limited life before they are digested or autolyse and then new coils are formed in their place. The fungus provides the sole source of nutrients during the first years of the life of all orchids — at least until the plants develop chlorophyll in their later stages.

A functionally similar type of mycorrhiza is found in the non-chlorophyllous *Monotropa* spp., but the fungal symbionts in these cases are either ectomycorrhizal species that are supplied with nutrients from a tree host (e.g. *Boletus* spp.) or they are tree pathogens such as *Armillaria* spp. (see Chapter 4). Labelled carbon can be supplied to the tree leaves as $^{14}CO_2$ and it moves down to the roots as labelled sucrose before it enters the mycorrhizal sheath and is found as labelled sugar alcohols or trehalose (see Chapter 6). Then the label is translocated along mycelial cords and is found as labelled sugars in *Monotropa*. This is an interesting twist on the common expression 'poacher turned gamekeeper' because the fungus is a poacher at one end but gamekeeper at the other.

Significance of mycorrhizas

Mineral nutrition

The different types of mycorrhizal association clearly have different roles. In the case of orchids and *Monotropa* the fungus supplies the plant with organic carbon and probably also with mineral nutrients. In most other cases the fungus depends on a plant for most or all of its carbohydrate requirements but can supply the plant with mineral nutrients from soil. Until recently it was assumed that the major role of mycorrhizal fungi was to supply plants with phosphate, because phosphorus is an extremely immobile element in soils. Even if phosphates are added to soil in soluble form they are soon immobilized as organic phosphates or calcium phosphate, etc. The extensive mycelial network of mycorrhizal fungi, coupled with their efficiency in phosphorus uptake (see Chapter 5), can help to overcome this limitation. This is probably the major role of the AM fungi in many natural and cropping environments, because inoculation with these fungi has been shown significantly to increase plant phosphorus levels in experimental conditions. However, in the more acidic soils of moorlands and in woodlands where the rates of decay of leaf-litter are relatively slow, the mycorrhizal fungi are now thought to play a much more significant role in obtaining nitrogen by means of their protease enzymes, and supplying nitrogen to the plants in the form of amino acids.

Water relationships

Mycorrhizal fungi can also benefit plants by tapping the soil water reserves. For example in an experimental system similar to that shown in Fig. 3.13 (p. 60), pine seedlings were inoculated with a mycorrhizal fungus and grown in a peat substrate until the fungus had formed mycelial cords that extended well beyond the root zone. Then the peat was allowed to dry to the point where non-mycorrhizal control seedlings died, but the mycorrhizal plants remained healthy because the mycelial cords could transport water from deeper in the peat. Such a role could be important in soils with poor water retention, such as mine spoil heaps where trees are planted (or colonize naturally) for land reclamation. It need not be restricted to cord-forming fungi because, as we noted in Chapter 7, all fungi can grow at water potentials below those that plants can tolerate. For this reason the AM fungi could be particularly important in desert communities, although this has received surprisingly little study. Bethlenfalvay *et al.* (1984) reported that many common desert plants, including cacti, are heavily mycorrhizal.

Protection against toxins and pathogens

On land reclamation sites the ectomycorrhizal fungi can protect trees from high concentrations of potentially toxic heavy metals, because these tend to be accumulated and immobilized in the mycorrhizal sheath. Also in these sites the mycelial networks contribute to soil stability by binding the soil particles. In a different context, there is substantial evidence that mycorrhizal fungi can help to protect plants against pathogenic attack. AM fungi have been shown to reduce the degree of infection by root pathogens but, conversely, can increase infection by leaf pathogens and plant viruses in experimental conditions. These are often likely to be indirect effects, stemming from the improvement of plant vigour and mineral nutrition. For example, phosphorus-deficient roots are especially susceptible to infection by *Pythium* spp. and by the take-all fungus; in the first part of this century wheat crops on the Canadian prairies were devastated by 'browning root rot' (*Pythium* spp.), but this problem almost disappeared when phosphate fertilizers became widely available.

Ectomycorrhizas can have a more direct role in plant protection. As an example, pine roots are protected against attack by *Phytophthora cinnamomi* if they have mycorrhizas of *Leucopaxillus cerealis* var. *piceina* (Marx, 1975). Most of this protection could be due to the 'mechanical' barrier of the mycorrhizal sheath, although it would be more appropriate to call this a 'non-host' barrier because *P. cinnamomi* is a parasite of plant tissues, not of the fungal sheath around the feeder roots. But, there might be an additional effect of a polyacetylenic antibiotic, diatretyne nitrile, that *Leucopaxillus* produces in culture and which strongly inhibits the growth of *P. cinnamomi*:

diatretyne nitrile:
$$HOOC–CH=CH–C\equiv C–C\equiv C–C\equiv N$$

Many other ectomycorrhizal fungi have been found to produce antibiotics in culture, and also to exhibit hyphal interference (see Chapter 11). These effects may be particularly relevant in Australia where *P. cinnamomi* has been introduced, causing major damage to the native jarrah (*Eucalyptus marginata*) forest community. In parts of Australia the jarrah trees show extensive 'dieback' and some of the understorey trees such as *Banksia* have been virtually eliminated (Podger, 1975). In Australia and New Zealand the spread of *P. cinnamomi* along roadways and water courses has also caused devastation to commercial forestry plantations, so that the unaffected forests have now been 'sealed off' — vehicles are not allowed to leave and the extracted timber is lifted out over barrier fences. A combination of factors seem to have contributed to the damage caused by *P. cinnamomi* in these regions, including severe phosphorus deficiency of the soils, because mycorrhizal development is restricted in the most severely deficient conditions. This leads to a cyclical scenario in which the trees grow poorly, so their small root systems cannot remove the water efficiently from the soil after periods of heavy rain, and there is little leaf-litter to capture and absorb the water. The periodic wet conditions then promote zoospore production by the pathogen, and the unprotected, non-mycorrhizal root tips are rapidly destroyed.

Mycorrhizas as key components of ecosystems

We end this chapter with a brief discussion of mycorrhizas in ecosystem function, because these fungi might play an even greater role than anything we have considered so far. The evidence is far from complete, but it suggests that mycorrhizal networks might link many of the plants within a habitat—even plants of different species because of the general lack of host-specificity of these fungi. For example, it is known that tree seedlings can be linked to a 'mother' tree or 'nurse' tree by a common mycorrhizal network, such that $^{14}CO_2$ fed to the leaves of a larger tree can be found as label in the roots and shoot tissues of nearby seedlings. The amounts of nutrients transferred in this way are probably too small to be significant for feeding of the seedlings, but at least the parent tree could support the establishment of mycorrhizal infection of the seedling roots, sparing the seedling some cost in terms of its limited photosynthetic capacity—the annual cost of maintaining mycorrhizas is estimated to be at least 10% of the total photosynthetic production of a tree.

Perhaps more important is the potential of the mycorrhizal network to retain and conserve mineral elements. An estimated 70–90% of ectomycorrhizal rootlets die and are replaced annually. If these rootlets were not interconnected by mycelia then they would be decomposed by the soil microbial community and at least some of the nutrients could be leached from the soil by rain-water. However, the mycorrhizal mycelial connections can help to retain mineral nutrients within the system, withdrawing them from the degenerating mycorrhizas to others that are still functioning. If this happens to any significant degree in mycorrhizal systems as a whole, then plants that are not tapped into the network would be at a significant disadvantage. The only available option would be to colonize disturbed habitats, and this is precisely where plants of the non-mycorrhizal families are commonly found. There is still much to learn about the role of mycorrhizas in natural community dynamics and ecosystem functioning (Allen, 1992).

References

Allen, M.F. (ed.) (1992) *Mycorrhizal Functioning*. Chapman & Hall, New York.

Armstrong, G.M. & Armstrong, J.K. (1981) *Formae speciales* and races of *Fusarium oxysporum* causing wilt diseases. In: *Fusarium: Diseases, Biology and Taxonomy* (eds P.E. Nelson, T.A. Toussoun & R.J. Cook), pp. 391–9. Pennsylvania State University Press, University Park, PA.

Beckman, C.H. & Talboys, P.W. (1981) Anatomy of resistance. In: *Fungal Wilt Diseases of Plants* (eds C.E. Mace, A.A. Bell & C.H. Beckman), pp. 487–521. Academic Press, New York.

Bethlenfalvay, G.J., Dakessian, S. & Pacovsky, R.S. (1984) Mycorrhizae in a southern California desert: ecological implications. *Canadian Journal of Botany*, **62**, 519–24.

Collmer, A. & Keen, N.T. (1986) The role of pectic enzymes in plant pathogenesis. *Annual Review of Phytopathology*, **24**, 383–409.

Cooper, R.M., Longman, D., Campbell, A., Henry, M. & Lees, P.E. (1988) Enzymatic adaptations of cereal pathogens to the monocotyledonous primary wall. *Physiological and Molecular Plant Pathology*, **32**, 33–47.

Deacon, J.W. & Scott, D.B. (1983) *Phialophora zeicola* sp. nov., and its role in the root rot-stalk rot complex of maize. *Transactions of the British Mycological Society*, **81**, 247–62.

Deacon, J.W. & Scott, D.B. (1985) *Rhizoctonia solani* associated with crater disease (stunting) of wheat in South Africa. *Transactions of the British Mycological Society*, **85**, 319–27.

Dodd, J.L. (1980) The role of plant stresses in development of corn stalk rots. *Plant Disease*, **64**, 533–7.

Esler, G. & Coley-Smith, J.R. (1984) Resistance to *Sclerotium cepivorum* in *Allium* and other genera. *Plant Pathology*, **33**, 199–204.

Gerik, J.S. & Huisman, O.C. (1985) Mode of colonization of roots by *Verticillium* and *Fusarium*. In: *Ecology and*

Management of Soilborne Plant Pathogens (eds C.A. Parker *et al.*), pp. 80–3. American Phytopathological Society, St Paul.

Gianinazzi-Pearson, V., Gollotte, A., Dumas-Gaudot, E., Franken, P. & Gianinazzi, S. (1994) Gene expression and molecular modifications associated with plant responses to infection by arbuscular mycorrhizal fungi. In: *Advances in Molecular Genetics of Plant–Microbe Interactions* (eds M.J. Daniels, J.A. Downie & A.E. Osbourn), pp. 179–86. Kluwer Academic, Dordrecht.

Harley, J.L. (1989) The significance of mycorrhiza. *Mycological Research*, **92**, 129–39.

Harley, J.L. & Smith, S.E. (1983) *Mycorrhizal Symbiosis*. Academic Press, London.

Hayman, D.S. (1970) The influence of cottonseed exudate on seedling infection by *Rhizoctonia solani*. In: *Root Diseases and Soil-borne Pathogens* (eds T.A. Toussoun, R.V. Bega & P.E. Nelson), pp. 99–102. University of California Press, Berkeley.

Honée, G., van den Ackerveken, G.F.J.M., van den Broek, H.W.J. *et al.* (1994) Molecular characterization of the interaction between the fungal pathogen *Cladosporium fulvum* and tomato. In: *Advances in Molecular Genetics of Plant–Microbe Interactions* (eds M.J. Daniels, J.A. Downie & A.E. Osbourn), pp. 199–206. Kluwer Academic, Dordrecht.

Komada, H. (1990) Biological control of fusarium wilts in Japan. In: *Biological Control of Soil-borne Plant Pathogens* (ed. D. Hornby), pp. 65–75. CAB International, Wallingford.

Lamb, C.J., Brisson, L.F., Levine, A. & Tenhaken, R. (1994) H_2O_2-mediated oxidative cross-linking of cell wall structural proteins. In: *Advances in Molecular Genetics of Plant–Microbe Interactions* (eds M.J. Daniels, J.A. Downie & A.E. Osbourn), pp. 355–60. Kluwer Academic, Dordrecht.

Marx, D.H. (1975) The role of ectomycorrhizae in the protection of pine from root infection by *Phytophthora cinnamomi*. In: *Biology and Control of Soil-borne Plant Pathogens* (ed. G.W. Bruehl), pp. 112–15. American Phytopathological Society, St Paul.

Mitchell, R.T. & Deacon, J.W. (1986) Differential (host-specific) accumulation of zoospores of *Pythium* on roots of graminaceous and non-graminaceous plants. *New Phytologist*, **102**, 113–22.

Nelson, E.B. (1987) Rapid germination of sporangia of *Pythium* species in response to volatiles from germinating seeds. *Phytopathology*, **77**, 1108–12.

Nelson, E.B. (1992) Bacterial metabolism of propagule germination stimulants as an important trait in the biological control of *Pythium* seed infections. In: *Biological Control of Plant Diseases* (eds E.C. Tjamos, G.C. Papavizas & R.J. Cook), pp. 353–7. Plenum Press, New York.

Podger, F.D. (1975) The role of *Phytophthora cinnamomi* in dieback diseases of Australian eucalypt forests. In: *Biology and Control of Soil-borne Plant Pathogens* (ed. G.W. Bruehl), pp. 27–36. American Phytopathological Society, St Paul.

Punja, Z.K. & Grogan, R.G. (1981) Mycelial growth and infection without a food base by eruptively germinating sclerotia of *Sclerotium rolfsii*. *Phytopathology*, **71**, 1099–103.

Ride, J.P. & Pearce, R.B. (1979) Lignification and papilla formation at sites of attempted penetration of wheat leaves by non-pathogenic fungi. *Physiological Plant Pathology*, **15**, 79–92.

Schneider, R.W. (1982) *Suppressive Soils and Plant Disease*. American Phytopathological Society, St Paul.

Simon, L., Bousquet, J., Levesque, R.C. & Lalonde, M. (1993) Origin and diversification of endomycorrhizal fungi and coincidence with vascular land plants. *Nature*, **363**, 67–9.

Staskawicz, B.J., Ausubel, F.M., Baker, B.J., Ellis, J.G. & Jones, J.D.G. (1995) Molecular genetics of plant disease resistance. *Science*, **268**, 661–7.

VanEtten, H.D., Soby, S., Wasmann, C. & McCluskey, K. (1994) Pathogenicity genes in fungi. In: *Advances in Molecular Genetics of Plant-Microbe Interactions* (eds M.J. Daniels, J.A. Downie & A.E. Osbourn), pp. 163–70. Kluwer Academic, Dordrecht.

VanEtten, H.D., Sandrock, R.W., Wasmann, C.C., Soby, S.D., McCluskey, K. & Wang P. (1995) Detoxification of phytoanticipins and phytoalexins by phytopathogenic fungi. *Canadian Journal of Botany*, **73**, S518–S525.

Wastie, R.L. (1962) Mechanism of action of an infective dose of *Botrytis* spores on bean leaves. *Transactions of the British Mycological Society*, **45**, 465–73.

Wastie, R.L. & Sankar, G. (1970) Variability and pathogenicity of isolates of *Colletotrichum gloeosporioides* from *Hevea brasiliensis*. *Transactions of the British Mycological Society*, **54**, 117–21.

Chapter 13

Fungal parasites of humans, insects and nematodes

In contrast to their major significance as parasites of plants, fungi cause relatively few diseases (**mycoses**) of humans and other warm-blooded animals. Nevertheless, these diseases are becoming important in acquired immune deficiency syndrome (AIDS) patients and with the use of immunosuppressants for treatment of leukaemia and in transplant surgery. Fungal pathogens can be life-threatening in these circumstances and we shall see in Chapter 14 that there are few really satisfactory drugs to control them.

Fungi more commonly attack insects, nematodes and other invertebrates, acting both as natural population regulators and holding promise for development of commercial biocontrol products. This chapter discusses the adaptations of fungi for parasitism of animals, focusing on the two main areas of practical interest: the mycoses of humans and the fungal parasites of insects and nematodes.

Mycoses of humans

Table 13.1 shows some of the major fungi that infect humans and other warm-blooded animals. They fall into essentially three groups.

1 The **dermatophytes** that cause ringworm, athlete's foot and other superficial mycoses. These fungi are very common, and parasitism is their natural mode of growth.

2 The fungi that grow as relatively harmless **commensals** on the mucosal membranes but which can, in appropriate conditions, become invasive. *Candida albicans* is the classic example of this.

3 A much larger and diverse group of **opportunistic pathogens** that can grow in the lungs and can invade the tissues of severely compromised hosts,

but which are found normally as saprotrophs in soil or on plant or animal remains.

The nomenclature of some of these fungi poses a problem. Typically, they grow as hyphae or yeasts in the body, but when cultured on agar they produce asexual sporing stages and they have been named according to these (deuteromycota). However, in recent years some of them have been found to be heterothallic, and when paired with strains of opposite mating type on agar they can produce sexual stages, usually cleistothecia of the ascomycota (see Chapter 1, Fig. 1.9). The generic names have been changed to reflect this (Table 13.1), but the diseases were based on the older names (e.g. cryptococcosis is caused by *Cryptococcus neoformans*, now called *Filobasidiella*). So, we will retain the old names for simplicity.

Dermatophytes

The dermatophytes, or ringworm fungi, are a clearly defined group of about 40 species, traditionally named according to their asexual stages *Trichophyton*, *Microsporum* and *Epidermophyton* (deuteromycota; Fig. 13.1). When mated, they produce a cleistothecial stage in the genus *Arthroderma* (ascomycota), but the natural role of the sexual stage and even of the conidia is largely unknown because these fungi seldom produce spores on an animal host.

The different species of dermatophyte show characteristic, although not absolute, associations with different animals and different body parts. For example, among the **anthropophilic** species (characteristically associated with humans) *T. interdigitale* usually grows between the toes, causing athlete's foot, *T. rubrum* causes a more intense, chronic infection of the foot, and *M.*

Table 13.1 Some fungi that cause mycoses of humans and other vertebrates.

Primary route of entry	Fungus	Sexual stage	Disease	Natural habitat
Skin	*Trichophyton* (24 species) *Microsporum* (16 species) *Epidermophyton* (2 species)	} *Arthroderma* (ascomycota)	} Ringworm, tinea, dermatomycosis	} Keratinized tissues, soil, wild and domesticated animals
	Pityriasis versicolor	None		
Wounds	*Phialophora,* *Cladosporium,* *Sporothrix,* etc.	Often none	} Subcutaneous mycoses; chromomycosis, sporotrichosis, etc.	} Saprotrophic in soil, on vegetation, etc.
Mucosa	*Candida albicans*	None	Candidosis, vulvovaginitis, thrush	Commensal on mucosa
Lungs	*Aspergillus fumigatus, A. flavus, A. niger,* etc.	None	Aspergillosis; lungs or invasive	Saprotrophic in soil, plant material
	Blastomyces dermatitidis	*Ajellomyces* (ascomycota)	Blastomycosis; lungs, skin lesions, bones, brain	Saprotrophic
	Coccidioides immitis	None	Coccidioidomycosis; lung, systemic	Saprotrophic in soil
	Cryptococcus neoformans	*Filobasidiella* (basidiomycota)	Cryptococcosis; lung, brain, meninges	Bird excreta, vegetation
	Histoplasma capsulatum	*Ajellomyces* (ascomycota)	Histoplasmosis; lung; rarely systemic	Bird and bat excreta
	Paracoccidioides brasiliensis	None	Paracoccidioidomycosis; lung, cutaneous, lymph nodes	Soil?
	Pneumocystis carinii	None	Pneumonia	Humans, animals

audouinii causes head ringworm. Other species are characteristic of animals (**zoophilic**) but infect humans through prolonged contact with infected animals; examples of this are *M. canis* on cats and *T. verrucosum* on cattle.

All the dermatophytes grow in the non-living tissues of the skin, nails and hair, often in the narrow zone just above the region where keratin is deposited. They seldom invade the living epithelium but they cause irritation, presumably because of their antigenicity, and this leads to scratching and the spread of infection. Mild wounding leads to an increased rate of cell division and keratinization of the skin, which favours these fungi by increasing the supply of substrates. On the other hand, an acute inflammatory response leads to cessation of keratinization which can sometimes cause the infection to resolve spontaneously. The main factors that promote the common dermatophytic infections of humans are high humidity caused by tight clothing, and mild

abrasion. Direct host-to-host transmission seems to be rare; instead, animals often are infected by using common scratching posts, and humans can be infected by using shared facilities such as towels and bathing places. The inoculum in these cases is shed flakes of skin or hair, and the dermatophytes survive well in these materials provided that the conditions are not wet enough to encourage bacterial growth.

In laboratory conditions many of the dermatophytes can degrade keratin, a protein that most other organisms cannot degrade because of the presence of numerous disulphide bonds between the sulphur-containing amino acids. The dermatophytes also grow in a characteristic way on keratinized materials, producing hyphal fronds in the planes of weakness between skin layers (Fig. 13.2) or producing 'perforating organs' consisting of columns of cells for penetrating the keratin-filled cortex of hair (Fig. 13.2). However, these distinctive growth forms are seen only on shed materials,

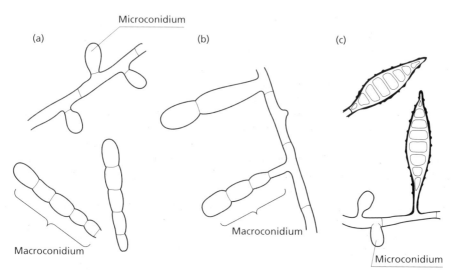

Fig. 13.1 Spore stages of dermatophytic fungi. (a) *Trichophyton* produces multicellular macroconidia up to 50 μm long, and small microconidia, about 4 μm. (b) *Epidermophyton* produces macroconidia with few septa, and no microconidia. (c) *Microsporum* produces spindle-shaped macroconidia with thick walls and small microconidia.

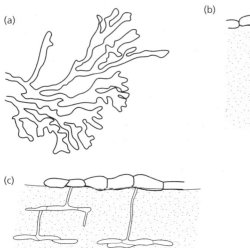

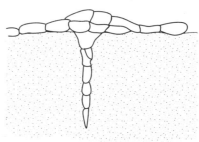

Fig. 13.2 Characteristic hyphal growth forms of dermatophytic fungi. (a) Flattened hyphal fronds growing in the planes of weakness in a layered substratum such as skin. (b) A 'perforating' organ growing into a physically resistant keratinized substratum such as nail or hair. (c) Hyphal fronds seen in transverse section of a layered substratum such as skin.

and seldom if ever when the dermatophytes are growing on the host. Even keratin may not be degraded until after the tissues are shed, and these fungi are generally poor competitors in their saprotrophic phase on shed tissues. Thus, the anthropophilic and zoophilic species seem to have become host-adapted by growing as persistent somatic colonies on a continuous supply of fresh substrate — the continuously regenerating skin, nails and hair. In this phase they show none of the characteristic features that they still retain from their presumably saprotrophic ancestors—asexual sporulation, sexual reproduction, keratin utilization and distinctive growth forms.

The dermatophytes are closely related to some saprotrophic (**geophilic**) species that grow on keratinized materials in soil. These soil fungi can be 'baited' by burying degreased hair, but they are seen relatively late in a succession. The initial colonizers are species of *Fusarium*, *Penicillium* and zygomycota that use the more readily available polysaccharides, mucopolysaccharides and proteins. Overlapping with these, or following them, are species of *Chaetomium*, *Humicola* and *Gliocladium*. Only then do the keratin-degrading *Athroderma* spp. appear and produce their cleistothecial sexual stages.

The control of dermatophytic infections is discussed in Chapter 14. Because these fungi grow superficially they can be treated by topical application of oxidizing agents (e.g. potassium permanganate) or antibiotics (e.g. nystatin, amphotericin B). For more serious infections, however, they are treated by oral administration of the antibiotic griseofulvin.

Candida albicans

C. albicans is a yeast which, in appropriate conditions, can change to a transitory hyphal form — a 'pseudomycelium' that forms further yeast cells by budding (see Chapter 4). It is diploid and has no known sexual stage. It is a very common commensal of humans, being found on the mucosal membranes of the mouth, gut or vagina of more than 50% of healthy individuals, and in most cases it causes no harm because its population is kept in check by a combination of the body's natural defences and by bacterial competition. However, *C. albicans* can become a significant problem if this balance is seriously disturbed. Then, it proliferates and invades the mucosa, causing local irritation, and in extreme cases it can grow systemically in the body, with fatal consequences. Thus, part of the key to understanding the role of *C. albicans* as a pathogen lies in the predisposing factors that enable the fungus to proliferate and become invasive.

C. albicans sometimes causes 'thrush' in newborn babies, growing on the mucosa of the mouth and throat and producing a white speckling from which the disease is named. This condition often is associated with delivery through an infected birth canal, and the fungus is able to proliferate on the mucosa of the neonate before a normal, balanced microbial population has developed in the mouth. *C. albicans* also causes vulvovaginitis, especially when women are pregnant or are using oral contraceptives; in these cases there is a strong hormonal predisposition, discussed later. One of the traditional and simple ways of alleviating the problem is by topical application of natural yoghurt containing a high population of lactic acid bacteria. *Candida* frequently causes stomatitis (inflammation of the buccal mucosa) of people who wear dentures. This is associated with several factors, including abrasion, adhesion to the dental plastic and probably the closed environment with lack of 'flushing' underneath the denture plate. *Candida* infection of the gut is often associated with prolonged antibacterial therapy, especially with tetracycline antibiotics which suppress the bacterial population. Stress can be an additional factor in *Candida* infections; for example, astronauts have been found to develop high populations of *C. albicans* during space flights. All these relatively mild, but distressing, infections involve localized invasion of the mucosa when *Candida* changes from a yeast to hyphal form. In extreme conditions, especially in the advanced stages of diabetes or cancer, *C. albicans* can invade and grow systemically in the body tissues; the source of these infections may be a contaminated catheter used for delivering fluids intravenously.

This catalogue of examples shows that *Candida* is essentially a non-specific pathogen in conditions that favour its development. However, none of the predisposing factors alone would seem to be sufficient to explain the onset of invasive growth, so attention has focused on the virulence determinants of *C. albicans* and how these might be influenced by the predisposing factors.

Virulence determinants of C. albicans

In vaginal infections and many other clinical cases the yeast cells adhere strongly to the epithelia, then they sprout germ-tubes which penetrate the tissues. This suggests that two of the principal virulence determinants may be the ability of the yeast to adhere and to undergo the dimorphic switch to a hyphal form. *In vitro* studies have confirmed that *Candida* adheres to the shed epithelial cells of the vagina, and that this adhesion is stronger on cells obtained from pregnant women. The two most common cell types in the vaginal epithelium are the intermediate cells and the superficial cells. *Candida* binds most strongly to the intermediate cells *in vitro*, and these cells predominate when there are high levels of progesterone (during pregnancy, at stages of the menstrual cycle and in oral contraception). However, progesterone seems also to have a direct effect, because *Candida* binds more strongly to the epithelial cells of non-pregnant women if progesterone is added *in vitro*.

These general correlations between *in vitro* adhesion and the course of clinical infections are supported by other studies. For example, *C. albicans* adheres to cells of the buccal cavity, to the methyl acrylate resin of dentures and to the surfaces of catheters. By using buccal epithelia or denture resin as 'model systems' *in vitro*, the adhesion of *Candida* was found to be enhanced strongly if the fungus had been grown on high levels of galactose, maltose or sucrose, rather than on glucose. This is associated with the presence of an additional fibrillar-floccular layer on the surface of the yeast wall, probably a mannoprotein. Nevertheless, the differential effect of sugars seems to be strain-related, because *Candida* strains from active infections show it frequently, whereas strains from asymptomatic carriers can show it much less often. Also, it may be a specific feature of *C. albicans* rather than of *Candida* spp. in general, because the adhesion of non-pathogenic *Candida* spp. and *Saccharomyces cerevisiae* is not altered markedly when they are grown on different sugars. In support of the studies on stomatitis, strains of *C. albicans* from active vaginal infections are found to adhere more strongly *in vitro* than do isolates from healthy people. This raises the possibility that venereal spread of vaginal candidosis might involve the transmission of strongly adherent strains.

In summary of the adhesion studies, there seems little doubt that adhesion is an important feature of pathogenesis, but it is also clear that adhesion is influenced by several interacting factors including strain variations, substrate availability, the type of surface presented to *Candida* and other factors such as progesterone. The adhesion of *Candida* is discussed in detail by Douglas (1995) and Calderone (1995).

The dimorphic switch is also central to the pathogenicity of *C. albicans*, because invasion of the mucosa is always achieved by hyphal growth. In fact, this distinguishes *C. albicans* from all other (non-pathogenic) *Candida* spp. in clinical samples, and it is used in a simple diagnostic test: the yeast is cultured on a selective medium, then a loopful of inoculum is transferred to a vial containing horse serum and incubated at 37°C for 4–5 h. Of all the *Candida* strains that might occur in clinical samples only *C. albicans* will sprout germ-tubes in these conditions, although several other *Candida* spp. from non-human sources can do so.

The rare cases of systemic spread of *C. albicans* in the blood and lymph tissues are always associated with severely predisposing factors such as leukaemia, advanced diabetes, prolonged corticosteroid therapy, etc. But, strangely, *C. albicans* seldom grows systemically in AIDS patients, who tend to develop other systemic fungal infections instead. *Candida* grows systemically as a yeast in the body fluids, but dimorphism plays an important role in one respect. If cells of *C. albicans* are mixed with white blood cells (macrophages and polymorphonuclear leucocytes) *in vitro* the yeast cells can be engulfed and destroyed, but some of them break out of the phagocytes by converting to hyphae and then produce a further population of yeasts. In these conditions the vigour of the host defence system will be crucial in containing an infection. In Chapter 14 we will see that the major drugs used to treat systemic candidosis (ketoconazole and related compounds) act synergistically with the host defences, because at even low concentrations these drugs suppress the transition from yeast to mycelial growth.

Opportunistic and incidental pathogens

Theoretically, any fungus that can grow at 37°C could be a potential pathogen of humans, but in practice the spectrum is much narrower than this (see Table 13.1). A few common saprotrophic species of *Phialophora*, *Sporothrix*, *Cladosporium* and *Acremonium* can infect wounds and cause localized, damaging, subcutaneous mycoses. Many examples of this are described by Kwon-Chung and Bennett (1992). A different spectrum of fungi characteristically infect through the lungs because their air-borne spores are small enough to reach the alveoli (see Chapter 9). These fungi are more significant causes of life-threatening systemic mycoses, and they are discussed below. Again, the treatment of these infections is considered separately in Chapter 14.

Aspergillosis

A. fumigatus, A. flavus and *A. niger* produce abundant air-borne conidia which can establish in the lungs of people with impaired respiratory function and then produce dense, localized saprotrophic colonies termed **aspergillomas.** Usually these remain non-invasive, surrounded by fibrous tissue of the host. Aspergillomas are quite common in poultry that are fed on moulded grain, and can occur in farm workers who regularly handle such materials. Infection of the lungs is incidental so far as the fungi are concerned because they grow as saprotrophs on plant organic matter (see Chapters 7 and 10) and they have no means of disseminating from the lungs to spread the infection to other hosts. *A. fumigatus* sometimes becomes invasive and grows systemically in the body, either from infections in the lungs of immunosuppressed people or after entry through surgical wounds. It is a significant problem in transplant surgery, when the patient's immune system is artificially suppressed.

Yeasts and dimorphic fungi

A characteristic group of yeasts and dimorphic fungi cause systemic infections of compromised individuals. These fungi include *Cryptococcus neoformans, Coccidioides immitis, Histoplasma capsulatum, Paracoccidioides brasiliensis, Blastomyces dermatitidis* and *Pneumocystis carinii.* The diseases tend to be geographically localized, reflecting specific habitat requirements of the fungi involved, and in some cases they are known to occur in wild animals which may be the natural reservoir for infection of humans. In all cases these infections are initiated when air-borne spores enter the lungs, and they either remain localized there, causing tissue disruption and cavitation, or they can progress to other parts of the body by spread of yeast cells in the circulatory system. People with impaired respiratory function can be particularly prone to lung infection, and people in the advanced stages of diabetes, leukaemia and immunosuppressive disorders can often develop the systemic disease. Yet, skin tests with antigens of some of these fungi indicate that a substantial proportion of the population in the endemic areas have been exposed to infection at some stage and

perhaps suffered only mild, flu-like symptoms before the infection spontaneously resolved. Thus, these fungi pose a significant and perpetual threat to a sector of the population who are immunocompromised, but they are not particularly damaging to the population at large.

Cryptococcus neoformans

C. neoformans is a haploid yeast, with a sexual stage in the genus *Filobasidiella* (basidiomycota; Fig. 13.3). It causes the disease called **cryptococcosis**, which is now the fourth most common life-threatening disease of AIDS patients in the USA (7–8% of cases) and is also relatively common in AIDS patients in Britain and the rest of Europe (3–6%) (Kwon-Chung & Bennett, 1992). Consistent with this, *C. neoformans* has a worldwide distribution as a saprotroph. It has been isolated frequently from the old, weathered droppings of pigeons on window-ledges in cities, although it does not compete well in wet droppings where bacteria can raise the pH to growth-inhibitory levels. It also occurs in the droppings of other birds and in soil enriched with these, but it does not infect the birds themselves.

In the past it was assumed that dried, powdery birds' droppings might represent the natural source of air-borne inoculum that enters the lungs. But, the discovery of a sexual stage with air-borne basidiospores of 1.8–3.0 μm diameter suggests that these spores might be a more common source of infection because they are the right size to be deposited in the alveoli (see Chapter 9). Unfortunately, the natural habitat of the sexual stage is unknown, and this is a hindrance to epidemiological studies. It does not seem to be bird excreta; instead, it is speculated that birds become contaminated during feeding — perhaps on vegetable matter, and the fungus then grows in their droppings.

C. neoformans is an unusual fungus because it has a thick, rigid polysaccharide capsule around the yeast cells. The capsule is an important virulence determinant, because naturally non-capsulate strains or mutants that lack the capsule are non-virulent. These strains are more readily ingested by leucocytes than are capsulate cells *in vitro*, but the addition of capsular components to acapsular cells can restore their resistance to

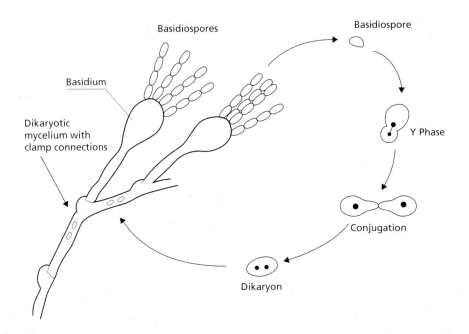

Fig. 13.3 *Cryptococcus neoformans* and its sexual stage *Filobasidiella neoformans*. The budding yeast cells (Y phase) conjugate and produce a dikaryotic cell which forms a mycelium on which inflated basidia develop. Meiosis in the basidia leads to production of four haploid nuclei which migrate to the tips of the basidia. Then chains of haploid basidiospores are budded from the tips of the basidia. These spores germinate to form a budding yeast phase.

phagocytosis. In addition, *C. neoformans* produces a phenoloxidase, which distinguishes it from all other species of *Cryptococcus*. This enzyme seems to be important for pathogenesis because it acts on phenolic compounds to produce melanin, which is deposited in the yeast walls, perhaps helping to protect the cells against oxidants in the host tissues. *C. neoformans* can grow in the lungs, causing either mild or chronic, persistent pneumonia, depending on the individual's degree of lung dysfunction from other causes. In fact, many people may have been exposed in this way in Britain, Australia and the USA, judging from random testing for skin reactions to cryptococcal antigens: a delayed hypersensitivity response indicates prior exposure. But, the most damaging phase of the disease occurs when chronic lung infection leads to 'silent' dissemination to the central nervous system, where the fungus shows a predilection for growth in the cerebral cortex, brain stem, cerebellum and meninges. This is invariably fatal if not treated, and it is common in AIDS patients, perhaps because a primary lesion in the lung has remained quiescent for many years and leads to secondary spread when the immune system is seriously impaired.

Coccidioides immitis

C. immitis is a dimorphic fungus with an unusual life cycle (Fig. 13.4). It grows as hyphae in its saprotrophic phase in soil, and produces thick-walled **arthroconidia** by fragmentation of the hyphae. These spores are air-borne and enter the lungs, where they swell and each grows to produce a large multinucleate **spherule**. Then, the protoplasm of the spherule is cleaved to produce **endospores**, which are released in the lung and produce further spherules. So, unlike most dimorphic fungi, *C. immitis* does not have a budding yeast phase.

C. immitis has a restricted geographical distribution, confined to the semi-arid or desert regions of south-western USA (California, Arizona), northern Mexico and isolated pockets of South America. It is a soil fungus, found in hot, sandy, alkaline conditions, consistent with its ability to grow at

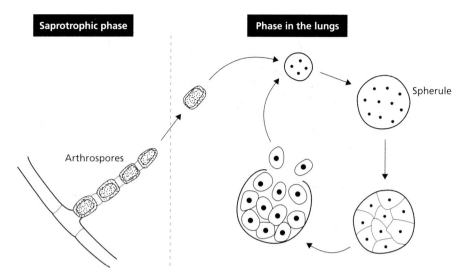

Fig. 13.4 Life cycle of *Coccidioides immitis*. In the saprotrophic phase the fungus grows as a mycelium which produces thick-walled arthrospores on short hyphal branches. The hyphal branches undergo synchronous septation to form a chain of compartments, then resources are withdrawn from every second compartment to supply the materials for wall thickening of the arthrospores. The arthrospores are wind-dispersed, enter the lungs and germinate to produce yeast-like swollen cells. As these grow, they continue to swell and undergo nuclear division, producing large mulitnucleate spherules. Then the cytoplasm cleaves around the nuclei to form uninucleate spores which are released when the spherule ruptures. The spores germinate to produce further spherules.

between 20 and 40°C and up to pH 9.0 in culture. Even so, it is not generally distributed in soil, but occurs in distinct localized patches, and there is evidence that it is most common near the burrows of soil-dwelling animals. Dogs, cats, rodents and other animals are susceptible to infection in the endemic areas, providing support for the view that wild animals and their carcasses may be a natural reservoir of infection. Also, within the endemic areas up to 50% of the human population can show a positive skin test to the fungal antigen, coccidioidin, obtained from broth cultures of the mycelial phase. Most of these people might have developed a mild pulmonary infection after exposure to dust storms which carry the arthroconidia. But, only a low proportion of people develop the disease, **coccidiomycosis**. This is characterized by pneumonia-like symptoms and a later (1–3-week) development of a blistering rash, representing a delayed response to the antigens. With time, the lung symptoms can develop further, and the fungus can disseminate to other tissues, causing lesions of the bones, the subcuta-neous tissues, the meninges and the major organs. This disseminated form of the disease is usually associated with immunosuppression resulting from transplant surgery, lymphoma, corticosteroid therapy or AIDS.

Transmission from infected to healthy people is extremely rare, but *Coccidioides* poses a major threat to laboratory workers because the arthroconidia are highly infective. There have been several reports of technicians and students being infected in this way, with at least one fatality (although not of a student!).

Histoplasma capsulatum

Histoplasma is a fungus with a 'typical' mould–yeast dimorphism: it grows as hyphae on laboratory media at 25–30°C, but as a budding yeast at 37°C on cysteine-rich media. The hyphae produce single conidia on the ends of short branches; some of them are large, warty **macroconidia** (8–15 μm) whereas others are **microconidia** (2–4 μm) which could be the main infective agents in the lungs.

The spores germinate to form a germ-tube which rapidly gives rise to a budding yeast phase in appropriate conditions. This yeast phase is almost invariably found throughout the course of infection; the hyphae are involved in the saprotrophic phase in dead body tissues or in natural substrates.

The fungus has a wide geographical distribution, including much of the eastern USA, most of Latin America, and parts of South-East Asia, Africa and Europe (e.g. Italy). It occurs naturally as a saprotroph in faecal-enriched soil around poultry houses, in bat droppings in caves, and also in starlings' droppings in towns and cities. The birds themselves are not infected, and the fungus is not found in their fresh droppings. However, bats have been found to be naturally infected, with intestinal lesions. Spores produced from faecal materials are the most likely source of infection of humans, because there are recorded instances of speliologists having developed the disease (**histoplasmosis**) after visiting caves. Skin tests using an antigen, histoplasmin, obtained from culture filtrates suggest that a high proportion of the human population has been exposed to infection in the endemic areas; one such study estimated that 20% of all the people in the USA had been infected at some time from air-borne spores. Even in mild infections the fungus has been detected in the urine, suggesting that it might commonly be generalized, although in the vast majority of cases the infection resolves spontaneously. This fungus, like *Coccidioides*, can be acquired in the laboratory — when students have examined cultures in the classroom — because its air-borne spores can readily enter the lungs and cause acute pulmonary histoplasmosis. However, chronic pulmonary infection is confined to people with lung dysfunction from other causes, and it can lead to dissemination to the other body tissues. This dissemination occurs by carriage of the yeast cells in the lymph system, and is most common in immunodeficient people or in the very young (less than 1 year) or older (50 years or more) sectors of the population.

Blastomyces, Paracoccidioides *and* Pneumocystis

B. dermatitidis and *Paracoccidioides brasiliensis* are two important causes of systemic mycoses in the specific geographical regions where they occur — North America and Africa in the case of *Blastomyces*, and South America in the case of *Paracoccidioides*. Both are dimorphic, growing as mycelia at lower temperatures and as budding yeasts at 37°C in body tissues. They behave essentially like *Histoplasma* in terms of pathology so they need not be considered further.

A more interesting case is *Pneumocystis carinii*, a yeast of worldwide distribution and very common as the cause of a virulent pneumonia in AIDS patients. It cannot be grown in pure culture, but from examination of infected material it seems to divide by binary fission, and it produces ascospores after the somatic cells fuse by hyphal outgrowths. It infects humans, rats, mice, horses and rabbits, but the strains from these sources show serological differences and are not always cross-infectible, suggesting that there are different biotypes. *P. carinii* was originally described as a protist (protozoan) and it has only recently been recognized fully as a fungus.

The opportunistic pathogens: summary

When we take all the opportunistic pathogens together, we see a number of features that defy simple explanation. At one extreme is *Pneumocystis* which cannot be grown in culture and so may have become adapted fully to a parasitic lifestyle. Yet, before the incidence of AIDS it was reported only rarely, suggesting that it is not a very efficient parasite. Some of the other pathogens, especially *Coccidioides* and *Histoplasma*, seem to be associated with wild animals in closed communities (rodents in burrows, bats in caves) so they might have a natural cycle of infection, growing on excreta and animal remains but also perhaps infecting the newborn, the old or the sick. In these cases, infection of humans is almost certainly incidental, resulting from periodic exposure to quite high inoculum levels. At the other extreme are fungi such as *Cryptococcus* which are associated with bird excreta and other, unknown natural materials. For these fungi the infection of humans or other warm-blooded animals seems entirely incidental, because they lack the one essential feature of a parasitic lifestyle which is a natural means of transmission from host to host. Nevertheless, we cannot dismiss any possibility because knowledge of the natural behaviour of all these fungi is still very sparse — an inevitable but unfortunate consequence of the emphasis on the health of humans, the **unnatural** host. These fungi have a spectrum of features that could be considered as virulence determinants, including dimorphism, the ability to grow at 37°C, capsular polysaccharides, preference for nutrient sources rich in organic nitrogen, spores that can be deposited in the lungs, and often a predilection for certain body organs or tissues in the disseminated phase of disease.

Insect-pathogenic fungi

Many fungi are specifically adapted to parasitize insects and other arthropods. Some common examples of these **entomopathogens** are given in Table 13.2 and shown in Fig. 13.5. We will deal first with the general aspects of their mode of parasitism (Charnley, 1989) and then consider some specific examples of their roles in natural or applied biocontrol of insect pests.

The infection cycle

The general infection cycle of insect-pathogenic fungi is summarized in Fig. 13.6. A spore attaches to the insect cuticle and germinates to form an appressorium, usually over an intersegmental region of the host. In *Metarhizium anisopliae* this behaviour closely parallels that of the rust fungi of plants described in Chapter 4, because the germ-tube recognizes the host topography or artificial surfaces with appropriate ridges and grooves. Penetration occurs by a narrow penetration peg beneath the appressorium, and involves the actions of cuticle-degrading enzymes. Many of these fungi produce lipases, proteases and chitinase in culture, although the latter is perhaps least important because the structural integrity of the cuticle is conferred mainly by disulphide bridges between the proteins. The penetration peg either penetrates both the epicuticle and procuticle or it penetrates only the epicuticle and then forms plates of hyphae between the lamellae of the procuticle, exploiting the zones of mechanical weak-

Table 13.2 Some common fungi that parasitize insects and arthropods.

Fungus	Hosts*	Distribution
Metarhizium anisopliae (deuteromycota)	Many: Lepidoptera, Coleoptera, Orthoptera, Hemiptera, Hymenoptera	Worldwide
Beauveria bassiana (deuteromycota)	All	Worldwide
Hirsutella thompsonii (deuteromycota)	Arachnida (mites)	Widespread
Cordyceps militaris (ascomycota)	Many larvae and pupae of Lepidoptera, some Coleoptera and Hymenoptera	Worldwide
Nomuraea rileyi (deuteromycota)	Larvae and pupae of Lepidoptera, Coleoptera	Worldwide
Paecilomyces farinosus (deuteromycota)	Many (Lepidoptera, Diptera, Homoptera, Coleoptera, Hymenoptera, Arachnida)	Worldwide
Verticillium lecanii (deuteromycota)	Several, especially aphids, scale insects	Widespread, tropics and sub-tropics
Entomophthora, Erynia and similar fungi (zygomycota)	Various, often host-specific, e.g. *Entomophthora muscae* on housefly, *Erynia neoaphidis* on aphids	Worldwide
Coelomomyces species (chytridiomycota)	Mosquitoes and midges; often host-specific	Common

* Lepidoptera—butterflies and moths; Diptera—flies; Homoptera—bugs; Coleoptera—beetles; Hymenoptera—wasps and bees; Orthoptera—grasshoppers, locusts; Hemiptera—sucking bugs; Arachnida—spiders and mites.

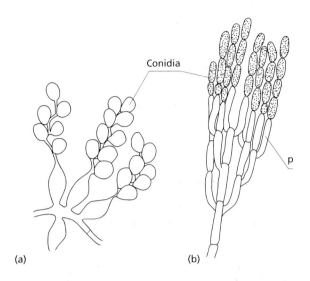

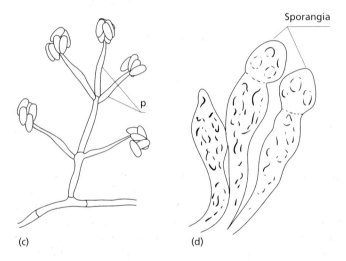

Fig. 13.5 Sporing structures of some common insect-pathogenic fungi. (a) *Beauveria bassiana* (deuteromycota); the conidia are formed alternately on an extending tip of a conidiophore. (b) *Metarhizium anisopliae* (deuteromycota); the conidia are formed in chains from phialides (p) at the tips of a branched conidiophore. (c) *Verticillium lecanii* (deuteromycota); the conidia are formed in clusters at the tips of phialides. (d) *Entomophthora* spp. (zygomycota); the single terminal sporangium functions as a single spore (see Chapter 9).

ness. Further penetration hyphae develop from these fungal plates.

The infection is aborted in these early stages if the insect moults. Otherwise, the fungus invades the epidermis and hypodermis, causing localized defence reactions. If these are overcome, then the hyphae either ramify in the insect tissues or, most frequently, produce swollen **blastospores** (yeast-like budded cells), **hyphal bodies** (short hyphal fragments) or protoplasts (*Entomophthora* and related zygomycota) that proliferate in the haemolymph (insect blood). This 'unicellular' phase of growth and dissemination leads to insect death, either by depletion of the blood sugar levels

or by production of toxins (see below). Then, the fungus reverts to a mycelial, saprotrophic phase and extensively colonizes the body tissues. Usually, at least some of the tissues are colonized before the insect dies—the fat body in particular. Finally, the fungus converts to either a resting stage in the cadaver or, in humid conditions, grows out through the intersegmental regions to produce conidiophores. In aphids infected by *Verticillium lecanii* (see Fig. 13.5) it is not uncommon to see conidiophores projecting from many parts of the body while the insect is still moving. In other host–parasite interactions the insect is killed more rapidly, before the body tissues are extensively

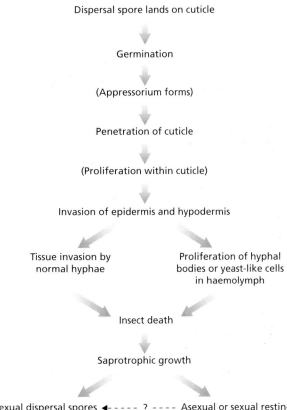

Dispersal spore lands on cuticle

Germination

(Appressorium forms)

Penetration of cuticle

(Proliferation within cuticle)

Invasion of epidermis and hypodermis

Tissue invasion by normal hyphae

Proliferation of hyphal bodies or yeast-like cells in haemolymph

Insect death

Saprotrophic growth

Asexual dispersal spores ◄ - - - - ? - - - - Asexual or sexual resting stage

Fig. 13.6 General infection cycle of an insect-pathogenic fungus; stages that do not always occur are shown in parentheses.

invaded, indicating the involvement of toxins. For example, both *Beauveria bassiana* and *M. anisopliae* produce depsipeptide toxins in laboratory culture, and they are active on injection into insects. The toxins of *M. anisopliae* (termed destruxins; Fig. 13.7) are thought to be significant in pathogenesis because infected insects die rapidly before there is extensive tissue invasion. The role of the toxin of *B. bassiana* (termed beauvericin; Fig. 13.7) is less clear because this fungus invades the tissues more extensively before the host dies, and the pathogenicity of strains is not always correlated with their *in vitro* toxin production. An alternative or additional role of the toxins may be to preclude bacterial growth in the cadaver during the saprotrophic phase, because the toxins have mild antibacterial activity.

Host range

Beauveria and *Metarhizium* have very wide host ranges, including hundreds of insects in the Orthoptera (grasshoppers), Coleoptera (beetles), Lepidoptera (butterflies and moths), Hemiptera (bugs) and Hymenoptera (wasps). As would be expected, these fungi grow readily in laboratory culture, making them attractive candidates as biocontrol agents. In contrast, the entomopathogenic zygomycota (*Entomophthora muscae* on house-flies, *Erynia neoaphidis* on aphids, etc.) typically are highly host-specific and they cannot be grown in axenic culture. *Coelomomyces psorophorae* also cannot be grown in culture; it is a member of the Lagenidales in the oomycota (see Chapter 1). and it has an unusual life cycle (Fig. 13.8). It parasitizes mosquito larvae but the zoospores produced from these larvae can only infect a copepod such as *Cyclops*. Similarly, the fused motile gametes released from *Cyclops* will only infect mosquito larvae, so this fungus has an obligate alternation of hosts in the haploid and diploid phases. This, and its inability to grow in culture,

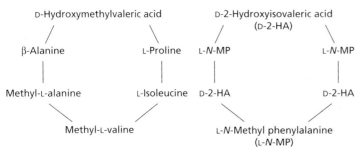

D-Hydroxymethylvaleric acid   D-2-Hydroxyisovaleric acid
                                              (D-2-HA)

β-Alanine   L-Proline   L-N-MP   L-N-MP

Methyl-L-alanine   L-Isoleucine   D-2-HA   D-2-HA

        Methyl-L-valine   L-N-Methyl phenylalanine
                                          (L-N-MP)

Destruxin B (*Metarhizium anisopliae*)   Beauvericin (*Beauveria bassiana*)

Fig. 13.7 Destruxin and beauvericin, two peptide toxins of insect-pathogenic fungi.

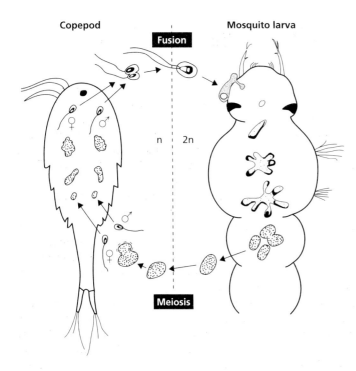

Fig. 13.8 Life cycle of *Coelomomyces psorophorae*. **Diploid phase** (2n): a motile zygote attaches to a mosquito larva, encysts, produces an appressorium and penetrates the host to form a weakly branched mycelium. This produces thick-walled resting sporangia which are released after the host dies. **Haploid phase** (n): meiosis occurs in the resting sporangia and haploid zoospores of different mating types are released. These infect a copepod (*Cyclops vernalis*) and form a thallus that eventually produces gametangia. The gametangia release motile gametes, which fuse either inside or outside the host. The resulting diploid zygote can only infect a mosquito larva. (Adapted from Whisler *et al.*, 1975.)

limits its potential for use in biocontrol of mosquitoes. But, it might be possible to manipulate the *Coelomomyces* population indirectly by promoting the populations of copepods so that there is an abundant source of inoculum for infection of mosquito larvae.

Epidemics caused by entomopathogenic fungi

Insect-pathogenic fungi, especially the zygomy-cota, can cause natural and spectacular crashes in the populations of their hosts. An example is shown in Fig. 13.9, where the population of the broad bean aphid, *Aphis fabae*, was reduced in middle and late season by a natural epidemic of *Erynia neoaphidis* and *Neozygites fresenii* (zygomycota). However, in all these cases the increase in pathogen population lags behind that of the host, because the relationship is density-dependent — the host must reach a critical density before the conditions are right for an epidemic. This means

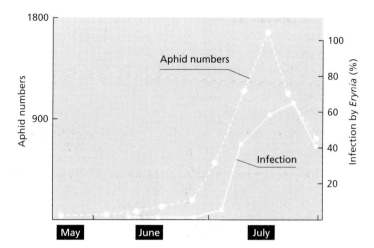

Fig.13.9 Numbers of aphids (*Aphis fabae*) in a broad bean crop in the course of a season in Britain (broken line) and per cent infection by the pathogenic fungus *Erynia* (solid line). (Based on data in Wilding & Perry, 1980.)

that there is always at least some degree of crop damage. Moreover, these epidemics are influenced strongly by environmental conditions, particularly humidity, because all the entomopathogens require a high humidity for sporulation and they also need high humidity for spore germination on an insect cuticle. Thus, natural epidemics are unpredictable and even commercially applied inocula of these fungi are prone to fail in practice unless the correct environmental conditions can be guaranteed.

The behaviour of the insect host also can limit the development of epidemics. As an example of this, Dobersky and Tribe (1980) found that *B. bassiana* was extremely common near the bases of elm trees that had died from Dutch elm disease in southern Britain (see Chapter 9). The fungus was isolated from 81% of all bark samples and 94% of all soil samples at the tree bases. It infects both the adult bark beetles and the larvae in laboratory conditions, indicating that it might be used as a biocontrol agent of the disease vector. However, the incidence of natural infection of the beetles in the dead trees was extremely low, perhaps because the opportunity for epidemic spread is limited by the many separate, non-merging tunnels that the beetles produce in the bark (see Chapter 9, Fig. 9.6).

Developments in practical biocontrol

Some of the limitations in natural biocontrol can be

overcome by using commercially produced inocula if these are formulated to be persistent and are applied in optimum conditions. Feng *et al.* (1994) reviewed the technology for producing *B. bassiana*, which is commercially attractive because of its wide host range. These authors report that about 10 000 t of spore powder formulation of *B. bassiana* are produced annually in China for treatment of $0.8–1.3 \times 10^6$ ha of forest and agricultural land. In North America and Western Europe, however, there are no commercially available formulations of *B. bassiana*, and few **mycoinsecticides** of any sort because field trials have given inconsistent control.

The preferred method of producing commercial inocula is by submerged liquid culture, but this raises a problem with *B. bassiana* and most other entomopathogens because they grow as thin-walled budding cells (blastospores) in these conditions, and the blastospores cannot tolerate drying and storage. For this reason, conidia are required for commercial inocula. They are produced either on solid food sources such as cereal bran or in two-stage systems where blastospores or mycelia are produced in liquid culture and then applied to solid matrices for production of conidia. The addition of osmotica to the culture medium can lead to an accumulation of compatible solutes in the spores (see Chapter 7) and enable these to germinate at somewhat lower humidity than they would otherwise require. The inoculum formulations can also contain mineral oil, helping to maintain the moisture needed for germination on the insect cuticle. Alternatively, the fungi can be entrapped in alginate beads with a suitable bulking agent such as kaolin. Many fungi survive drying in these conditions, protected by the matrix of the bead, and they sporulate from the surface of the bead when it is remoistened.

Verticillium lecanii

V. lecanii (deuteromycota) is one of the most successful commercial biocontrol agents of insects. It is used primarily in glasshouses for the control of aphids on potted chrysanthemums and (using a different strain) for control of whitefly on cucumbers. For these purposes it is produced as fermenter-grown conidia because it is one of the relatively few fungi that produce conidia readily in liquid culture. It occurs naturally as a parasite of aphids and scale insects in the subtropics, because it requires relatively warm conditions (greater than 15°C) for infection. Like all the entomopathogenic fungi, it needs a high relative humidity during the germination and penetration phases, but then the humidity can be reduced without affecting its parasitism. All these conditions make this fungus ideal for use on potted chrysanthemums, one of the most important year-round horticultural crops. These plants are 'blacked-out' with polythene sheeting for part of each day to give short day lengths that initiate flowering, and this also raises the humidity for infection by *V. lecanii*. A single spraying of conidia just before blacking-out can be sufficient to give season-long control of the important aphid *Myzus persicae*, but it was found to be less effective for two minor aphid pests of chrysanthemum, *Macrosiphoniella sanborni* and *Brachycaudus helichrysi*. Hall and Burges (1979) showed that this was not related to inherent differences in susceptibility of the three aphid species. Instead, it is explained by their behavioural differences. *M. persicae* tends to feed on the undersides of leaves where the humidity is higher, and it is more mobile than the other aphids on chrysanthemum, feeding for short times and then moving on because chrysanthemum is not its preferred host plant. As mentioned earlier, *V. lecanii* often sporulates on the body while an aphid is still alive, so the infected individuals of *M. persicae* can spread the infection to other individuals of

this species feeding in the same locations on the crop.

Gillespie and Moorhouse (1989) discuss further examples of the use of insect-pathogenic fungi in biocontrol, but we end this account with an interesting digression. *V. lecanii* not only infects insects but also colonizes the sporing pustules of rust fungi (e.g. *Uromyces dianthi*, carnation rust) on leaf surfaces, and it parasitizes the fungal spores in experimental conditions. To some degree this behaviour is strain-dependent, but three of the eight strains obtained from sporing pustules could also parasitize aphids, suggesting that the fungus is relatively non-specialized as a parasite and might even be used as a dual control agent. It is interesting to note that fungi and insects have several features in common, including the presence of chitin in the wall or cuticle, trehalose and mannitol as major carbohydrates, and lipids and glycogen as major energy storage compounds (see Chapter 6). So, the ability of a fungus to parasitize both types of host is perhaps not surprising.

Fungi on nematodes

Nematodes (eelworms) are small animals, usually a few millimetres long. They are extremely common in soil, animal dung and decomposing organic matter; their populations in European grasslands, for example, are estimated to range from $1.8–120 \times 10^6$ m^{-2}. Many of the nematodes are saprotrophs that feed on bacteria or other small organic particles, but some are parasites of animals including humans (e.g. *Trichinella spiralis* which invades human muscle tissue), and some are parasites of crop plants — the root-knot nematodes (*Meloidogyne* spp.), cyst nematodes (e.g. *Heterodera*, *Globodera* spp.) and various ectoparasitic and burrowing nematodes. The chemicals that can be used to control parasitic nematodes in living plants or organic matter are extremely toxic and thus environmentally undesirable. For this reason there has been much interest in the **nematophagous** (nematode-feeding) fungi and other nematode parasites.

Nematophagous fungi are common in organic-rich environments, and they include representatives of all the major fungal groups (Table 13.3). Here we will consider the three major types with different adaptations for feeding on nematodes: (i) the 'predatory' nematode-trapping fungi; (ii) the endoparasitic fungi; and (iii) the parasites of nematode eggs or cysts. Barron (1977) gives an extensive account of all these fungi.

Nematode-trapping fungi

The nematode-trapping fungi are predatory species which capture their prey by specialized devices: adhesive hyphae (*Stylopage* and *Cystopage*; zygomycota), adhesive nets (e.g. *Arthrobotrys oligospora*; Plate 13.1, facing p. 210), short adhesive branches (e.g. *Monacrosporium cionopagum*), adhesive knobs (e.g. *M. ellipsosporum*; Plate 13.1), non-constricting rings, and constricting rings that are triggered to contract when a nematode enters them (e.g. *Dactylella brochopaga*). More than one type of mechanism can be found in different species of a genus.

All these fungi are considered to be essentially saprotrophic, but to exploit nematodes as an additional source of nutrients. Consistent with this, they grow readily on a range of organic substrates in laboratory culture, and they even include some wood-decaying basidiomycota. Also, some of them (e.g. *A. oligospora*) coil round the hyphae of other fungi in culture, indicative of mycoparasitic behaviour (see Chapter 11) or a nutritionally opportunistic lifestyle. Nevertheless, their specialized trapping devices indicate that they are adapted to exploit nematodes. In some cases they produce the traps during normal growth in culture. But, in other cases (e.g. *A. oligospora*) the traps are produced only in the presence of nematodes or nematode diffusates, and this can be mimicked by supplying small peptides or combinations of amino acids such as phenylalanine and valine.

All the nematode-trapping fungi have an essen-

Table 13.3 Examples of major types of nematophagous fungi.

Fungus	Grouping	Infective unit
CHYTRIDIOMYCOTA		
Catenaria anguillulae	Endoparasite	Zoospore
OOMYCOTA		
Nematophthora gynophila	Endoparasite	Zoospore
Myzocytium humicola	Endoparasite	Adhesive zoospore cyst
ZYGOMYCOTA		
Stylopage, Cystopage spp.	Predators	Adhesive hyphae
DEUTEROMYCOTA		
Arthrobotrys oligospora	Predator	Adhesive nets
Monacrosporium cionopagum	Predator	Adhesive branches
Dactylaria candida	Predator	Adhesive knobs
		Non-constricting rings
Dactylella brochopage	Predator	Constricting rings
Drechmeria coniospora	Endoparasite	Adhesive conidia
Hirsutella rhossiliensis	Endoparasite	Adhesive conidia
Verticillium chlamydosporium	Egg parasite	Hyphal invasion
ASCOMYCOTA		
Atricordyceps (sexual stage of	Endoparasite	Non-adhesive conidia
Harposporium oxicoracum)		
BASIDIOMYCOTA		
Hohenbuhelia (sexual stage of	Endoparasite	
Nematoctonus pachysporus)	Predator	Adhesive conidia
Pleurotus ostreatus	Toxin producer	Adhesive traps
		Toxic droplets

tially similar mode of parasitism. The nematode is captured by an adhesive, then the fungus penetrates rapidly by means of a narrow penetration peg, swells to form an infection bulb in the host, and hyphae grow from this infection bulb to fill the nematode body and absorb its contents. Usually, this phase is completed within 1–3 days, then the hyphae grow out of the dead nematode to produce further capture organs or to sporulate.

The initial adhesion is almost instantaneous and effectively irreversible. Even when the nematode thrashes wildly to free itself from the fungus, the trapping organ will break from the hyphae and remain attached to the nematode, then initiate infection. Yet, the trapping organs are not 'sticky' in the general sense of the term, because they do not accumulate soil debris, etc. Instead, the adhesive is a lectin-like material that binds strongly to specific saccharide components of the nematode surface. This has been studied in detail for *A. oligospora* and also for the endoparasitic species *Drechmeria coniospora* (Tunlid *et al.*, 1992). The ability of *A. oligospora* to trap the saprotrophic nematode *Panagrellus redivivus* is annulled in the presence of *N*-acetylgalactosamine, presumably because this compound binds to the recognition site on the fungal lectin. *Panagrellus* is known to have *N*-acetylgalactosamine components on its surface, because it binds to commercially available lectins (e.g. wheat germ agglutinin) that recognize this sugar derivative. A glycoprotein with this binding property has been isolated from *A. oligospora*, but it is found only on the surface of the traps, not on the normal hyphae so it presumably is the product of a differentiation-specific gene. Other fungus–nematode combinations have also been investigated for binding specificity by using sugars to try to block adhesion. These studies indicate that *Arthrobotrys conioides* has a lectin that recognizes α-D-glucose or α-D-mannose residues, *Monacrosporium eudermatum* has a lectin for L-fucose, and *Drechmeria coniospora* has a lectin for sialic acid. So, the adhesion can be to some degree nematode-specific. For example, *Monacrosporium ellipsosporum* does not capture *Xiphinema* spp. (which feed on the root tips of many plants) but these nematodes are captured by other species of *Monacrosporium*, *Arthrobotrys* and *Dactylaria*. However, Tunlid *et al.* (1992) cautioned against simple interpretations based on lectin-binding studies, because ultrastructural studies suggest that the adhesive on fungal traps might change when nematodes become attached to it: more adhesive material might be released or there might be a rearrangement of the adhesive so that different binding sites are exposed in it.

The trapping of nematodes by some wood-decay fungi has focused attention on the role that nematodes might have as supplementary nitrogen sources, overcoming the critically low nitrogen content of wood (see Chapter 10). The wood-rotting 'oyster fungus' *Pleurotus ostreatus* not only forms adhesive traps but also produces droplets of toxin from specialized cells, and nematodes immobilized by this toxin are then invaded and digested. However, the emphasis on nematode trapping as a means of nitrogen capture has perhaps been exaggerated. It seems better to regard these fungi as nutritional opportunists. In this respect it is relevant that *A. oligospora* forms traps in response to nematode diffusates, but it is suppressed by high glucose or phosphate concentrations, suggesting that nematode trapping by this species may be an adaptation to nutrient limitation in general.

Endoparasites and egg parasites

The endoparasitic fungi are quite different from the trapping fungi because they seem to depend on nematodes as their main or only food source in nature, even though many of them can be grown in laboratory culture. Consistent with this, they show a strong density dependence on their hosts (Jaffee, 1992). The zoosporic fungus *Catenaria anguillulae* (chytridiomycota) is perhaps the least specialized example because it grows on several types of organic material in nature, including liver fluke eggs. Also, its zoospores do not settle easily on moving nematodes in water films, and instead it accumulates at the body orifices of immobilized or dying nematodes (Plate 13.2, facing p. 210). This contrasts with *Myzocytium humicola* (oomycota) which also produces zoospores but they encyst soon after release and then germinate to produce an adhesive bud which attaches to a passing nematode. The other endoparasites also produce adhesive spores—for example, *Hirsutella rhossiliensis* (Plate 13.3, facing p. 210) and *Drechmeria coniospora*, which seem to attract nematodes by chemotaxis, helping to ensure that attachment occurs. Then, the adhered spores germinate rapidly and the hyphae fill the host, killing it within a few days. Finally, the hyphae grow out through the host wall and produce a further batch of spores. From a single parasitized nematode, *Hirsutella* can produce up to 700 spores, and *Drechmeria* is reported to produce up to 10 000 spores.

The parasitic efficiency of these endoparasites is so high that they can have a significant impact on host populations in appropriate conditions. The best example of this is the role of *Nematophthora gynophila* (oomycota; Fig. 13.10) in controlling the cereal cyst nematode, *Heterodera avenae*, on oat crops in Britain (Kerry *et al.*, 1982). The female nematode penetrates the root just behind the root tip and lodges with her head inside the endodermis, where the host cells respond by swelling into nutrient-rich 'giant cells' from which the nematode taps the host nutrients. As the female grows her body distends into a lemon shape which ruptures the root cortex so that her rear protrudes from the root. Then she is fertilized by wandering males (it happens even in nematodes!) and her uterus fills with eggs which develop into larvae. At this stage the larval development is arrested, the female dies and her body wall is transformed into a tough, leathery cyst which can persist in soil for many years, making these nematodes difficult to control by conventional means. The nematode population was found to increase progressively when oats were grown repeatedly in a site, but then spontaneously declined to a level at which the nematode no longer caused significant damage. Investigation of these 'cyst–nematode decline' sites revealed a high incidence of parasitism of the females by *N. gynophila*, coupled with parasitism of the eggs (i.e. the sacs containing the individual arrested larvae) by another fungus, *Verticillium chlamydosporium* (deuteromycota). *Nematophthora* infects the females, presumably from zoospores, and fills most of their body cavity with thick-walled resting spores (oospores) so that the cyst, if formed at all, contains relatively few eggs but a large number of fungal spores. *Verticillium* has a different role from this—it is a facultative parasite of nematode eggs, which it destroys after they have been released into soil.

Although this natural biocontrol is highly effective, it has not led to the widespread use of *N.*

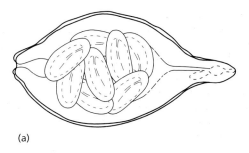

(a)

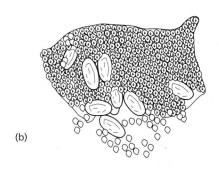

(b)

Fig. 13.10 Diagrammatic representation of the cereal cyst nematode *Heterodera avenae* and the parasitic fungus *Nematophthora gynophilia*. (a) Mature healthy female nematode containing embryonated eggs; the female body will subsequently develop a leathery cuticle and become a cyst that persists in soil. (b) A cyst filled with oospores of *N. gynophila* which parasitizes the female nematode; the eggs themselves are not infected. (Based on a photograph in Kerry & Crump, 1980. *Transactions of the British Mycological Society* **74**, 119–25.)

gynophila in biological control. Part of the reason is that this fungus cannot be grown in axenic culture —it is an obligate parasite of nematodes—and in any case it is effective only in soils wet enough to favour zoospore activity. The egg parasite *V. chlamydosporium* can be cultured easily but is less effective as a biocontrol agent, especially when acting alone.

References

Barron, G.L. (1977) *The Nematode-destroying Fungi*. Canadian Biological Publications, Guelph, Ontario.

Calderone, R.A. (1995) Recognition of endothelial cells by *Candida albicans*: role of complement-binding proteins. *Canadian Journal of Botany*, **73**, S1154–9.

Charnley, A.K. (1989) Mechanisms of fungal pathogenesis in insects. In: *Biotechnology of Fungi for Improving Plant Growth* (eds J.M. Whipps & R.D. Lumsden), pp. 85–125. Cambridge University Press, Cambridge.

Deacon, J. W. & Saxena, G. (1997) Orientated zoospore encystment and cyst germination in *Catenaria anguillulae*, a facultative endoparasite of nematodes. *Mycological Research*, **101**, 513–22.

Dobersky, J.W. & Tribe, H.T. (1980) Isolation of entomogenous fungi from elm bark and soil with reference to ecology of *Beauveria bassiana* and *Metarhizium anisopliae*. *Transactions of the British Mycological Society*, **74**, 95–100.

Douglas, L.J. (1995) Adhesin–receptor interactions in the attachment of *Candida albicans* to host epithelial cells. *Canadian Journal of Botany*, **73**, S1147–53.

English, M.P. (1965) The saprophytic growth of non-keratinophilic fungi on keratinised substrata, and a comparison with keratinophilic fungi. *Transactions of the British Mycological Society*, **48**, 219–35.

Feng, M.G., Poprawski, T.J. & Khachatourians, G.G. (1994) Production, formulation and application of the entomopathogenic fungus *Beauveria bassiana* for insect control: current status. *Biocontrol Science and Technology*, **4**, 3–34.

Gillespie, A.T. & Moorhouse, E.R. (1989) The use of fungi to control pests of agricultural and horticultural importance. In: *Biotechnology of Fungi for Improving Plant Growth* (eds J.M. Whipps & R.D. Lumsden), pp. 55–84. Cambridge University Press, Cambridge.

Hall, R.A. & Burges, H.D. (1979) Control of aphids in glasshouses with the fungus *Verticillium lecanii*. *Annals of Applied Biology*, **102**, 455–66.

Jaffee, B.A. (1992) Population biology and biological control of nematodes. *Canadian Journal of Microbiology*, **38**, 359–64.

Kerry, B.R., Crump, D.H. & Mullen, L.A. (1982) Studies of the cereal cyst nematode, *Heterodera avenae*, under continuous cereals, 1975–1978. II. Fungal parasitism of female nematodes and eggs. *Annals of Applied Biology*, **100**, 489–99.

Kwon-Chung, K.J. & Bennett, J.E. (1992) *Medical Mycology*. Lea & Febiger, Philadelphia.

Tunlid, A., Jansson, H.-B. & Nordbring-Hertz, B. (1992) Fungal attachment to nematodes. *Mycological Research*, **96**, 401–12.

Whisler, H.C., Zebold, S.L. & Shemanchuk, J.A. (1975) Life history of *Coelomomyces psorophorae*. *Proceedings of the National Academy of Sciences, USA*, **72**, 693–6.

Wilding, N. & Perry, J.N. (1980) Studies on *Entomophthora* in populations of *Aphis fabae* on field beans. *Annals of Applied Biology*, **94**, 367–78.

Chapter 14
Prevention and control of fungal growth

In this final chapter we deal with the major approaches for controlling fungi. This serves to link the topics of previous chapters and put them in a practical context. We shall cover four broad themes:

1 control by management of physical and biological factors;
2 the use of fungicides, with special reference to plant disease control;
3 the major antifungal antibiotics;
4 control of the pathogens of humans.

However, we will not always adhere rigidly to these distinctions because sometimes it is appropriate to compare the different approaches to a problem.

Management by control of physical and biological factors

Here we cover various aspects of non-chemical control, including management of the environment and manipulation of microbial agents. These approaches depend on a thorough understanding of the 'target' fungus and the external factors that influence it. Some of the approaches are obvious and we will not dwell on them. Suffice to say that, in general, these approaches should be considered before all others because often they are the cheapest, safest and most durable. We saw the background to this in Chapter 7 when we considered the environmental conditions for fungal growth. The following practical examples illustrate some approaches.

Prevention of food spoilage by environmental and biological means

All food products need to be stored so as to min-

imize spoilage. In theory this could be done by controlling any single variable such as temperature. In practice, however, it can be cheaper and less damaging to the product if a combination of less extreme treatments is used. We saw in Chapter 7 that the moulding of stored grain is influenced by the combination of temperature and water content.

The rotting of fresh produce like fruits and vegetables is a major problem, and it prevents some of the more exotic products (mangoes, papaya, etc.) from being widely and cheaply available on world markets. Low-temperature storage is costly and potentially damaging to fruit tissues, so the storage of apples and pears, for example, is achieved by a combination of cool temperatures and elevated levels of carbon dioxide in **controlled atmosphere storage**. This combination not only limits the growth of the major fruit-rot fungi but also delays fruit ripening so that fruits are less susceptible to attack. Without controlled atmosphere storage, 'one rotten apple in a barrel' can cause all the fruit to rot because ethylene is generated as a wound response and it triggers the ripening of the other fruits. Commercial bananas are shipped green and then ripened artificially by exposure to an ethylene-generating chemical. The green fruit not only survives handling and shipping (e.g. 2–3 weeks from the Central Americas to the European market) but also arrests the growth of *Colletotrichum musae* (banana anthracnose) from latent infections on the skin (see Chapter 12). The role of latency and quiescence in plant pathology was reviewed by Williamson (1994). Nevertheless, in practice it is necessary to take account of all potential problems, so bananas usually are sprayed with fungicides before shipping to prevent finger rot caused by *Phytophthora* spp. which can colonize rapidly through the fruit stalks that have been trimmed with cutting knives in the washing and packing sheds.

Wounding provides portals for entry of many other fruit-rot pathogens and is a major cause of market losses. Some minor wounds can be protected by waxing (commonly done for apples) or fungicides (for bananas and other fruits where the skin is not eaten). Alternative biocontrol methods are now being explored in many countries. Many harmless organisms have been shown experimentally to colonize fruit wounds and exclude pathogens from them. Yeasts are especially promising because they can be produced easily in fermenters and would meet with little consumer resistance — they are present on fruit surfaces in any case. Of several yeasts that have been tested for biocontrol, strains of *Pichia guillermondii* were among the best for control of apple rots caused by *Botrytis cinerea* and *Penicillium*. Most yeasts probably exert their effect by competing for nutrients at the wound sites, but *P. guillermondii* also binds to the pathogen hyphae *in vitro* and damages them, perhaps by release of a β-1,3-glucanase which it is known to produce in culture. Unfortunately, *P. guillermondii* can grow at 37°C and this could markedly reduce its acceptability to consumers. So, *Candida* spp. that do not grow at this temperature and that occur naturally on fruit surfaces are being explored instead. It may not be necessary to use live organisms at all, because the cell wall fractions of many yeasts, including *Saccharomyces cerevisiae*, can act as powerful elicitors of plant defence mechanisms. Also, chitosan obtained by chemical deacetylation of crab shell has been found to disturb the growth of several fungi *in vitro*, causing excessive branching, wall alteration and cytoplasmic disorganization. This perhaps occurs by interference with wall enzymes such as chitin deacetylase, which could be especially important for zygomycota such as *Rhizopus stolonifer* (which rots several soft fruits) because of the high chitosan content in their walls (see Chapter 2). Many of these experimental treatments might not alone give reliable control of post-harvest diseases, but they might be combined for effective control (Wilson & Wisniewski, 1989).

Sanitation, quarantine and other strategies

There is a long history of using biological knowledge to avoid problems. We will consider briefly some examples.

• Crop rotation is one of the most effective means of controlling soil-borne plant pathogens that survive in crop residues. In warm, moist soil conditions, a 1-year break from cereal crops is enough to reduce the soil inoculum of take-all fungus (*Gaeumannomyces graminis*) to non-damaging levels, provided that weed hosts such as rhizomatous grasses are eliminated. A 2-year break is similarly effective for control of *Pseudocercosporella herpotrichoides* (eyespot of cereals).

• Various crop management practices can be altered to avoid disease. For example, in many countries stinking smut of wheat (*Tilletia caries*) is seed-transmitted and must be controlled by seed-applied fungicides, but in the Pacific north-west of the USA it is mainly soil-borne. If winter wheat is sown early it reaches the most susceptible stage for infection while the soil moisture and temperature are unsuitable for germination of the smut spores and so the disease is avoided. Other simple and effective disease avoidance practices include the liming of soil to prevent serious clubroot disease of crucifers (*Plasmodiophora brassicae*), and maintenance of soil phosphate levels to avoid serious damage by *Pythium* in cereals. As noted earlier, browning root rot of wheat (*Pythium graminicola*) in the North American prairies was virtually eliminated when phosphatic fertilizers became widely available.

• Meteorological forecasting has been highly effective for controlling air-borne pathogens that require specific conditions for sporulation. New Zealand operates a facial eczema warning system for *Pithomyces* (see Chapter 11) so that farmers can remove their sheep from affected pastures in the danger period. Control of potato blight (*Phytophthora infestans*) has for long relied on weather reports; the early work on this disease identified 'Beaumont periods' when appropriate conditions of temperature and humidity persist for a number of days, favouring sporulation of the pathogen, so that farmers can most effectively time their fungicide sprays to protect crops from infection.

• For management of major gene resistance against powdery mildew of cereals (*Erysiphe graminis*), two epidemiological approaches have been advocated. The first involves breeding multi-lines of individual crop cultivars, i.e. 10 or so lines that are genetically identical except for possession of different resistance genes. When the mixture is sown in a field it presents a mosaic of plants with different susceptibilities to any one race of the pathogen, so that even if one of the 10 lines is heavily diseased the others will be unaffected. This results in more than the expected 90% yield because of compensatory growth by the healthy plants when there is reduced competition from the susceptibles. But, more than this, it also slows the development of disease by various mechanisms, including interception of inoculum and possible systemic acquired resistance when plants are challenged by an incompatible race of the pathogen. The alternative suggestion is to mix crop cultivars within a field to achieve the same effects, but this suffers from the potential problem of their different maturity dates.

• Sanitation can be a highly effective control or avoidance strategy, illustrated in the case of the mycoparasites *Trichoderma harzianum* and *Pythium oligandrum*. These fungi have much potential for control of plant pathogens (see Chapter 11) but they can also be a problem in commercial mushroom production by antagonizing the mycelium of *Agaricus bisporus*. A distinctive strain (or strains) of *T. harzianum* has become a problem in British mushroom sheds, apparently because the spores contaminate the trays of compost from soil or old compost that has not been removed thoroughly from the floors. *P. oligandrum* similarly has caused serious cropping losses in a few mushroom-production units.

• The growth of fungi such as *Amorphotheca resinae* and *Paecilomyces varioti* in aviation fuel-storage tanks has been particularly difficult to control. They grow on the *n*-alkanes of aviation kerosene (see Chapter 5) and cause problems by blocking filters and corroding the walls of the storage tanks when they produce acids as metabolic by-products. At least one air-crash has been attributed to *A. resinae* in the 1960s, before the seriousness of this problem was recognized. Theoretically, these fungi should not grow in the fuel because they require water. But, it is almost impossible to prevent water from seeping into storage tanks or condensing during changes of air

temperature, especially in aircraft tanks which are subjected to extreme temperature variations during flight. Then, the fungi grow at the fuel–water interface, where there is sufficient aeration for utilization of the alkanes (see Chapter 6). The problem is now controlled by a combination of measures — lining of the tank walls with rubberized materials to prevent corrosion, regular checking and cleaning of filters, and addition of biocides to the fuel. These biocides include mixtures of organoborates, or ethylene glycol monoethyl ether, which has the advantage of acting also as an antifreeze.

• Quarantine is one of the mainstays for preventing the spread of pathogens, but is becoming increasingly difficult to maintain with the increase in international travel and transport. Its value often is appreciated only in retrospect. The major epidemics of Dutch elm disease in North America and Western Europe have been linked to introductions of elm timber that was not de-barked to remove the beetle vectors (see Chapter 9). The strain of *Cryptococcus neoformans* that is increasingly being isolated from acquired immune deficiency syndrome (AIDS) patients has been traced to tropical origins. The same strain was discovered to grow on certain types of eucalypt tree in Australia, and the regions outside of Australia where the clinical strain occurred corresponded to regions where these types of tree had been imported—an association that could not have been foreseen for quarantine purposes. The spread of Panama disease of bananas (*Fusarium oxysporum*; see Chapter 12) to most banana-growing regions of the world has almost certainly been due to transport of planting material—bananas are established from 'suckers' (see Plate 12.2) at the base of the stem, because commercial bananas are sterile triploids, unable to be established from seeds. The pathogen is thought to have its centre of origin in the south-east Asian peninsula and to have been disseminated by traders in the last century. The development of technologies for raising sterile tissue-culture plantlets (actually derived by meristem culture) should overcome such problems in the future. More recent and embarrassing examples can be cited. A serious root-rot disease of raspberry canes caused by *Phytophthora fragariae* var. *rubi* is found now in most production areas of the world. It seems to have been spread in vegetative propagating material (canes) of new, improved cultivars. Consistent with spread of this pathogen from a single source in recent times is the finding that restriction fragment length polymorphisms (RFLPs; see Chapter 8) are virtually identical in pathogenic strains across the world (Stammler *et al.*, 1993). The same was true in the past for the spread of *P. fragariae* var. *fragariae*, the cause of red core disease of strawberry plants; and for a particular form (biovar 3) of the crown gall bacterium, *Agrobacterium tumefaciens*, on grapevines. This pathogen has been discovered to grow as a systemic, symptomless endophyte, so in the past it would have escaped detection in 'visibly clean' vegetative propagation material.

Biological and integrated control

Biocontrol has been discussed at many points in this book, so here we review it in broader terms. It can be defined as the practice by which, or process whereby, an organism that is undesirable is controlled by the activities of others. This definition covers both natural and purposeful biocontrol, and it accommodates the various practical approaches that are used to achieve control:

• purposeful introduction of an organism to control another;
• purposeful exploitation and management of naturally occurring biocontrol systems;
• purposeful change of some environmental factor so as to promote the activities of natural agents;
• any combination of the above, because often it is necessary to change some environmental factor to favour an introduced biocontrol agent.
We shall consider these approaches in turn.

The first approach (purposeful introduction) has been discussed in this book already. Examples are the use of *Phlebiopsis gigantea* to control root rot of pine (see Chapter 11); the use of hypovirulence to control chestnut blight (see Chapter 8); the use of mycoparasites or other organisms as seedling protectants (see Chapter 11, and see later); the control of *Fusarium* wilt by inoculation with nonpathogenic *Fusarium* strains (see Chapter 12); the control of fruit rots; the protection of leaves against ice-nucleation active bacteria (see Chapter 11); and the use of *Verticillium lecanii* and other insect pathogens (see Chapter 13).

The purposeful exploitation of naturally occurring biocontrol was seen in relation to decline of cereal cyst nematode (see Chapter 13), the exploitation of soils naturally suppressive to *Pythium* diseases of cotton (see Chapter 11), the natural protective role of mycorrhizas (see Chapter 12) and the management of take-all patch disease of turf involving *Phialophora* (see Chapter 11). In most cases this disease can be avoided simply by maintaining the turf pH at a level that favours *Phialophora*. In all these cases, management is the key issue: when the natural control system has been identified, it can be managed rationally.

The third approach, involving change of some environmental factor to promote the activities of natural biocontrol agents, may seem the least defensible use of the term biocontrol. Some people prefer the term **integrated control** for this, but in any case it has a strong biological component. For example, in Chapter 11 we noted that the use of a fungicide to control *Pithomyces chartarum* (facial eczema of sheep) probably has a biological component because the fungicide changes the microbial balance in favour of the competitors of *Pithomyces*. The same principle is applied widely in glasshouses, where soil is pasteurized to 65–70°C by steam–air mixtures (**aerated steam**) rather than by treatment with steam alone (100°C). This is not only cheaper but also more effective, as shown by the experimental results in Table 14.1. In this experiment, trays of soil naturally contaminated with the seedling pathogen *Pythium ultimum* were treated at different temperatures then sown with pepper seedlings, and a small inoculum of another pathogen, *Rhizoctonia solani*, was placed in one corner of each tray, to simulate a surviving pocket of inoculum that would be found just beyond a treated area in a commercial glasshouse. The area of diseased plants in each box was then assessed. Treatment of the soil for 30 min at any temperature (61, 70 or 100°C) eliminated the resident *Pythium* population, but after treatment at 100°C the soil was sterile and highly favourable for spread of *Rhizoctonia*, whereas soil treated at 61°C suppressed this pathogen. The reason was that *Bacillus* spp. survived exposure to this temperature as dormant spores and then grew rapidly to fill the biological 'vacuum'. It is well known that seedling pathogens such as *Rhizoctonia, Pythium*

Table 14.1 Disease caused by *Rhizoctonia solani* when introduced into one corner of trays of soil treated for 30 min at different temperatures to eradicate a natural soil population of *Pythium* and then densely seeded with pepper (*Capsicum*). (Data from Olsen & Baker, 1968.)

Temperature (°C)	Disease caused by *R. solani*		Area of disease caused by resident *Pythium* (cm²)
	Area (cm²)	Linear spread (cm)	
No treatment	None	0.3	103
100	253	18	0
71	65	9	0
60	3	2	0

and *Fusarium* spp. rapidly recolonize sterilized soils, and that pasteurization prevents this in practice. In the experiment above, the cheapest treatment was the best, because it was based on biological principles.

Chemical **fumigants** also can be used to eradicate pathogens from glasshouse or field soils, and they have differential effects on the fungal population. For example, the mild chemical fumigant, carbon disulphide, has been used as a pre-planting treatment to eradicate *Armillaria mellea* from small woody fragments in old peach orchards after most of the diseased root systems were removed mechanically. Its success cannot be explained by fumigant action alone because the chemical cannot penetrate completely into woody roots in which the pathogen survives. Instead, the fumigated soil is recolonized rapidly by antagonistic *Trichoderma* spp., which then invade the roots and displace the pathogen weakened by the fumigant. Figure 14.1 shows this effect in controlled experimental conditions, but using the stronger fumigant methyl

bromide to eradicate *Armillaria* from root pieces. *Trichoderma* spp. colonized fumigant-treated roots naturally from soil inoculum and displaced the pathogen, whereas *Trichoderma* could not displace it from untreated roots.

In the warmer parts of the world, solar heating (**solarization**) beneath thin polyethylene sheeting can be used as an alternative to chemical control. Over a 20–30-day period until the film is degraded the soil temperature can reach over 40°C at the surface and 30°C or more at even 50-cm depth. The spores of most soil-borne pathogens can survive these temperatures *in vitro*, but in moist, solar-heated soil they are virtually eliminated by the activities of competitors and antagonists that flourish at the higher temperatures (Katan, 1987).

Chemical control

Fungitoxic chemicals play a major role in crop protection, in controlling the life-threatening pathogens of humans and in preventing decay of

Fig. 14.1 Displacement of *Armillaria mellea* from pieces of woody root by *Trichoderma* colonizing the roots from natural soil. (a) Roots colonized by *Armillaria* were treated with methyl bromide vapour for 3 h before burial in soil. This led to rapid displacement of the 'weakened' *Armillaria* by *Trichoderma*. (b) Roots colonized by *Armillaria* were exposed to air for 3 h before burial. The pathogen persisted in the roots and was not displaced by *Trichoderma*. (The fumigation treatment alone (without burial in soil) did not kill *Armillaria* so it could be isolated from 100% of roots.) (From Ohr *et al.*, 1973.)

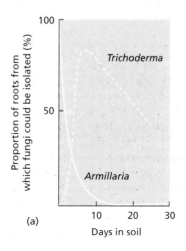
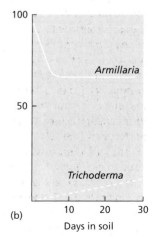

277

various materials. But, the choice of chemical depends on the specific needs. For example, for the preservation of structural timbers, fence posts and telegraph poles the chemical must be persistent and non-specific, so these materials are treated with tar oils such as creosote, other fungitoxic phenolic compounds or copper–chrome–arsenate. For cosmetics, ointments, etc., the preservatives must be non-toxic, non-irritant and compatible with the product formulation; low-molecular-weight alcohols, and esters of *p*-hydroxybenzoic acid (parabens) are used frequently in these cases. For control of plant or human pathogens the essential requirement is that the chemical has a high degree of **selective toxicity** so that it kills the pathogen but does not damage the host.

Selective toxicity is relatively easy to achieve for a bacterial pathogen, because peptidoglycan is found only in the walls of prokaryotes, and protein synthesis occurs on 70S ribosomes. This explains why most of the antibiotics used in medicine act on either bacterial walls or protein synthesis. It is much more difficult to control a eukaryote in the tissues of another eukaryote, and so research has aimed at identifying the unique structural or metabolic features of fungi that could be specific targets for antifungal agents. Figure 14.2 shows that several of these targets are used currently for control of plant or human diseases, but others remain to be exploited. We will see examples of this in the following sections. Further details of many of the chemical control agents can be found in Trinci and Ryley (1984).

Fungicides in plant disease control

The term fungicide is used in a broad sense in plant disease control, for any compound that kills

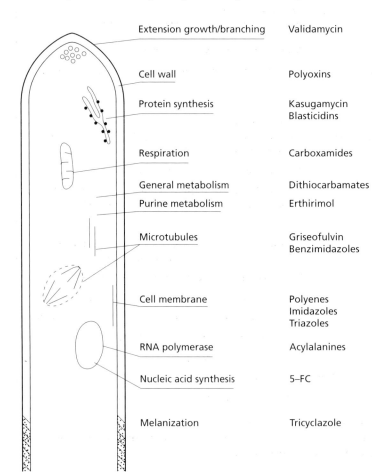

Extension growth/branching	Validamycin
Cell wall	Polyoxins
Protein synthesis	Kasugamycin Blasticidins
Respiration	Carboxamides
General metabolism	Dithiocarbamates
Purine metabolism	Erthirimol
Microtubules	Griseofulvin Benzimidazoles
Cell membrane	Polyenes Imidazoles Triazoles
RNA polymerase	Acylalanines
Nucleic acid synthesis	5–FC
Melanization	Tricyclazole

Fig. 14.2 The main targets currently exploited for chemical control of fungi that cause plant and human diseases. Some examples of the chemical control agents are shown. (Based on an illustration in Trinci & Ryley, 1984.)

or inhibits a fungus without killing it (in which case, strictly, it is a fungistatic compound). Almost all these compounds are chemically synthesized, because antibiotics are too expensive or are reserved for pharmaceutical use. Fungicides must be approved for use by a regulatory body, which also specifies the crops on which they can be used and any 'withholding period' between the last application and crop harvesting. In Britain the approved products are updated annually in the *UK Pesticide Guide*. The current list covers over 100 active ingredients, and numerous formulations of these. Table 14.2 lists only a few representative examples, with a selection of the diseases concerned, by way of illustration. The corresponding chemical structures are shown in Fig. 14.3. For his-

torical and practical purposes, they are divided into three groups: (i) inorganic fungicides; (ii) organic contact or protectant fungicides; and (iii) systemic fungicides.

Inorganic fungicides

Inorganic fungicides were the first to be used: elemental sulphur in 1846, copper in 1882 and mercury around 1920. The value of sulphur was always limited by its insolubility, but powdered sulphur was used widely for control of powdery mildew diseases because their mycelia and spores develop on the leaf surface, supported by haustoria in the underlying cells (see Chapter 12). Sulphur is still available as a combined fungicide

Table 14.2 Some representative fungicides and examples of their use for disease control*.

Type of compound	Mode/site of action	Examples of usage
INORGANIC PROTECTANT		
Sulphur	Many, general	Powdery mildews (grape, hops, etc.)
Copper	Many, general	Potato blight; downy mildews (grape, hop, etc.); apple scab
ORGANIC PROTECTANT		
Dithiocarbamates		
Maneb, Zineb, etc.	Thiol enzymes	Potato blight; leaf diseases of cereals, horticulture
Thiram	Copper/respiration	Seed treatment (damping-off), foliar diseases (*Botrytis* of strawberry, horticulture)
Dinocap	Respiration	Powdery mildew of fruit, flower crops
Iprodione	Nucleic acids	Foliar for *Septoria* of cereals; *Botrytis, Alternaria, Sclerotinia* on many crops; *Rhizoctonia* on potato; turf diseases
ORGANIC SYSTEMIC		
Benzimidazoles	Microtubules	Many diseases; cereal eyespot, *Botrytis* on beans, turf diseases, fruit storage rots, powdery mildews in horticulture
Carboxin (in mixtures)	Mitochondria	Seed treatment for rape, cereals; foot rots, smut, ear blight
Imazalil	Sterol synthesis	Powdery mildew of cucumber, rose, ornamentals
Triforine	Sterol synthesis	Powdery mildew of barley, apple, hop, cucumber, ornamentals
Ethirimol (in mixtures)	Nucleotides	Seed treatment for barley seedling blights, smuts
Tricyclazole	Melanin synthesis	Rice blast (*Magnaporthe grisea*)
Fosetyl-aluminium	Phosphate metabolism	*Phytophthora* diseases
Acylalanines (metalaxyl, etc.)	RNA synthesis	*Phytophthora* and *Pythium*; downy mildews

* All examples are illustrative, not comprehensive. Products must be approved officially for each specific use (see *UK Pesticide Guide*, updated annually, or equivalent national publications).
RNA, ribonucleic acid.

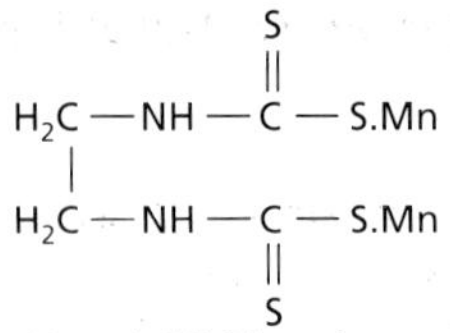

Maneb (**dithiocarbamate**)

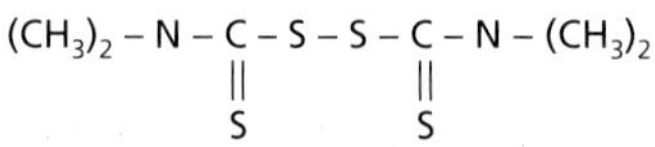

Thiram (**dimethyldithiocarbamate**)

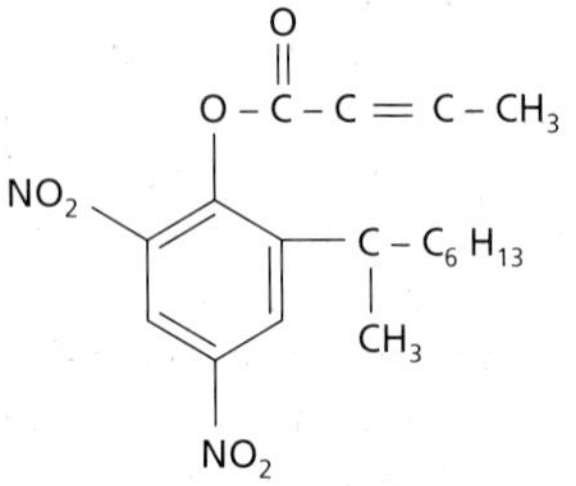

Dinocap (**dinitrophenol**)

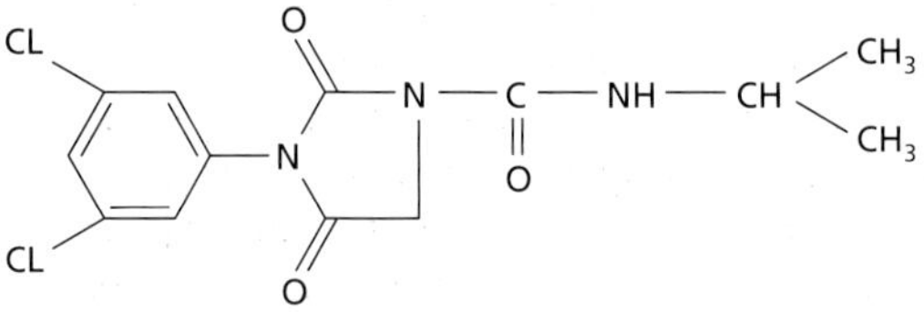

Iprodione (**dicarboximide**)

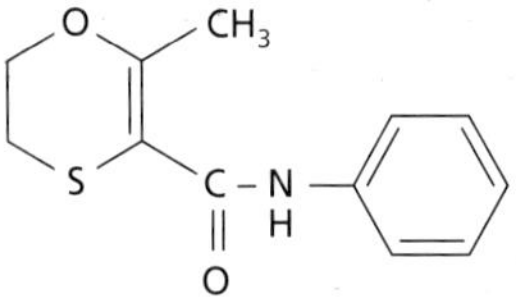

Carboxin (**carboxamide**)

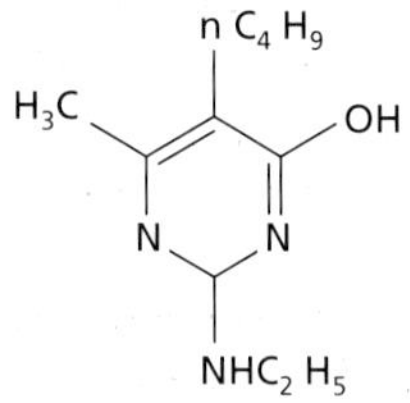

Ethirimol (**2-aminopyrimidine**)

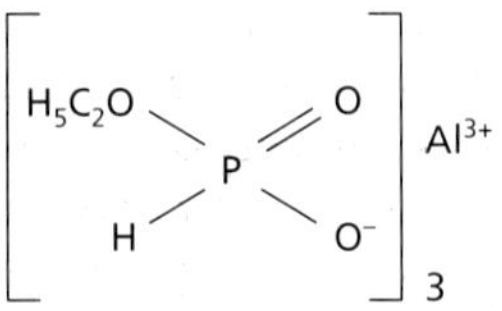

Fosetyl aluminium

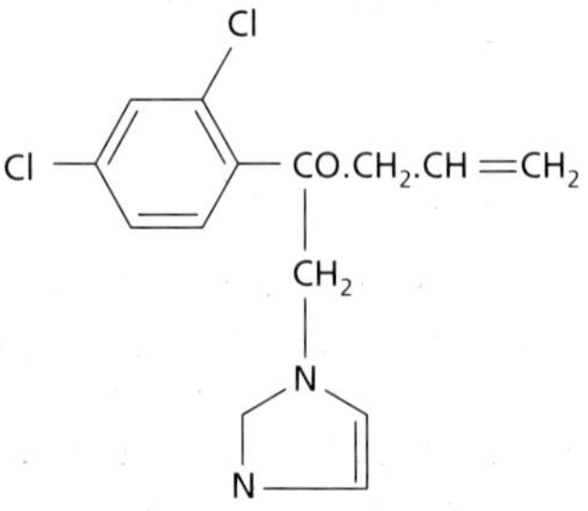

Imazalil (**imidazole**)

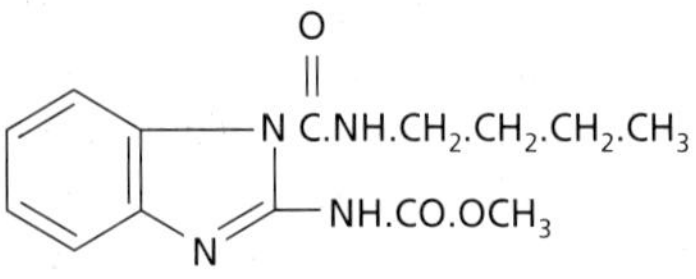

Benomyl (**benzimidazole**)

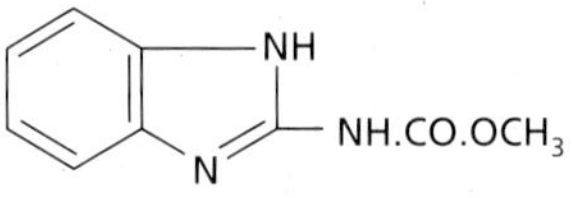

Carbendazim (**MBC**)

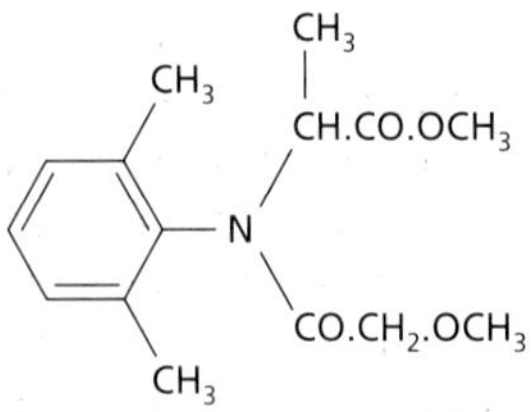

Metalaxyl (**acylalanine**)

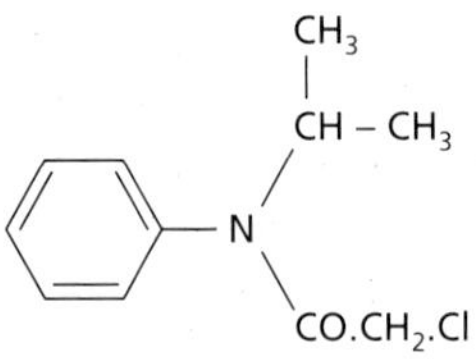

Propachlor (**herbicide**)

Fig.14.3 Structures of some representative fungicides used for plant disease control. The herbicide Propachlor is shown for comparison with the fungicide Metalaxyl.

and foliar feed, and is one of the few fungicides available to 'organic' growers. Copper-based fungicides have been used extensively for their broad antifungal and antibacterial spectrum. They were spectacularly successful against downy mildew of grapevines (*Plasmopara viticola*), for which Burgundy mixture and Bordeaux mixture were developed. Usually, the copper fungicides are complexes of copper sulphate with lime (Bordeaux mixture) or they are based on copper hydroxide or copper oxychloride. They control many leaf and fruit diseases, but are potentially phytotoxic. They are used widely in developing countries because of their low cost. Mercury always had restricted usage because of its acute mammalian toxicity. Its use finally was banned in Britain in April 1992 but much earlier in many other countries.

All these metal ions interfere with basic metabolic processes, so fungi do not develop resistance to them easily. In the relatively few cases where resistance does occur it usually is due to reduced uptake of the toxic metal.

Organic contact (protectant) fungicides

Organic fungicides were developed in the 1930s and quickly replaced the inorganic fungicides. They are termed protectant or contact fungicides if they act only near the site of application, to protect the plant surface or control an established infection. Thus, they need to be applied over the whole surface of the plant, and reapplied to protect new growth. They are, however, durable fungicides and many of them are still in use today. Fungi do not develop resistance to them easily because they interfere with basic metabolic processes and often have multiple sites of action in the cell.

The **dithiocarbamates** (maneb, zineb, mancozeb, etc.; Fig. 14.3) are used to control many leaf diseases. All of these compounds have thiol groups and they inhibit respiration non-specifically by interfering with sulphur-containing enzymes. However, thiram (Fig. 14.3) may act more specifically by forming a complex with copper and then disrupting the enzyme complex (pyruvate dehydrogenase complex) that converts pyruvic acid to acetyl-coenzyme A (CoA) (see Chapter 6). Dinocap (Fig. 14.3) is a dinitrophenol-type compound which uncouples respiration from adenosine triphosphate (ATP) synthesis; it acts specifically on powdery mildews. The **dicarboximides** such as iprodione (Fig. 14.3) are thought to interfere with nuclear division; although they affect many fungi, they do not inhibit oomycota. They are relatively unusual among the protectant fungicides because pathogens develop resistance to them quite readily *in vitro* and then show resistance to other members of the group (cross-resistance). This could indicate a single-site mode of action or a common uptake or detoxification mechanism. Nevertheless, only moderate levels of resistance are encountered in field conditions, probably because resistance is coupled with a reduction of virulence and thus of environmental fitness.

Systemic fungicides

Systemic fungicides are absorbed by plants and then distributed internally, even to sites remote from the site of application, so the development of these compounds in the 1960s revolutionized crop protection. It was no longer necessary to cover a plant completely, and these fungicides could eradicate even existing, deep-seated infections; moreover, they would be transported into new growth to protect this from subsequent infection. They are used widely today, both as sprays for control of shoot diseases and as seed-applied treatments. However, with few exceptions they travel only upwards, in the xylem, and not downwards in the phloem, so they cannot be used to control the major soil-borne pathogens. Their use in practice has also revealed another problem: the active ingredients are not general toxicants but, instead, have highly specific modes of action on a single step in an enzymic pathway or by binding to a specific component of the fungal cell. Because of this, fungi can develop resistance to them easily, often by a single-gene mutation which alters fungicide binding to the target site. A striking example of this was seen when the **acylalanine** fungicides (see later) were first introduced for control of potato blight (*Phytophthora infestans*); in some areas the pathogen developed resistance within the first year, so these compounds were withdrawn immediately from the market and only released again when mixed with contact fungicides to help extend their useful life.

Despite these problems, the systemic fungicides have become the mainstay of crop protection, and they also reveal much about the unique metabolic features of fungi. We consider some of the major types below.

Benzimidazoles (Fig. 14.3)

The benzimidazole group were the first systemic fungicides to be introduced — thiabendazole in 1964, benomyl in 1967, and then others such as carbendazim (1970) and thiophanate-methyl. Ironically, thiabendazole was developed and released in 1962 as an antihelminthic, and only later was its fungicidal role discovered. This empirical approach to fungicide development is still the most successful—thousands of chemicals synthesized by chemists are put through routine assays (Shepherd, 1987). Thus, benomyl and carbendazim were discovered and patented separately after thiabendazole, but now we know that they have similar or identical modes of action: they are converted to benzimidazole-2-yl-carbamate (MBC) which binds to spindle microtubules and blocks mitosis. They also affect tip growth, either as a result of blocking mitosis or by affecting the cytoplasmic microtubules which are common in the hyphal tip (see Chapter 3).

Davidse (1986) reviewed the modes of action and biological impact of these fungicides. They are selectively antifungal, with little or no effect on plant and animal cells despite their mode of action on microtubules. The reason for this was shown *in vitro* because MBC binds strongly to the extracted β-tubulin of fungi, preventing this from forming a dimer with α-tubulin and therefore preventing the self-assembly of microtubules when the tubulin dimers attach end to end. MBC does not bind strongly to the tubulins of higher animal or plant cells, nor of the oomycota. Instead, the tubulins of these organisms bind to another antimitotic agent, colchicine, which does not affect the fungi.

The benzimidazoles are used successfully to control a large number of pathogens, but some fungi (e.g. *Alternaria*) are naturally tolerant of them, as we saw in relation to facial eczema in Chapter 11. This is probably due to reduced uptake through the wall or membrane. In practice, the benzimidazole-sensitive fungi also can

develop tolerance (resistance) if the fungicides are used repeatedly. It is caused by point mutations at various sites in the β-tubulin gene, sometimes causing reduced binding of the fungicide to β-tubulin and sometimes affecting the interaction of β-tubulin with α-tubulin or with the tubulin-associated proteins, all of which are necessary for functional microtubules. The mutants often show reduced environmental fitness, probably because any alteration of an important functional molecule such as β-tubulin involves some loss of efficiency. This might help to explain why the benzimidazoles still are used widely, more than 30 years after they were first introduced.

Sterol synthesis inhibitors

The main fungal sterol, **ergosterol**, is a major component of the fungal plasma membrane (see Chapter 2), whereas cholesterol has a corresponding role in higher animals, and the cholesterol-like phytosterols in plants and oomycota. All sterols are produced by the same basic pathway (see Chapter 6) in which mevalonic acid condenses to form isoprene units, then several of these units join together and the molecule undergoes cyclization and various substitutions. However, a key step in the fungal pathway to ergosterol is the removal of a methyl group from the carbon-14 (C-14) position of the sterol precursor (see Fig. 14.8), and this does not occur in the plant and animal sterols. This step is catalysed by an enzyme, C-14 demethylase, which has an iron-containing cytochrome as its coenzyme.

Several systemic fungicides act specifically on this enzyme, and therefore are selectively toxic to fungi — they block demethylation so that the fungus cannot synthesize its normal sterols and its membrane function is impaired. The structures of these fungicides are quite diverse but their common feature is the possession of a heterocyclic ring attached to a large lipophilic group (see imazalil in Fig. 14.3). The lipophilic part of the molecule is thought to bind to the demethylase enzyme, while a nitrogen in the heterocyclic ring associates with the iron of the coenzyme. We return to this later when we consider some of the antimycotic drugs used for human therapy.

Some of the sterol demethylase inhibitors have a wide spectrum of activity against fungi, but some act specifically against powdery mildew fungi, perhaps reflecting differences in fungicide uptake by the different fungal groups. None of them affects the oomycota such as *Pythium* and *Phytophthora*, which do not synthesize their own sterols.

Other systemic fungicides

There are several systemic fungicides that act more or less specifically on particular fungal groups. For example, the **carboxamide** fungicides such as carboxin (Fig. 14.3) act against basidiomycota (rusts, smuts, *Rhizoctonia solani*) by interfering with energy generation; they inhibit the step in the tricarboxylic acid (TCA) cycle where succinate is dehydrogenated to produce fumarate (see Chapter 6). The **2-aminopyrimidine** fungicides such as ethirimol (Fig. 14.3) act specifically against powdery mildew fungi by inhibiting an enzyme (adenosine deaminase) involved in purine metabolism. In this case, fungicide-tolerant mutants often show no alteration of the enzyme which could explain their resistance, but are thought to circumvent the inhibition by obtaining purines from the host plant.

The **acylalanines** (e.g. metalaxyl; Fig. 14.3) act specifically against oomycota by inhibiting the nuclear ribonucleic acid (RNA) polymerase of these fungi and thus blocking RNA synthesis. These fungicides are interesting for several reasons. Unlike most other systemic fungicides, they can move downwards in the phloem to at least some degree, although in practice they are usually drenched into soil for the best control of *Pythium* or *Phytophthora* in the root zone. Also, they are closely related structurally to the systemic herbicides such as propachlor (Fig. 14.3); only relatively small differences in the substituent groups on these molecules determine whether they are systemic fungicides, systemic herbicides or inactive molecules. As a final point, tolerance of the acylalanine fungicides can develop rapidly in practice, even though the oomycota are diploid. The tolerance is thought to involve a mutation in the RNA polymerase gene, so that even a single mutation on one chromosome can enable the fungus to grow. This resistance also is very stable in the absence of continuing selection (i.e. when the fungicide is no longer used), and sexual crossing between 'partly resistant' strains can generate

'full' homozygous resistant progeny. There is now widespread resistance to the acylalanines in European populations of *P. infestans*. The development of this resistance has coincided with the widespread appearance of the A2 mating type of *P. infestans* in Europe, whereas previously only the A1 mating type was found (both mating types are common in Mexico where *P. infestans* is thought to have originated on the native *Solanum* spp.). Presumably the development of resistance was favoured by sexual crossing.

Fosetyl-aluminium (Fig. 14.3) is another fungicide that acts specifically against oomycota, and it has the remarkable property of being fully phloem-mobile. So, it can be applied to the shoots and it moves into the roots, making it extremely valuable for control of root-infecting *Phytophthora* spp. One of its principal roles has been as a trunk-injection treatment of avocado trees; a single annual application can give excellent control of *Phytophthora cinnamomi* on the roots. Fosetyl-aluminium is remarkable also for its mode of action. It shows little or no activity against oomycota in laboratory culture, and this led initially to the view that it might act by inducing host resistance. There may be some truth in this, but subsequent work showed that fosetyl-aluminium decomposes readily in plants to yield phosphorous acid and this simple mineral acid can suppress the growth of oomycota in culture media of low phosphate content.

The **melanin synthesis inhibitors** such as tricyclazole have no effect on fungal growth in culture, but they can interfere specifically with the infection process, especially when fungi infect from melanized appressoria (see Chapter 4). These fungicides have found a limited application in the control of *Colletotrichum* spp. and especially of *Magnaporthe grisea* (the cause of rice blast). Wolkow *et al.* (1983) showed that, for fungi that infect directly through the leaf cuticle from melanized appressoria, the melanization is essential for 'focusing' of the penetration peg. Normally, this peg arises from a narrow region of the appressorium where the wall is non-melanized in contact with the host. If the fungus cannot synthesize melanin then there is a broad zone of contact with the host surface and the attempted penetration often fails. Melanin-deficient mutants of these pathogens show a similar inability to penetrate the host. The reason is perhaps that melanin strengthens the appressorium wall so that the considerable osmotic pressure in the appressorium is focused at the penetration point, as we saw in the case of rice blast in Chapter 4. Unfortunately, the main target pathogens, including rice blast, can rapidly develop resistance to these fungicides in practice, because the fungicides act on one enzymic step in the melanin biosynthetic pathway.

Systemic fungicides: the new era

New systemic fungicides are still coming onto the market, and in this respect their specific, single-site modes of action are an advantage—it is doubtful that any new compound that acts as a general toxicant would ever be approved for use by the regulatory and environmental authorities. However, the vulnerability of these compounds to the emergence of pathogen resistance has caused a change in the approach to chemical control. The new 'resistance-management' strategies involve various approaches: the formulation of systemic chemicals with protectant fungicides (e.g. carbendazim with mancozeb for control of many leaf diseases of cereals); the alternating use of systemic chemicals with different modes of action; and the coupling of systemic fungicides with non-chemical control methods such as plant cultivar resistance or, in the near future, biocontrol agents.

The discovery of systemic acquired resistance to plant pathogens (see Chapter 12) has led recently to the development of novel chemical control agents which have no effect on the fungi but act by inducing the host-resistance response. One such compound is based on **salicylic acid**, an intermediate of the plant phenylpropanoid pathway that is activated during local acquired resistance as a prelude to systemic acquired resistance.

Antibiotics in plant disease control

There are many antifungal antibiotics (Table 14.3), but only a few are selective enough to be used for disease control. For example, cycloheximide (Fig. 14.4) cannot be used despite its marked antifungal action because it blocks protein synthesis on the 80S ribosomes of all eukaryotes. Some other antifungal antibiotics are reserved for medical use, discussed later, although streptomycin (Fig. 14.4) has been used in the past to control some powdery mildew pathogens (see Chapter 4). The few antibiotics that have been used recently for plant disease control were developed specifically for this

Table 14.3 Some antifungal antibiotics.

Compound	Produced by	Fungi affected	Site/mode of action
Cycloheximide	*Streptomyces griseus*	All	Protein synthesis
Griseofulvin	*Penicillium griseofulvum*	Many (not oomycota)	Microtubules
Polyenes (nystatin, filipin, amphotericin B)	*Streptomyces* spp.	Many (not oomycota)	Cell membrane
Polyoxins	*Streptomyces*	Many (not oomycota)	Chitin synthesis
Validamycin A	*Streptomyces*	Some plant pathogens	Morphogen
Blasticidin-S	*Streptomyces*	Some plant pathogens	Protein synthesis
Kasugamycin	*Streptomyces*	Some plant pathogens	Protein synthesis
Streptomycin	*Streptomyces*	Oomycota	Calcium?
Antimycin A	*Streptomyces*	Many	Respiration
Patulin	*Penicillium, Aspergillus*	Many	Respiration
Pyrrolnitrin	*Pseudomonas* spp.		
Pyoluteorin	*Pseudomonas* spp.		
Gliotoxin	*Trichoderma virens*		
Gliovirin	*T. virens*		
Viridin	*T. virens*	Various *in vitro*	Implicated in biocontrol of plant pathogens
Trichodermin	*Trichoderma*		
6-Pentyl-α-pyrone	*Trichoderma*		
Suzukacillin	*Trichoderma*		
Alamethicine	*Trichoderma*		

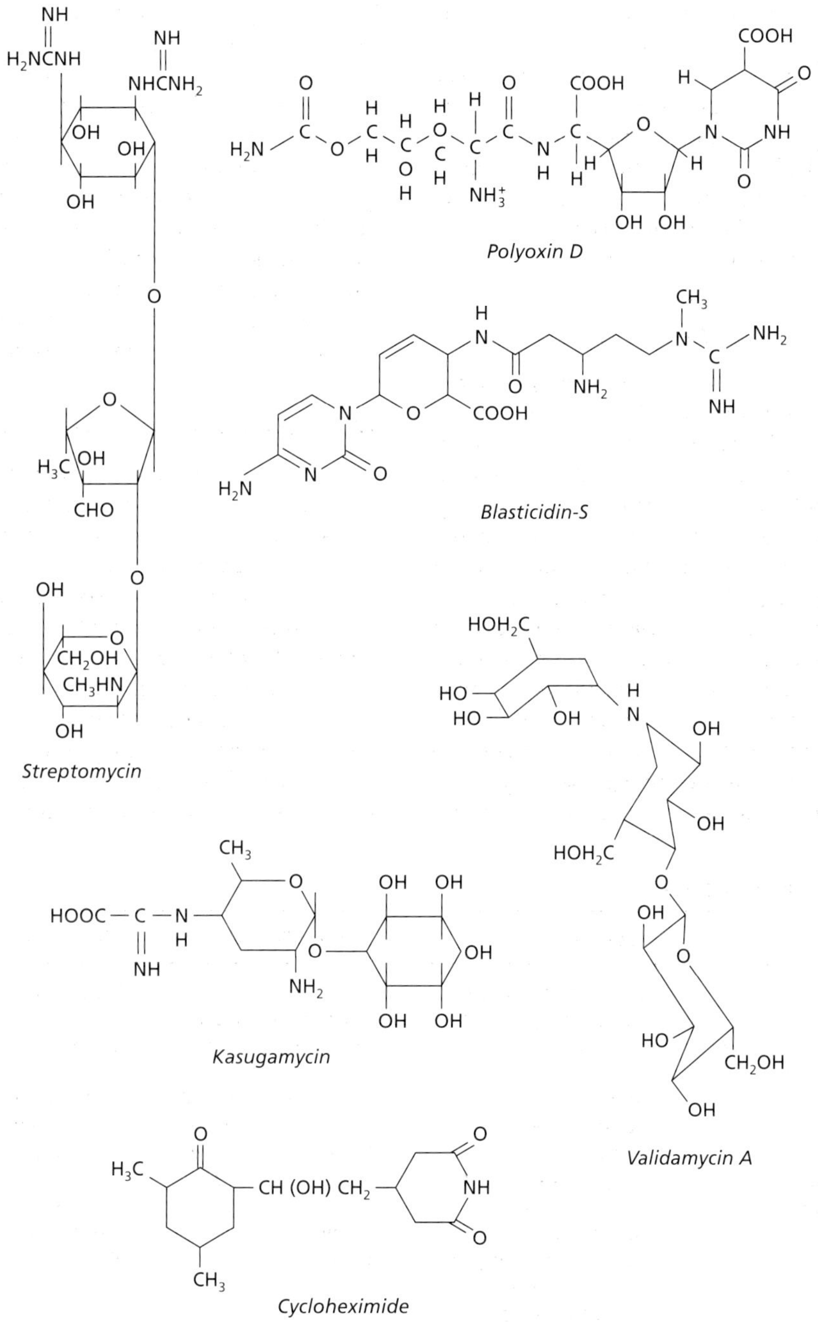

Fig.14.4 Structures of some antifungal antibiotics, including those used for plant disease control.

purpose in Japan. We consider them briefly below, and we also consider some further examples of the roles of antibiotics in biocontrol of plant diseases.

Polyoxin and nikkomycin

Polyoxin D (Fig. 14.4) is a specific inhibitor of chitin synthesis. It binds strongly to the active site of chitin synthase, competing with the normal substrate, uridine diphosphate (UDP)-*N*-acetyl-glucosamine (see Chapter 6). In fact, the structure of polyoxin closely resembles that of a sugar nucleotide. Nikkomycin is a similar compound that acts in the same way. The unique mode of action of these compounds raised the hope that they might have significant roles in controlling fungi and insects—the two major groups of chitin-containing organisms—while having no effect on higher animals or plants. *In vitro* they inhibit the growth of many fungi, including plant pathogens and the fungi that cause mycoses of humans. They have been used in practice to control some plant diseases, especially sheath blight of rice (*Rhizoctonia*) and black spot of pear (*Alternaria*) in Japan. However, the pathogens develop tolerance of them rapidly in field conditions, owing to the selection of mutant strains that show reduced antibiotic uptake. They are completely ineffective for disease control in animal hosts, because they are taken into fungal cells through membrane proteins used normally for the uptake of dipeptides, and the abundance of small peptides in animal tissues blocks uptake of the antibiotics.

Blasticidin, kasugamycin and validamycin (Fig. 14.4)

These three antibiotics have found limited roles in Japanese agriculture. Blasticidin-S inhibits protein synthesis in bacteria and a few fungi, by binding to the larger subunit of 70S and 80S ribosomes. It was developed for control of rice blast (*M. grisea*) but later it was replaced by another protein synthesis inhibitor, kasugamycin, which has lower mammalian toxicity. Resistance has developed to both of these compounds in field conditions: it is based on reduced uptake of blasticidin, and reduced binding of kasugamycin to the ribosome. Validamycin A is used to control several diseases caused by *Rhizoctonia*, including seedling diseases, black

scurf of potato and sheath blight of rice, but it is inactive against most fungi and bacteria. It has no effect on *Rhizoctonia* in rich culture media, but it causes abnormal branching and cessation of growth in weak media. It is known to inhibit the synthesis of myo-inositol in *R. solani*, indicating an important role of this compound for normal growth and pathogenicity of this fungus.

Antibiotics implicated in biocontrol

The roles of antibiotics in interfungal interactions were discussed in Chapter 11, so here we consider only the practical applications of these compounds in disease control. Several new biocontrol products are undergoing field trials or have been marketed in the last 5 years. One of these, marketed as GlioGard, consists of small, air-dried alginate beads containing fermenter-grown biomass of *Trichoderma virens* (=*Gliocadium virens*) and wheat bran as a food base (Mintz & Walter, 1993). This product acts primarily, if not exclusively, by production of the antibiotic gliotoxin (see Chapter 11). The dried granules are mixed with soil-less rooting media in glasshouses and left for several days after the medium is moistened, before seedlings or rooted cuttings are transplanted into the beds. Timing of these operations is crucial, because the fungus must grow on the wheat bran substrate and produce gliotoxin which diffuses into the rooting medium, protecting the plants from pathogenic attack. If there is a delay in planting then the antibiotic levels start to decline and the degree of protection is reduced considerably.

Rhizosphere bacteria seem to represent an abundant source of antifungal compounds. Table 14.4 shows the results of 'non-selective' screening (on trypticase soy agar) of the dominant bacteria from a range of plants by a commercial company (Leyns *et al.*, 1990). The most common bacteria with antifungal properties included *Pseudomonas* spp. (some of which produce the antibiotics pyrrolnitrin, pyoluteorin and phenazines), *Xanthomonas maltophilia* (unknown antibiotics), *Bacillus subtilis* and other *Bacillus* spp. (some of which produce iturins, mycosubtilins, bacillomycin, fengymycin, mycobacillin and mycocerein) and *Erwinia herbicola* (which produces herbicolins). Some of these bacteria are used in commercial biocontrol formulations. For example, *B. subtilis* is

Table 14.4 Bacteria with antifungal properties isolated by 'non-selective' methods from the root zone of crop plants*. (Adapted from Leyns *et al*., 1990.)

Crop	Number of plants examined	Number of dominant bacterial isolates	Isolates with antifungal properties (%)
Sugarbeet	1550	6780	38
Maize	503	1508	27
Soybean	450	1139	21
Sunflower	450	1119	22
Barley	36	175	92
Grape	36	231	100

* Trypticase soy agar was used for isolation because it yielded the highest numbers of bacteria and did not select specifically for known types of bacteria.

marketed for seed treatment of cotton and soybeans, to protect against damping-off fungi. Others have been implicated in natural biocontrol in field conditions, a classic example being the role of fluorescent pseudomonads (*P. fluorescens* and *P. putida*) in suppressing the take-all disease of cereals (*Gaeumannomyces graminis*).

When wheat is grown in successive years in a field the level of take-all disease can build up progressively, reaching a peak in the third or fourth year (depending on site and seasonal factors). But, the level of disease can decline spontaneously in the following years, so that wheat can be grown continuously with acceptable yields provided that sufficient nitrogen is supplied (Fig. 14.5). This **take-all decline** or take-all suppressiveness is known to be caused by micro-organisms because it is lost if the soil is pasteurized or treated with chemical fumigants (as experimental treatments), and suppressiveness can be transferred to a non-

suppressive soil if this is pasteurized (to reduce its microbial population) and then supplemented with 1–5% of suppressive soil. A similar suppressiveness can be induced experimentally by growing wheat seedlings for 3–4 weeks in pots of take-all infested soil, then removing the plants and resowing repeatedly. After the third or fourth 'crop' of wheat the soil is highly suppressive. This does not happen if wheat is sown successively in the absence of take-all, nor if take-all inoculum is added in the absence of plants. So, the suppressive organisms must proliferate on the take-all infected roots.

Several organisms might contribute to this suppressiveness, but fluorescent pseudomonads are considered to play a major role, especially the fluorescent pseudomonads that produce **phenazine-1-carboxylic acid**. These antibiotic-producing pseudomonads are found to colonize the take-all lesions, exploiting the nutrients

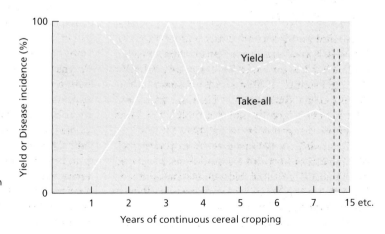

Fig. 14.5 Diagrammatic illustration of take-all decline, a common phenomenon in continuous cereal cropping.

released from the damaged tissues, and they give significant control of take-all infection when applied to cereal seeds in experimental conditions. One strain of *P. fluorescens* (strain 2–79) has been investigated intensively by workers in Washington State, USA. The purified phenazine antibiotic was shown to inhibit *G. graminis* at concentrations less than $1\,\mu g\,ml^{-1}$ *in vitro*, and seedlings grown from seeds coated with *P. fluorescens* (strain 2–79) were found to have 27–43 ng antibiotic per 1 g of root and closely adhering soil (Thomashow *et al.*, 1990). This was considered sufficient to reduce the growth by *G. graminis*, based on the *in vitro* studies. When transposon mutagenesis was used to disrupt the antibiotic gene then strain 2–79 was much less effective in controlling the pathogen on roots, but the full biocontrol activity was restored when a cosmid vector containing the phenazine gene was inserted in the mutants.

This important work raises the prospect of using *P. fluorescens* as a biocontrol agent in practice, so that wheat might be protected during the peak take-all years. However, to date this has run into the problem that might have been predicted from studies on the ecology of the rhizosphere (see Chapter 10). In field conditions strain 2–79 of *P. fluorescens* was found to grow well on the root tips of autumn-sown wheat when the bacterium was applied to seeds, but its population declined behind the root tips and also when the wheat crop went through the winter months. It flourished again in the spring, but now on take-all lesions that had developed in the meantime (Weller, 1983). Thus, strain 2–79, like all micro-organisms, is suited to a particular niche—in this case the zones of highest nutrient exudation near the root tips and in disease lesions. It can delay the infection by *G. graminis*, and perhaps also antagonize the fungus in lesions so that less inoculum remains for a following wheat crop. But, it cannot protect the whole of the root system for the whole season. This is true of biocontrol agents in general—they need to be applied to specific sites at specific times, such as to seeds for protection during the short phase when a plant is most susceptible to seedling pathogens. For season-long control it may be necessary to use mixtures of organisms which complement one another by being adapted to different niches (Deacon, 1991).

Control of mycoses of humans

The control of fungal pathogens of humans presents difficulties greater than any we have considered so far. Not only is it necessary to control one eukaryote within the tissues of another, but also the host often is extremely ill, with severely impaired defence systems (see Chapter 13) and low tolerance of potentially toxic compounds. We should remind ourselves that even the antibiotics that control bacterial infections do not work alone; they act in conjunction with the host's defences.

There are very few antimycotic agents that can be used successfully in humans. We consider them below and then discuss the potential for developing further antimycotic drugs.

Griseofulvin

Griseofulvin (Fig. 14.6) is produced by the fungus *Penicillium griseofulvum*. It is administered orally to control serious dermatophytic infections of the skin, nails and hair, because it accumulates specifically in the keratinized tissues where these fungi grow. However, it is fungistatic rather than fungicidal, so that the treatment must be prolonged until the infected tissues have sloughed off.

The history of development of griseofulvin is interesting. It was discovered in the 1960s by the ICI company during screening of fungal metabolites for plant disease control. The *in vitro* test was a germination assay on spores of the onion pathogen, *Botrytis allii*, and griseofulvin was identified as a 'curling factor' which caused the germ-tubes to grow in curls or spirals. Because of this morphogenetic effect and the finding that it affected only chitin-walled fungi, it was thought to act on the fungal walls. Consistent with this, it was found to cause gross thickening of the walls, with heavy deposition of glucans. However, this is now known to be a secondary effect of the primary mode of action on microtubules. In cell-free systems griseofulvin prevents the assembly of α- and β-tubulin into tubulin dimers, and perhaps also affects the tubulin-associated proteins. So, it has similar effects to the benzimidazole fungicides in being an antimitotic agent (on the spindle microtubules) and disrupt-

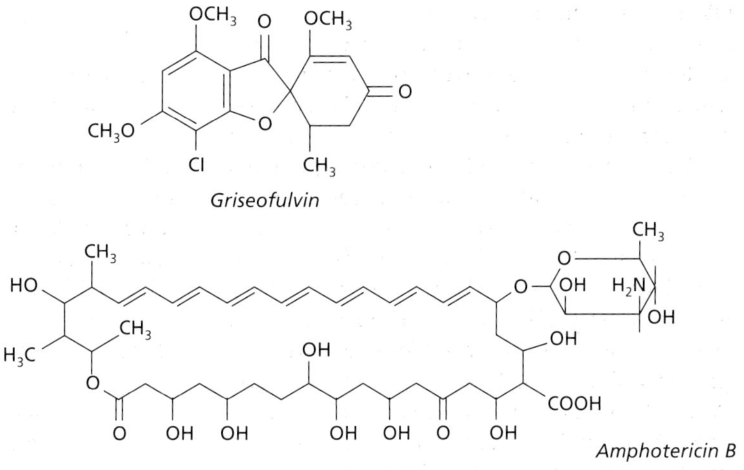

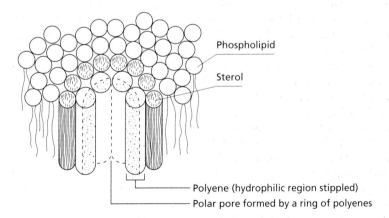

Fig. 14.6 The antibiotics griseofulvin and amphotericin B, used for treatment of mycoses of humans.

ing cellular transport processes (via the cytoplasmic microtubules). However, it acts at a different site from the benzimidazoles because mutants that develop resistance to the benzimidazoles are not resistant to griseofulvin, and vice versa. Although resistance to the antibiotic can occur quite frequently *in vitro*, it has not been a major problem in clinical practice.

Polyene macrolide antibiotics

The polyene macrolide antibiotics are produced by *Streptomyces* spp. Their typical structure is exemplified by amphotericin B (Fig. 14.6). An aminosugar group is attached to a large lactone ring with a series of double bonds on one side (hence the term polyene) and a series of hydroxyl groups on the other side. So, the molecule has both a hydrophobic and a hydrophilic face. By X-ray crystallography the molecule is seen to have a rod-like structure, about 2.1 nm long in the case of amphotericin B, similar to a membrane phospholipid. The polyenes insert in the cell membrane by associating with sterols (the hydrophobic face)

Fig. 14.7 Proposed mode of action of amphotericin B and similar large polyene antibiotics on the cell membrane. The antibiotic has strong affinity for the fungal membrane sterol, ergosterol, causing ergosterol molecules to be repositioned in the membrane, so that the polyene-sterol complexes form polar pores. Note that only one layer of the phospholipid bilayer of the membrane is shown. (Based on a diagram in Gale *et al.*, 1981.)

289

and are thought to cause rearrangement of the sterols so that a group of eight polyene molecules forms a ring with the hydrophilic faces in the centre (Fig. 14.7). Thus, they form a polar pore through which small ions (potassium, hydrogen) can pass freely, disrupting the cell's ionic control.

Almost all eukaryotic cells have sterols in the membranes, although some oomycota do not produce them or even need them for growth. So, the polyene antibiotics would not, at first sight, seem to be selective antifungal agents. However, there is a large range of polyenes that differ in the ring size and presence or nature of the aminosugar group, and they show preferential binding to different sterols. Three of these compounds—amphotericin B, nystatin and pimaricin — have a much higher affinity for ergosterol than for cholesterol, the mammalian membrane sterol, so they are selectively antifungal. Nystatin and pimaricin can be applied topically as creams to control *Candida* infection of the vagina, etc., or as powder formula-

tions to control athlete's foot. Amphotericin B has less mammalian toxicity than these others and also is water-soluble, so it can be given intravenously for control of systemic mycoses of humans. It is the drug of choice for many of these infections, including *Candida, Aspergillus, Blastomyces, Coccidioides* and *Histoplasma* in some cases. It has serious side effects, including damage to the kidneys, but these infections almost inevitably would be fatal unless treated.

Azole drugs

The azole drugs are ergosterol synthesis inhibitors similar to those used for plant disease control, but with structures designed specifically for treatment of systemic mycoses of humans. They act by blocking demethylation at the C-14 position, during the pathway of ergosterol synthesis from the precursor **lanosterol** (Fig. 14.8). Depending on whether these drugs have two or three nitrogens in the five-

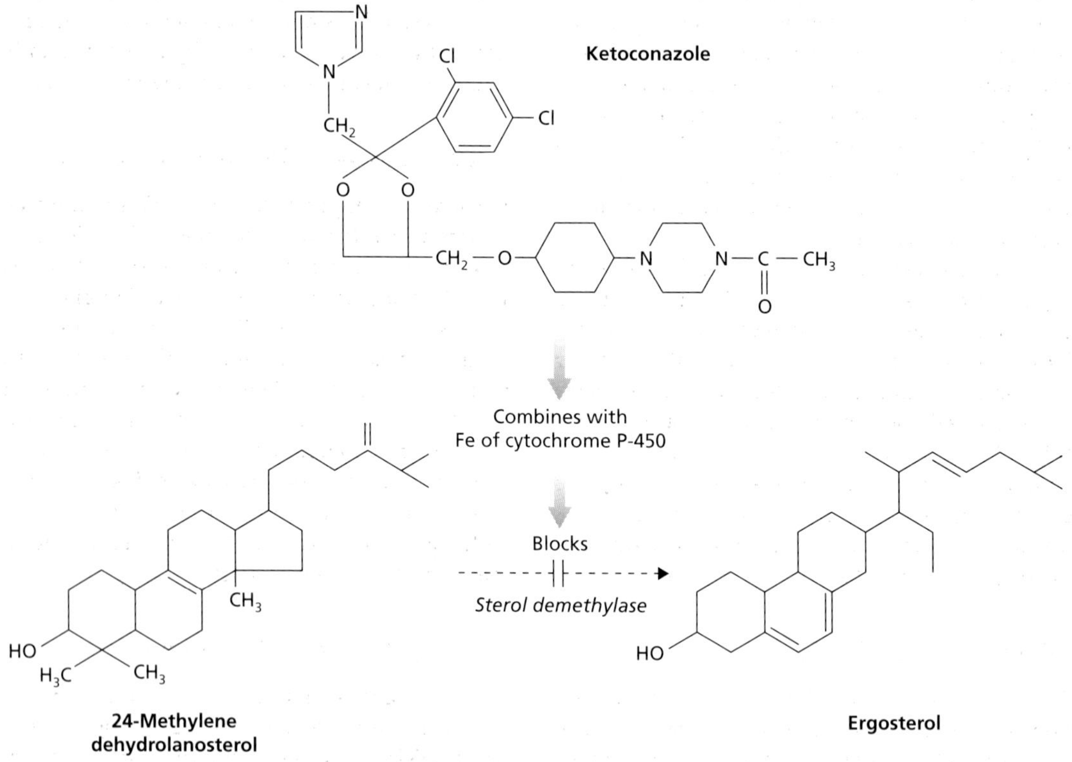

Fig.14.8 Structure and mode of action of the antifungal drug, ketoconazole, which blocks the demethylation step in the synthesis of ergosterol from sterol precursors.

membered heterocyclic ring, they are termed imidazoles or triazoles. Thus, **ketoconazole** (Fig. 14.8) is an imidazole, whereas the more recent compound, **fluconazole**, is a triazole. The azole drugs are used commonly to control pulmonary and systemic infections by *Candida*. By blocking the demethylation step of the sterol pathway, they cause the depletion of ergosterol so that fungi incorporate other sterols in the membrane. The effect of this has been studied in a 'model system' by exploiting the fact that *Saccharomyces cerevisiae* can be grown anaerobically but then needs to be supplied with sterols because many of the basic biosynthetic pathways do not operate in anaerobic conditions (see Chapter 6). The cells grow normally when supplied with ergosterol, but they become leaky and cannot assemble the wall in the normal way if supplied with lanosterol, presumably because the membrane function is impaired. In this respect it will be recalled that the major wall-synthetic enzymes, chitin synthase and glucan synthetase, are integral membrane proteins (see Chapter 3) so presumably any change in the membrane composition could affect these enzyme functions. In *Candida*, one of the effects of the azole drugs is to prevent the phase transition from yeast to hypha. As we noted in Chapter 13, this is important for invasion of tissues and for allowing *Candida* to break out of macrophages that have engulfed the yeast cells.

The azoles are selectively antifungal because the typical fungal sterols are demethylated whereas the mammalian sterol (cholesterol) is not so. However, the azole drugs can affect other processes in mammalian cells because they act on **cytochrome P-450**, the sterol demethylase cofactor (Fig. 14.8) which is also involved in the synthesis of testosterone. There is evidence that fungi are developing resistance to some of the azoles in clinical practice, equivalent to the resistance that has developed in agriculture.

5-Fluorocytosine (5-FC)

5-FC is a fluorine-substituted nucleoside, developed originally as an antitumour agent, but it has an additional role as an antimycotic agent for treatment of systemic *Candida* and *Cryptococcus*. Its mode of action is shown in Fig. 14.9. It is taken up by the **cytosine permease** of the cell membrane,

then deaminated to **5-fluorouracil** which, through a series of steps, is incorporated into RNA in place of uracil, causing impaired RNA function. It also impairs deoxyribonucleic acid (DNA) synthesis because one of the products of 5-fluorouracil inhibits the enzyme **thymidylate synthase** in the pathway that generates thymidine nucleotides.

5-FC is selectively toxic to fungi because mammalian cells have a low rate of uptake of the molecule and have little or no ability to convert it to 5-fluorouracil because they lack the enzyme cytosine deaminase (Fig. 14.9). However, 5-FC affects only a small number of fungi, notably *Candida*, *Cryptococcus* and *A. fumigatus*, and even these fungi easily develop resistance to it in clinical practice. The cause of this resistance can be studied *in vitro* by supplying cells with the various intermediates along the pathway shown in Fig. 14.9. Most resistant strains from clinical specimens are found to be deficient in uridylate pyrophosphorylase activity, but a few lack a cytosine permease or deaminase. Because of the rapid development of resistance, 5-FC is not used alone but in conjunction with other drugs, especially amphotericin B, with which it has a significant synergistic effect.

The special case of *Pneumocystis carinii*

We noted in Chapter 13 that *P. carinii* is an unusual fungus which for many years was classified as a protozoan. The principal means of controlling it is with the drug pentamidine which is thought to act on DNA. It is also sensitive to the antibacterial agents trimethoprim and sulfomethoxazole, which disrupt the folate pathway involved in transfer of groups from the amino acid serine to a range of other compounds. *P. carinii* is not affected by the antifungal polyenes or azoles because it has cholesterol, not ergosterol in its cell membrane. Although *P. carinii* is placed taxonomically within the fungi, it is still a most peculiar organism.

New targets for chemical control

There is continuing need to find new chemical control agents for fungal pathogens in agriculture and medicine. One of the current approaches is to use targeted screens in which specific features of fungi are identified (Georgopapadakou & Walsh, 1994). If chemicals can be found that interfere with

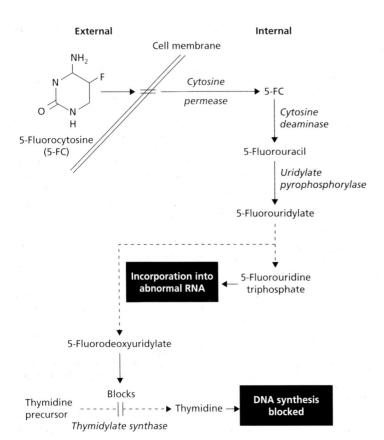

Fig. 14.9 Mode of action of 5-fluorocytosine which blocks deoxyribonucleic acid (DNA) synthesis and leads to production of defective RNA.

these features *in vitro* or in simple bioassays then they should be selectively antifungal and can be tested further for suitability in host systems. Some of these potential targets are listed below. The list is by no means complete but it serves to emphasize the several ways in which fungi are a biochemically unique group of organisms.

• Further stages of the ergosterol biosynthetic pathway, including the steps where already known compounds, the **morpholines**, block reduction or isomerization after the demethylation step.

• β-1,3-glucan synthesis: the **papulacandins** and **echinocandins** are known to inhibit this process, and the echinocandins in particular look promising for control of *P. carinii*.

• Elongation factor 3, which is involved in protein synthesis by fungi but not involved in mammalian protein synthesis.

• Fungal carbohydrates: the central roles of trehalose and polyols in fungal metabolism are clear

areas of difference from most other organisms, with several potential target enzymes.

• Polyamine synthesis: **polyamines** (spermidine, putrescine) are known to be essential for fungal growth, although their roles are still unclear; in some fungi they seem to be synthesized solely by a route involving ornithine decarboxylase, which is a potential target for control (Walters, 1995).

• Differential uptake systems: in several cases the differential toxicity of a compound, or development of resistance to it, is related to uptake systems of fungi, suggesting that these might be targets worthy of further study.

Conclusion: the challenges of modern mycology

Fungi impact on almost every aspect of our daily lives, in direct and indirect ways. They damage crops, spoil foodstuffs, cause disease, provide food and flavourings and industrial products, etc.,

maintain the
rganic matter.
ogy are both
nd, fungi are
al sources of
stimated that
ngi have yet
worth, 1991),
as untapped
conserved in
challenges in
their relation-
tion mechan-
w pathogenic
re major chal-
gical relation-
levelop fungi
organisms to
ions occur in
uite different
cology'. There
ils of fungal
which can
drug design,
o the genetic
of fungi that
loiting fungi,
ental manage-
thus insepar-
d therein lies
he breadth of
n all the pres-
g agencies are

gicides: mech-
. *Annual Review*

gy in the devel-
oil-borne plant

pathogens. *Biocontrol Science and Technology*, **1**, 5–20.

Gale, E.F., Cundcliffe, E., Reynolds, P.E., Richmond, M.H. & Waring, M.J. (1981) *The Molecular Basis of Antibiotic Action*, 2nd edn. Wiley, London.

Georgopapadakou, N.H. & Walsh, T.J. (1994) Human mycoses: drugs and targets for emerging pathogens. *Science*, **264**, 371–3.

Hawksworth, D.L. (1991) The fungal dimension of biodiversity: magnitude, significance and conservation. *Mycological Research*, **95**, 641–55.

Katan, J. (1987) Soil solarization. In: *Innovative Approaches to Plant Disease Control* (ed. I. Chet), pp. 77–105. Wiley, New York.

Leyns, F., Lambert, B., Joos, H. & Swings, J. (1990) Antifungal bacteria from different crops. In: *Biological Control of Soil-borne Plant Pathogens* (ed. D. Hornby), pp. 437–44. CAB International, Wallingford.

Mintz, A.S. & Walter, J.R. (1993) A private industry approach: development of GlioGard™ for disease control in horticulture. In: *Pest Management: Biologically Based Technologies* (eds R.D. Lumsden & J.L. Vaughn), pp. 398–404. American Chemical Society, Washington.

Ohr, H.D., Munnecke, D.E. & Bricker, J.L. (1973) The interaction of *Armillaria mellea* and *Trichoderma* spp. as modified by methyl bromide. *Phytopathology*, **63**, 965–73.

Olsen, C.M. & Baker, K.F. (1968) Selective heat treatment of soil and its effect on inhibition of *Rhizoctonia solani* by *Bacillus subtilis*. *Phytopathology*, **58**, 79–87.

Shepherd, M.C. (1987) Screening for fungicides. *Annual Review of Phytopathology*, **25**, 189–206.

Stammler, G., Seemuller, E. & Duncan, J.M. (1993) Analysis of RFLPs in nuclear and mitochondrial DNA and the taxonomy of *Phytophthora fragariae*. *Mycological Research*, **97**, 150–6.

Thomashow, L.S., Weller, D.M., Bonsall, R.F. & Pierson III, L.S. (1990) Production of the antibiotic phenazine-1-carboxylic acid by fluorescent *Pseudomonas* species in the rhizosphere of wheat. *Applied and Environmental Microbiology*, **56**, 908–12.

Trinci, A.P.J. & Ryley, J.F. (1984) *Mode of Action of Antifungal Agents*. Cambridge University Press, Cambridge.

Walters, D. (1995) Inhibition of polyamine biosynthesis in fungi. *Mycological Research*, **99**, 129–39.

Weller, D.M. (1983) Colonization of wheat roots by a fluorescent pseudomonad suppressive to take-all. *Phytopathology*, **73**, 463–9.

Williamson, B. (1994) Latency and quiescence in survival and success of fungal plant pathogens. In: *Ecology of Plant Pathogens* (eds J.P. Blakeman & P. Williamson), pp. 187–207. CAB International, Wallingford.

Wilson, C.L. & Wisniewski, M.E. (1989) Biological control of postharvest diseases of fruits and vegetables: an emerging technology. *Annual Review of Phytopathology*, **27**, 425–41.

Wolkow, P.M., Sisler, H.D. & Vigil, E.L. (1983) Effect of inhibitors of melanin biosynthesis on structure and function of appressoria of *Colletotrichum lindemuthianum*. *Physiological Plant Pathology*, **22**, 55–71.

Index